Television, Social Media,
and Fan Culture

Television, Social Media, and Fan Culture

Edited by Alison F. Slade, Amber J. Narro,
and Dedria Givens-Carroll

LEXINGTON BOOKS
Lanham • Boulder • New York • London

Published by Lexington Books
An imprint of The Rowman & Littlefield Publishing Group, Inc.
4501 Forbes Boulevard, Suite 200, Lanham, Maryland 20706
www.rowman.com

Unit A, Whitacre Mews, 26-34 Stannary Street, London SE11 4AB

British Library Cataloguing in Publication Information Available

Library of Congress Cataloging-in-Publication Data

Television, social media, and fan culture / edited by Alison F. Slade, Amber J. Narro, and Dedria
Givens-Carroll.
p. cm.
Includes bibliographical references and index.
ISBN 978-1-4985-0616-8 (cloth : alk. paper) -- ISBN 978-1-4985-0617-5 (electronic)
1. Television viewers--Social aspects. 2. Television programs--Social aspects. 3. Online social net-
works--Social aspects. 4. Fans (Persons)--Social aspects. I. Slade, Alison, 1977- editor. II. Narro,
Amber J., editor. III. Givens-Carroll, Dedria, 1960- editor.
PN1992.55T46 2016
302.23'45--dc23
                                    2015034009

∞™ The paper used in this publication meets the minimum requirements of American
National Standard for Information Sciences Permanence of Paper for Printed Library
Materials, ANSI/NISO Z39.48-1992.

Printed in the United States of America

Contents

Introduction   ix
    Alison F. Slade

1 The New Network: How Social Media Is Changing—And
   Saving—Television   1
   Ryan Cassella

2 Spoiler Alert: Understanding Television Enjoyment in the
   Social Media Era   23
   Benjamin Brojakowski

3 Rhetorical Strengths and Limitations of Interactivity for
   Activism in the Stewart and Colbert Universe   43
   Christopher A. Medjesky

4 Fandom Communication in a Mediated Age: The Use of Twitter
   and Blogs for Dissent Practices among National Basketball
   Association Fans   63
   Corey Jay Liberman, Michael Plugh, and Brian Geltzeiler

5 What Types of #SportFans Use Social Media?: The Role of
   Team Identity Formation and Spectatorship Motivation on Self-
   Disclosure during a Live Sport Broadcast   89
   *Shaughan A. Keaton, Nicholas M. Watanabe, and Brody J.
   Ruihley*

6 The Online Community: Fan Response of *Community*'s
   Unlikely Fifth Season   109
   Matthew R. Collins and Danielle M. Stern

7 Game(s) of Fandom: The Hyperlink Labyrinths That
   Paratextualize *Game of Thrones* Fandom   127
   Garret Castleberry

8 Be Original: Examining Fan Comments on A&E's *Duck
   Dynasty* Facebook Page After the Robertson Suspension   147
   Michel M. Haigh

9 "The Parents Have the Dream, but the Children Are in the
   Nightmare": Digital Interactivity, *Toddlers & Tiaras* Viewers,
   and Social Networking Sites   161
   Leandra H. Hernandez

10 Zombie Fans, Second Screen, and Television Audiences:
   Redefining Parasociality as Technoprosociality in AMC's
   #Talking Dead   183
   Sabrina K. Pasztor and Jenny Ungbha Korn

11 Memes, Tweets, and Props: How Fans Cope When Shows Go
   Off the Air   201
   Alane Presswood and Steven Granelli

12 So Are the Days of Our Tweets: An Examination of Twitter Use
   by American Daytime Serials and Their Fans   219
   Marsha Ducey

13 *Army Wives* Connect: Lifetime Viewers' Everyday Lives and
   Fandom Converge in Online Communities   235
   Darcey Morris

14 "Butter," Facebook, and Paula Deen: Examining Fans' Use of
   Social Media in Crisis   261
   Michel M. Haigh and Shelley Wigley

15 Fans Can Be Journalists Too: A Look at Fan Interaction with
   HBO's *The Newsroom*   277
   Julia E. Largent and Jason Roy Burnett

16 It's Bigger on the Inside: Fandom, Social Media, and *Doctor Who* 295
   Krystal Fogle

17 Television Inspired Cosplay and Social Media   317
   Laura Kane and William Loges

18 Who Killed @TheLauraPalmer?: Twitter as a Performance
   Space for *Twin Peaks* Fan Fiction   335
   Kathryn L. Lookadoo and Ted M. Dickinson

19 Fifty Years of "The Man from U.N.C.L.E.": How the Ever-
   Changing Media Sustained and Shaped One of the Oldest Fan
   Communities   351
   Cynthia W. Walker

20 Managing Multiscreen   373
  Dan Faltesek

Index   395

About the Contributors   397

Introduction

Alison F. Slade

Prior to the advent of the Internet, fans relied on social gatherings in small groups to meet and discuss their favorite programs. Eventually these types of interactions grew into larger fan societies thanks to the Internet and evolving communication technology (e.g., ComicCon, Star Trek conventions). Social media has changed the face of fan cultures and how fans interact with and about the medium of television. Social media has brought about a revolution in fan culture, from fan uprisings to save programs to entire groups and pages dedicated to mourning lost programs and characters. This anthology seeks to explore how fans use social media in regard to television programming, characters, narrative and any and all types of interactions, as well as how television uses social media for engaging fan cultures.

This anthology offers a critical look at television programming, fandom, and social media from a variety of lenses. The first two chapters offer an introductory glimpse into the history and impact of social media and television programming. In chapter 1, Ryan Cassella analyzes the dynamic interaction between the television creator and the consumer to demonstrate how social media is changing the programming landscape in specific ways previous technologies could never dream of maintaining for consumers. Casella applies a revised version of gatekeeping theory for the media producer and argues the interactivity and flexibility of social media will increasingly distort lines between producer and consumer, thus forever shifting the power dynamic between the two.

In chapter 2, Benjamin Brojakowski continues the introduction to this research of television programming, fans, and social media by exploring the linear history of television as it relates to consumer enjoyment. Historically, television has offered the viewer an escape, a journey into a fantasy world unparalleled by other forms of media. Social media has enhanced the enjoy-

ment in the viewing experience by allowing fans the opportunity to enhance
their interactions with the medium of television. Brojakowski investigates
these historical and developing trends in viewer enjoyment via social media
by explaining the use of hashtags, live tweets, and interactive experiences
through digital and mobile applications and indicates how social media fan
users can avoid spoiling their experiences with their favorite programs.

The next chapters in the anthology embark upon the task of analyzing
specific programs and social media interactions and impact. Christopher A.
Medjesky begins this discussion in chapter 3 through a rhetorical lens by
examining the interactivity through electronic and social media as it is used
in the Jon Stewart and Stephen Colbert universe. Medjesky explores interactivity as a rhetorical tool in the Stewart and Colbert programs Rally to Restore Sanity and/or Fear by examining Twitter, mobile applications, merchandising, fan sites, online forums, and moments on the television shows
themselves to illustrate how these online activities can lead to offline participation in civic engagement.

In chapter 4, Corey Jay Liberman, Mike Plugh, and Brian Geltzeiler turn
their attention to the world of televised sports and discuss the use of social
media for purposes of dissent (or disagreement). Specifically, Liberman,
Plugh, and Geltzeiler utilize content analysis of two major NBA blog outlets
(*ESPN*'s "True Hoop Network" and *Yahoo*'s "Ball Don't Lie"), as well as
fandom tweets, in order to understand how fans use these platforms to produce communication and content regarding players, coaches, management,
and owners within the National Basketball Association. They examine how
these platforms are used to express dissent from the sports fan base.

Chapter 5 continues the discussion of televised sports, fandom, and social
media by examining the convergence of team identity formation and self-
disclosure via social media during live sport broadcasts. Shaughan A. Keaton, Nicholas M. Watanabe, and Brody J. Ruihley identify the major factors
involved with identity formation and self-disclosure, including fans' motives
for viewing sports and how they psychologically select and commit to specific teams. By analyzing social media discussions within sports fan online
forums, Keaton, Wattanabe, and Ruihley want to explore how sports fans use
social media during live broadcasts and what types of disclosures these fans
are making online.

In chapter 6, Matthew R. Collins and Danielle M. Stern examine how,
through agency, fans actively participated in the saving of the television
program *Community*. Utilizing two Web sites, the A. V. Club and the *Community* subreddit /r/community, this study analyzed the online fan discussions to explore how online fans react after a successful "save our show"
campaign and establishes how fans use their agency to save a show via the
Internet.

Similarly, Garret Castleberry's chapter 7 focuses on examining *contrasting fandom ontologies*[1] within online fan forums surrounding the popular HBO series *Game of Thrones.* By examining both television recappers and their subjective examinations of the series and the fan-based WinterIsComing.net, this study investigates how these modes of Internet fandom mirror the franchise property they reflect and how producers of the program do not have total control over fan communities and online forums. In essence, these fan communities contribute to cultural production and help to create legitimate and imitative archives.

In chapter 8, Michel M. Haigh explores the Facebook fan interactions surrounding the controversial interview by *Duck Dynasty* family patriarch Phil Roberston in December 2013. Haigh's study coded fan comments posted to *A&E*'s *Duck Dynasty* Facebook page immediately following Robertson's suspension from the program as well as the comments made after he was reinstated by *A&E*. The chapter discusses how the "us" versus "them" scenario could play out on the *Duck Dynasty* Facebook page and examined Facebook posts for an "us" versus "them" theme to see if fans felt a connection with the Robertson family.

Leandra H. Hernandez explores the ways in which fans interact with *Toddlers & Tiaras* via social media pages and the ways in which the *Toddlers & Tiaras* show interacts with its fans. In chapter 9, Hernandez critically explores what sorts of discourses are occurring about *Toddlers & Tiaras* on Tumblr, Facebook, and Twitter, and whether or not online discourse topics vary according to Web site and argues that fans utilize these online platforms to express their dedication to watching the series and, more importantly, debate the moral and ethical issues associated with "forcing" young girls into pageantry.

In chapter 10, Sabrina K. Pasztor and Jenny Ungbha Korn take on a study of *Talking Dead*, the companion, interactive talk show to the hit AMC program *The Walking Dead*. Pasztor and Korn use *Talking Dead* to illustrate this pioneering format of fan interactivity and the relationship among online social media and contemporary fandom regarding a hit television program. The chapter further explores the issue of redefining parasociality into a new form of technoprosociality.

In chapter 11, Alane Presswood and Steven Granelli utilize analysis of Twitter and Facebook to explore how fans cope with the cessation of a favored show and the loss of a significant, yet fictional presence in their lives—and how those behaviors consequentially affect networks and producers. Presswood and Granelli investigate four popular and cancelled programs, *Breaking Bad, Community, Firefly,* and *Buffy the Vampire Slayer*, to determine what happens within those active fan bases after their character-fan relationships are abruptly ended by networks.

Soap operas have long been standard fare for television programming throughout its history, and there exists a large body of research regarding soap operas and their fans. The advent of social media has had an impact on soap opera fans. In chapter 12, Marsha Ducey examines the relationship between soap operas and fans on Twitter, including how the soap opera producers use social media to engage fans and promote programming and fan discussions of each program.

In contrast to the dramatic flair present in soap operas, some programming seeks to mirror true life experience more closely. The Lifetime program *Army Wives* can offer viewers a glimpse into the real-life experiences and family dynamics of members of the U.S. military. In chapter 13, Darcey Morris analyzes the Lifetime sponsored *Army Wives* discussion boards in terms of female media consumers, the themes the real Army wives focused on within the online forums, and how the Lifetime network strategically utilizes the online forums for rebranding and targeting viewers' attentions, fandom, and behaviors.

In chapter 14, Michel M. Haigh and Shelley Wigley investigate the Paula Deen controversy and fan response via Facebook. Following Deen's fall from grace, Deen's supporters and fans rallied behind the beloved *Food Network* star, taking social media by storm to express their love and support for Deen and their discontent with the Food Network's decision. This chapter is focused on fan discourse on Facebook during the Paula Deen media crisis.

In chapter 15, Julia E. Largent and Jason Roy Burnett offer the reader an opportunity to explore how social media allows fans the chance to utilize platforms to perform as fictional characters from the popular HBO program *The Newsroom*. Drawing from aspects of identity theory, Largent and Burnett analyze fictional Twitter character accounts penned by fans and view fan involvement through the lens of performative identity.

Krystal Fogle focuses on the Internet arts site Pinterest as an outlet for fans to merge content and supply commentary for the BBC fan favorite *Doctor Who*. In chapter 16, Fogle uses rhetorical analysis to illustrate how *Doctor Who* fandom has progressed throughout television history, how these fans build group identity and perform online, and how fans have taken to Pinterest to bond over ancillary content. *Doctor Who* fans provide new texts via online communities like Pinterest and allow fans to use fantasy themes to speculate on plot and character development.

Correspondingly, throughout history fans have employed a variety of methods to interact with one another and to express their love for characters and programs. Cosplay, or costume play, is one method of fan interaction to show appreciation for characters and programming. In chapter 17, Laura Kane and William Loges investigate the world of cosplay through social media interactions between producers and fans. New media platforms, specifically Facebook and YouTube, have allowed producers the opportunity to interact with fans in a back-and-forth relationship and provide fans with new, never before seen or heard details of the programs. Fans can use this information in their cosplay, and this chapter analyzes these relationships through the encoding and decoding theoretical framework proposed by Edward Hall.

Kathryn L. Lookadoo and Ted M. Dickinson also explore participatory fan culture in their chapter 18 case study of the 2014 *Enter the Lodge* project, an online production of fan-fiction based on the 1990s cult classic *Twin Peaks*. The case study offers an analysis of how fans use Twitter to interact with other fans and participate in fandom to continue the narrative of the cancelled program *Twin Peaks*.

The final chapters in the anthology offer one final look to the past and how social media and programming will continue on into the future. In chapter 19, Cynthia W. Walker discusses the fifty-year history of fandom associated with *The Man From U.N.C.L.E.* and how social media has played a role in keeping this strong and active fan base of a historical program alive and well in the online fandom universe. Walker provides historical and current evidence of how media has sustained one of the oldest fan communities in our culture.

Finally, in chapter 20, Dan Faltesek brings the anthology to a close with a discussion and analysis on the future of social media usage and multiscreen advertising technologies. The chapter focuses on the fragmented media audience and the tactics media advertisers are faced with as the media system is glutted with choices for the media consumer. Faltesek argues media campaigns today assume a fragmented audience, and utilizes industry sales texts from Microsoft Advertising and texts from *Advertising Age* to explore multiscreen management and strategic manipulation of fandom communities.

Social media will continue to play a role in the fandom and fan culture surrounding not only television programming of the past, but the new programming of the future. This anthology discusses a wide variety of programming and the role social media plays within fan culture. It will be interesting to see how future scholarship beyond this anthology will continue to broaden the depth of research regarding fandom, social media, and television culture.

NOTE

1. Tom Gruber's shortened definition of *ontology* as a "specification of a conceptualization" suffices nicely for my brief use of the term (Gruber, Tom. 1992. "What Is an Ontology?" Retrieved online from Stanford's listserve. http://www-ksl.stanford.edu/kst/what-is-an-ontology.html.

Chapter One

The New Network

How Social Media Is Changing—And Saving—
Television

Ryan Cassella

Since its mass media debut more than half a century ago, television's continuous evolution has lacked a true revolution. As a dominant media fixture in the industrialized world, scholars and critics have continually put forth bold claims that fail to materialize each time TV is paired with a new high-tech companion. From the remote control, to the VCR, to the DVR, television's extensive string of technical developments has had varying degrees of long-term success. In the end, all of these technological advances have fallen well short of sparking a media upheaval that has promised—among many things—an open television content interface, a fully empowered television viewer, and an industry that caters to the tastes and wants of the individual.

Today, a new wave of technology is widely speculated to be television's long-awaited "next big thing." The interactive platform of Web 2.0 is making a profound impact on both the personal and professional lives of people around the world. Social networks—whether in the form of a celebrity Instagram feed or a revolutionary political tweet—are transforming modern communication. With over 75 percent of Internet users engaging with social media channels such as Pinterest and Facebook, it is no surprise that the television industry has embraced a technological movement that—unlike the many highly touted contraptions before it—is living up to the critical hype.[1]

In the following chapter, both the television creator and consumer are analyzed to demonstrate how social media will meet the bold predictions that deflated its technological predecessors. Applying a revised gatekeeping theory for the media producer perspective, I argue that social media's interactive nature and technological flexibility will continue to blur the line between maker and consumer as both parties adapt to produce a more collaborative and interactive television experience. Additionally, while a television creator must listen and react to his or her audience to be successful in a modern media setting, the unique communicative climate generated by social media allows media professionals to "retrain" their audiences as well. In this way, social media is reinventing the power dynamic between viewer and producer.

THE CONNECTED TELEVISION: A BRIEF HISTORY

Over the course of its technological evolution, television has passed through three critical "waves." TV's first wave came in the mid-twentieth century, when the device became the media hub for households in industrialized nations. Following the "natural life cycle model" proposed by Lehman-Wilzig and Cohen-Avigdor,[2] television raced through the initial stages of market penetration, growth, and maturation. During this period, there was a clear divide in the consumer-producer relationship, as a small group of media professionals created all of the content for a rapidly expanding baby boomer consumer base. Television's second wave came with the arrival of cable and satellite content providers. At this juncture, TV entered the "defensive resistance" stage of Wilzig and Avigdor's media life cycle model.[3] While the consumer-producer dynamic remained fairly rigid due to technological limitations, connectivity became vital to an industry fighting for consumers with more viewing options than ever.

Now, television is in the middle of its third wave, a "new media evolution" of connectivity that has given the user unparalleled choice in their entertainment options.[4] While the traditional means of watching television via a TV set remains the dominant practice for audiences, an increase in users seeking out alternative means for watching content points to a new era for the medium.[5] The span of consumer needs met by a "connected" television is unparalleled: the medium's content has never been more diverse or engaging, the user's exposure is supported by a variety of convenient viewing mediums, and the social context is one of an engaged and collaborative media network. It is in such a fertile and vibrant media environment that social media has grown.

THE CONNECTED CREATOR: A REVISED GATEKEEPING PERSPECTIVE

In television's "golden age," the content creator was a gatekeeper in the most traditional sense. The producer-consumer relationship during broadcast TV's mid-twentieth-century reign was a one-way line of communication, where a small group of content creators were in complete control of mass media content. As cable and satellite providers entered the television industry, the power dynamic between the content creator and media consumer began to shift. TV's "third wave" has continued to chip away at the media creator's isolated gatekeeping position. Now that the television viewer has altered the power dynamic with mass media conglomerates and professional content creators, the media professional's gatekeeper role has given way to a more collaborative, two-way media relationship. For social media, this progressive mass media environment makes platforms like Twitter and Facebook a promising middleman between the connected creator and empowered consumer.

As the modern media consumer becomes more of a "broadcaster," media creators are looking to optimize the "openness and transparency" of social media channels.[6] For example, Heineken's media-savvy sponsorship of the U.S. Open showcased the power of social platforms in connecting television programming to its viewers. Utilizing Twitter to post video highlights of matches, the Heineken brand worked seamlessly with ESPN's and CBS's live coverage of the event.[7] Tennis fans used Twitter to give their opinions on the coverage of their favorite stars, while media creators used the "immediacy" of the interactive channel to promote their live content offerings. Heineken's successful media campaign embraced the changing dynamic between producer and consumer and used the tools provided by social media to reach viewers in a more personalized fashion.

While today's television creator is no longer an omnipotent media gatekeeper, they now have the opportunity to be something even more valuable—the omniscient communicator. Jacobson's analysis of the news program, *The Rachel Maddow Show*, gives evidence of this evolving media creator.[8] Monitoring the program's official Facebook page for user activity, Jacobson found social media content influenced the show's television content.[9] This study supports the idea that social media channels are creating new lines of dialogue for the empowered television consumer and the media creator is listening intently. The professional media producer will be tasked with continuing the conversation with an increasingly involved audience via numerous social media channels.

THE SOCIAL MEDIA REVOLUTION IN TELEVISION

Twitter, Facebook, Foursquare, Pinterest, Shazam—the list of social media platforms in today's digital media environment is astounding and ever changing. For television content creators, finding ways to utilize a variety of these platforms for programming is an ongoing experiment. For CBS's daytime television lineup, Twitter was utilized, for example, as a promotional vehicle for viewer contests and giveaways.[10] The cable network SyFy took the Twitter platform one step further by not only using it as a promotional tool, but also as a way to broaden the social network of its online gaming community.[11] In a variety of television environments, TV professionals are making a critical creative jump by adopting social media channels to extend the reach of a television product.

The growth of social media usage in television shows no sign of slowing down, but implementation remains a trial-and-error process.[12] HGTV became an early adaptor of Pinterest, a photo sharing social media hub, when the network noticed a heavy pattern of traffic from its home page to user Pinterest pages. *Jimmy Kimmel Live* created new connections with fans via Shazam, an interactive social channel that works with popular music.[13] Examples such as these are evidence that social media channels can strengthen and grow a television audience base when they are utilized in a creative and dynamic manner. The formation of "brand communities"—where a television program is only one aspect of a comprehensive media experience—is rapidly becoming a television professional's long-term goal.[14]

As the value of social media in television becomes increasingly apparent, established social media platforms have begun to compete for the highly coveted space that works alongside original programming: the second screen. Facebook and Twitter, in particular, have made an aggressive push to be television's supplementary media destination.[15] Facebook shares user data with big networks, as industry leaders attempt to make specific connections between social media activity and TV consumer behavior.[16] Meanwhile, Twitter has joined forces with Nielsen to create a revamped TV ratings system that incorporates social media activity.[17] Media outlets will also rely on their own social platforms in the ongoing competition for the second screen. Nat Geo, for example, has focused on creating interactive "brain games" that build organically from its television content without the use of a Twitter or Facebook platform.[18] While today's fragmented audience base may not resemble the sedentary viewers of television's golden era, the goal of the media professional remains the same—strive to create "event TV."[19] Not every program boasts the built-in allure of live content, but a robust second screen experience can make any piece of media a must see event in real time.

BEYOND THE SECOND SCREEN

In his book *The Thank You Economy*, Gary Vaynerchuk discusses the vast potential of social media in all industries. Future survival for any business, Vaynerchuk argues, will hinge on the ability to establish personal connections with individuals via social channels.[20] The look of social media may change—particularly with consumers moving more toward mobile devices—

but the personalized connections established by social channels drive TV's future.[21] In fact, the next generation of media consumers is growing up in the on-demand age where multiplatform viewing is second nature.[22] For the media professional, this generational divergence in media consumption behavior will be an additional hurdle, as television programming will have to meet the needs of different age groups with very different expectations of what their television experience is supposed to be.

The continued growth of social media platforms, both nationally and globally, points to a bright future for the numerous outlets of social television.[23] Social platforms will continue to be part of the early stages of program planning and a critical channel of communication for the media professional during the entire creative process. Real-time feedback via social media is a resource that can make TV programming an adaptable medium, while enhanced consumer information can drive modern television advertising. At the helm, there must be the new media gatekeeper—one that realizes the media landscape has changed forever and is dedicated to using social media to build a better product.

RESEARCH AND METHODS

Through a series of in-depth, semi-structured interviews with social media leaders in television, this chapter looks at pertinent issues facing the TV industry today:

- What is the role of social media in television?
- How can social media be used to enhance the television viewer's experience?
- What does the future hold for the relationship between TV and social media?

The chapter also seeks to better understand the trial and error process of developing and implementing social platforms alongside traditional media. Via the perspective of television's new media gatekeeper, the themes of adaptation and connectivity are central:

- Which social media platforms are best suited for television's third wave?
- What content creates the most engaging second screen experience?
- Where will television professionals find their ideal social network for the sought-after "virtual circle" of media self-sufficiency?

Starting with his professional connections within the industry, the author utilized snowball sampling to get a comprehensive take on social media in television. Because the individuals needed for the research constituted a very specific group, networking to increase the study's sample size was critical to finding a variety of perspectives that cover different television content and formats. The study used eleven in-depth interviews with the goal of reaching "theoretical saturation" in the data.[24] Interviews were conducted over the phone with the researcher taking detailed notes on a series of industry-based questions and averaged a length of forty-two minutes. The research subjects held a variety of professional titles—all of which were directly related to the development and execution of social media for their respective television networks. The networks represented by interviewees covered a variety of content including live sports, children's programming, and reality television.

The study's qualitative data was analyzed using a customized coding approach. Elements of the Grounded-Theory Development were adopted for the study; however this work's objective did not culminate in a formal theory.[25] After developing the study's questionnaire from the six broader research questions above and gathering responses, the textual data was unitized by question number and interview subject. Category coding fell into industry and media themes that are laid out in the following results section. The coding process allowed for a linear presentation of the data that was used in place of the study's original research questions. Overall, the coding process was fairly seamless as interview subjects were well-versed in the interview topics and expressed their ideas and opinions in a clear manner.

RESULTS AND DISCUSSION

Social Media and TV: The Approach

While all interview subjects endorsed the idea that social media should be a user-focused endeavor, the tactical approaches they described varied greatly. One major philosophical rift between the interviewees was how social media should be applied in television programming. Some research respondents believed social media was a tool for interaction, while others viewed it as a vessel for information. Those in the interactive camp believed consumers needed to be engaged by social content in order to continue the conversation from the network's original programming. Subject A described the process as a "two-way conversation that offered a unique point of view." When the audience felt they were a part of an ongoing dialogue about their favorite show or character, he reasoned, a media creator could extend his or her brand beyond the confines of a weekly episode or live event.

For those focused on delivering information to their audience via social media, providing "insider" content was viewed as a way to forge stronger bonds with viewers. Subject F described it as being the "eyes and ears for a fanbase when they can't be there." To a degree, the debate between social media's use as an information diffuser versus an interactive tool harkens back to an earlier technological battle in television where the old guard's one-way line of communication butted heads with the new consumer's two-way line of communication. Whichever technique the research group chose, all participants agreed that social media in television should be anchored by a user-centric approach where the audience was always the first thing on the minds of TV's social media producers.

Another important variable in television's approach to social media is timing—are social platforms being used before, during, or after a program? Those espousing the merits of social network use before content airs described a process where word-of-mouth gained significant momentum on social networks, leading to "event television" moments.[26] One interviewee referenced the success of the television movie *Sharknado* as a prime example of social media generating significant pre-program interest for an otherwise insignificant piece of media. "Free advertising" was the term used by Subject C who interpreted social channels as a convenient platform for putting one's audience to work. In the viral culture of the digital era, he reasoned, consumer buzz could spread rapidly and increase interest before TV content was released.

Those championing the in-show use of social media saw it as a way to make the audience part of the show. The inclusive nature of social networks, Subject B noted, allowed consumers to feel like they were a "part of something bigger," thereby make a popular media event "something out of the ordinary." Several research participants mentioned the reality show *The Voice*—a singing competition program—as a show that harnessed social's power during its airing in a meaningful and organic way. Subject H mentioned recent experiments with Pinterest and Snapchat as effective social platforms for digital content during her network's shows. Overall, user interaction was the primary goal of TV producers creating social content during a program's airing.

Meanwhile, interviewees who cited the benefits of post-show social media activity frequently mentioned *Late Night with Jimmy Fallon*. Subject D described the show as the "gold standard for a model of content delivery that is outrageously good." The program's success at distributing novel and engrossing content after a nightly broadcast through social channels was viewed as vital to the show's commercial and critical success. More original content is landing in digital spaces outside of the traditional television network model as programs embrace the value of a robust multiplatform presence. While subjects proclaimed the merits of social media's use in each of the three timeframes, a majority agreed the greatest success was usually found by having a comprehensive social presence that covered the before, during, and after. AMC was referenced as a network that successfully utilized social media in all three time periods by surrounding its hit shows with engaging digital content, including a post-show driven by social activity.

When leveraging a program's content for social platforms, several media producers mentioned the unavoidable link between the media producer and marketer. As "stewards of a brand and pop culture," Subject E described the relationship between media maker and promoter as vital to a program's success in the digital era. When both production and marketing departments at a television network "responded to the conversation in social," he argued, a network had a better opportunity to turn social media activity into additional viewers as well as strengthen its ties with current fans. In both the production and marketing worlds, social media may not replace the core business of television, but it is an increasingly indispensable tool for connecting to one's audience.

While seen as a transformative device by most interview participants, several media professionals felt a successful approach to social media in television always began with a foundation in sound storytelling. In turn, a robust social channel supported and enhanced the "engageable moments" media professionals strived to create. Terms such as "organic" and "seamless" were used frequently by interviewees in their description of a successful marriage between TV and social. As Subject B added, a media creator cannot build additional content for social channels with the primary goal of making content "viral": "You can create a place [for content], but it doesn't mean people will go there." Maintaining a commitment to high quality, authentic content was the limit of a media professional's control. Beyond that, it was up to the viewing public to decide how far digital content could be stretched.

For many, social channels were a digital playground where a successful storyteller had the room to experiment and take calculated risks. Subject I stated that "at the end of the day, it's entertainment—if you can do it in more than one way, all the better." From this vantage point, it is clear that a gifted media professional can enhance a narrative with social. However, without the creative and technical skills that successful TV professionals have honed in their own industry, the bright possibilities within social media cannot be accessed. For those talented creators looking for the newest channels to expand their story, the competition can be daunting, as social media is a portal available to a myriad of gifted storytellers waiting to spin their own tale. As Subject K summarized, "The story is everywhere, if you're not telling the story, someone else will."

Social Media and TV: The Application

Incorporating the strategic approaches described by the study's participants began with today's two dominant social media platforms: Facebook and Twitter. For many subjects, the number of Facebook members—well over one billion—was too large of an audience size to ignore. No matter the content or audience a network was targeting, having a Facebook presence was viewed as a given in the digital age. For Twitter, its own impressive army of users was bolstered by its TV-friendly interface. Subject H called Twitter an "instant medium" that could supplement a television experience in real time. In distinguishing the best way to use the two dominant social platforms, Facebook was deemed the "less is more" medium, while Twitter was a "volume-based" asset. For producers involved in live programming, Twitter was a necessary tool for capturing audience feedback and engagement. Facebook, on the other hand, was utilized more for building long-term campaigns with multimedia content and engaging fans between program airings. Aside from the creative and networking advantages Facebook and Twitter offered TV producers, the application of social platforms also had a more cursory purpose. Like businesses in a variety of industries, television networks that ignored social integration ran the risk of looking dated and out of touch with the modern consumer.

Using Facebook and Twitter as ways to connect with consumers has evolved for media creators—especially with regard to live content. The first step of having a social media presence often came in the form of television graphics. A great deal of contemporary programming now has embedded Twitter hashtags to promote social conversations, while the "lower third" region of screens is home to social "tickers" or "crawls" that provide timely user feedback and additional program information. Now, some contemporary programming is moving from supplementary graphics to a physical presence within programming. Several producers mentioned the implementation of "Twitter walls" into their live programming. By putting social media channels on prominent display, show creators are anchoring programs around social media, which a show's talent and viewers are both compelled to address.

To stimulate the competitive side of an audience, producers have employed on-air trivia games and other interactive challenges for their digital fans. Subject B described a socially driven game of tic-tac-toe that was a hit with his audience. Several sports producers referenced debate shows, where stoking the competitive fire of viewers via social channels was a very fruitful strategy. A few subjects mentioned programs that went as far as dedicating entire segments to social media. In these instances, programming was driven by social "calls to action," where fans submitted questions or multimedia content for a show's talent or hosts. From a minor visual presence to being a major part of a program's content, the process of social integration was on full display in the study's data.

In the new media landscape of convergence, media professionals are more comfortable than ever in handing a show's reigns to its audience. Subject B took the socially entrenched show segment one step further by

building parts of a program around user created content and feedback. Another show creator had a successful segment where fans used social channels
to share pictures and videos related to a common theme. Subject I noted that
programs revolving around user-content must take the time to recognize the
fans responsible for their valuable role. Making an on-air mention of a viewer's Twitter or Instagram handle can strengthen that "emotional" bond between a network's brand and its audience, while giving a viewer credit for his
or her work can be a strong incentive for future fan submissions.[27]

Collecting and organizing the constant stream of social submissions in a
productive way has been a challenge for networks. A majority of subjects
described a network Web site or smartphone app that functioned as a forum
for uploading and curating social content. One producer even described a
microsite linked to the network's main page that had the sole purpose of a
social media hub. Another interviewee mentioned Tumblr as a centralized
location for interacting with network talent and posting multimedia content.
Television-specific social platforms Viggle and tvtag—which reward consumers for "checking in" to their favorite programming—were also referenced
as potential digital homes for a network's vast amount of social activity.
Regardless of a network's primary home for its social media, having a strong
digital presence hinged on knowing one's audience. Subject K added there is
no "one size fits all in social" as each audience desires a unique digital
experience and each social platform delivers the experience in a different
way. The ongoing relationship between TV and social is a work in progress
for media creators who continue to view social platforms as a testing ground
for television's digital future.

Social TV: A Trial and Error Process

As media creators strive to maximize social media's potential in the world of
TV, several challenges have arisen. From a technical standpoint, integrating
social content has been a steep learning curve for production professionals.
Whether it was representing it graphically or making it a part of a show's
physical set, social media adds another layer to the production process. Subjects who were a part of the early social adoption population discussed the
increased pressures on graphic artists, show directors, and other personnel
faced with the additional responsibility of social media implementation. For
program's relying on social activity to drive the conversation, the overall
quality of viewer content was a challenge as well. Subject B mentioned the
lack of production value in user-created content and the need to create "social plants" in circumstances where engrossing media couldn't be obtained.
In these dire situations, media professionals created social profiles to pose as
regular viewers and contributed content to keep a sluggish program moving.
As social becomes more of a commonplace element in contemporary TV,

improved technology and a larger amount of a network's financial resources are helping relieve the production issues that come with user-based digital content.

Creative control was another major challenge for media professionals in the digital age. Allowing an audience to be part of the conversation can lead to negative comments and other "trolling" behaviors. Described as a "snark fest" by Subject J, media professionals can find it difficult to navigate through feedback that can turn negative in a hurry. Having staff members that filter social feedback for TV programs only tends to be part of the problem, as social users are free to continue creating negative content on the social channel of his or her choice. Often, a creative professional's best avenue of response is in the content they create. Network talent also posed a problem for media creators when they released controversial content on social media that was recycled by an engaged fan base. The controversial media could be as simple as an opinionated tweet that generated emotional responses from the audience or a more severe item that used politically incorrect or coarse language.

Ceding a level of control to one's audience takes a leap of faith, but the networking potential of social promoted by media writers like Vaynerchuk[28] is more than enough to win over a stubborn show runner or ease the pain of a talent's ill-advised tweet. As Subject E observed, coworkers who were unsure about social media's clout in television were often swayed by the rapidly improving data supporting social's impact on viewer habits and actions. Moreover, sharing control of content is not the same thing as relinquishing control. For example, Subject H worked for a network that initially outsourced its social media efforts. After lackluster returns, the network took ownership of its social activity and refocused its content in a far more effective manner by listening to the needs of its niche user base and making them a part of the social conversation.

Legal and privacy issues were an obstacle for new media professionals as well. Subject K worked on youth programming and described the care that went into her social strategies. The highly used "calls to action" to drive social traffic were not an option due to the legal constraints of the Children's Online Privacy Protection Act (COPPA). Any avenue of social engagement had to be prudently crafted as direct "sells" to an audience of minors were prohibited. Subject C recalled a social campaign where his audience was asked to involve their friends and family in multimedia submissions. The promotion received very little content and the network learned that even highly engaged fans had their limits when it came to the amount of personal information they were willing to share. As one producer reasoned, in most instances, a program's worst-case scenario with social media was an approach that did not generate a significant level of user activity or interaction.

However, legal and privacy issues raise the stakes even higher for media professionals responsible for the integrity of their brand.

Returning to the user-centric focus of the social media producer, several interviewees described situations where social content failed to connect with its audience. Subject D recalled an incident where a social "call to action" was rebuffed by an audience that felt the network had strayed from its core content. The social content involved a fashion critique of a star athlete's wardrobe. The producer viewed it as a learning experience and a tough reminder that media outlets should "know [their] lane" when delivering digital content. Subject F detailed a time period where her network added a new line of content to its regular programming and received resistance from fans on digital platforms. In order to win over her audience while expanding the network's user base simultaneously, the producer collaborated with the digital team attached to the incoming content and found common ground between both consumer groups.

Several subjects described scenarios in which social campaigns lacked the proper incentives to spur user activity. In one situation, a successful in-show sweepstakes built around a program's talent spawned another socially driven giveaway that failed to have the same impact. Subject A was involved in both campaigns and posited that the unsuccessful approach may have lacked a desirable reward or a strong connection to on-air network content. Subject G reinforced this idea that social content needed to have some kind of "payoff or payback" that "made it about [the audience]." The payoff could be in the form of a physical prize or rewarding content, as long as it resonated with the viewer and properly incentivized him or her to further engage with the network's programming. When executed properly, a network's digital content achieved the "virtual circle" where the producer fully integrated the emerging digital world of social media with original programming. [29]

A CREATOR AND THE AUDIENCE: AN ONGOING EVOLUTION

For the media creator of the digital era, social media is part of the daily routine. Its instantaneous nature makes feedback a constant and weekly social metric reports are part of the growing data set a TV producer uses to evaluate a network's programming. Working closely with the industry's growing digital creative teams is the present reality for TV professionals. The study's subjects described their evolving role as "audience friendly" content creators, who used social channels to make viewers a part of the creative process. Several media professionals referenced the collaborative "brand community" phenomenon, where media creators and consumers work together to build "cultural capital." [30] User loyalty was a prized commodity for interview subjects in an increasingly fragmented media world. If a more personal connection with an audience kept them in the fold, media creators were willing to put in the time and effort to make it work.

In attempting to turn social activity into increased viewership, subjects detailed the process of making social part of the larger storytelling process. Subject C spoke of "B" characters on a hit television series that were given a digital space to grow and flourish with alternative story lines. Another media professional argued that social platforms were an ideal space to capture more personal stories that connected to a network series or program. Subject D described a feel good story about a small town athletics team that could be linked to a national sporting event. In this way, the "emotional contagion" that motivates audiences to seek out certain types of media can be bolstered by social's contribution to a larger story arc.[31] Often airings of mass media can lack a personal touch and social channels are aiding networks in addressing the critical emotional component of good storytelling.

Audience feedback on social channels also functioned as a free focus group for show runners. Subject C admitted to storylines—even a show's entire creative development—being altered by social media feedback. Drawing on the incentive of fan involvement, a network even ran a successful social campaign where the audience chose the ending to a show's episode. Another interviewee referenced a notable production colleague who conversed with his audience on social channels while his shows aired for real time feedback. Similar to the sociological musings of Malcolm Gladwell in *The Tipping Point*, the digital specialist mentioned the importance of connecting with "social influencers" via social media.[32] These members of the new digital focus group are more socially active than the average media consumer and can have a significant level of authority over their peers. On the opposite end of the spectrum, there are far more passive audience members. Subject J described them as individuals who "just want to sit and relax [and] be told a story." Beyond the feedback on program content, Subject C mentioned the professional validation that came through digital channels as well. Having users complement a program's use of social media was confirmation for the social director that his efforts on the digital front were making a difference.

Modern media creators were willing to give consumers a voice in the creative process, but they have also found professional empowerment in social media as well. The current uncertain state of the traditional television market has led to less risk-taking in program development. On digital platforms, the media professional still has a place to exercise creative urges that may otherwise not see the light of day. Additionally, the two-way communication afforded by social channels is the new media gatekeeper's personal portal into the hearts and minds of their audience. Television consumers can—to a certain extent—be conditioned by a media professional's use of social for content promotion and distribution. When fans adopt the social channels utilized by a program and make it a part of their media viewing routine, the modern media creator alters the power dynamic in a substantial way. An empowered TV audience may be the reality for the foreseeable future, but social channels can also be used by media creators to curb the sometimes overwhelming momentum of user control.

The theme of content expansion in TV's use of social media also covers the business end of the industry. The level of audience engagement that social channels can provide is an inherent power that brands want to capture. Recent business deals, such as Instragram's partnership with Omnicom[33] and Tumblr teaming with Viacom[34] show the economic potential of popular social platforms. As one subject argued, because television and its potential sponsors are dealing with the same challenges of reaching an increasingly fragmented audience, media professionals can empathize with brands while using their increasingly sophisticated social metrics to win them over on the power of social media. In an effort to broaden the impact of "dynamic marketing" with social media, Subject C described the process of integrating a brand into a show's narrative. For one sponsorship campaign, the audience was given a glimpse into what a show's characters did after leaving their jobs. The characters used a certain automobile for their post-work adventures which, in turn, sponsored the digital shorts. Just as stories and characters can be extended with digital channels, the success of social campaigns support the idea that brands can broaden their reach with the same captivating content.

In her discussion of this topic, Subject K described an eye-opening tour of a major food chain's headquarters. Within the company walls was a well-outfitted production group capable of creating fresh and timely media content. The interviewee saw this as the future of branding, where companies with no media background would build their own personal content houses. For TV professionals, future financial success will depend on meeting the heightened demands of brands that understand the incredible power of unique and engaging content. Negotiating the visions of external sponsors in addition to internal network sales teams is a major challenge, one subject noted. However, as Subject E pointed out, "Our job is to package our [content's] assets to give a brand what they want," while also proving the effectiveness of campaigns built on social media. The current state of social branding on television is still in a developmental phase, yet numerous subjects saw the relationship improving with time as advertisers and television professionals move toward a more cohesive vision of social media's critical place in both industries' financial survival.

THE FUTURE OF SOCIAL AND TV

Unsurprisingly, the TV professionals working closest with social media do not see the relationship dissolving anytime soon. Subject I went as far to say that "social media will be synonymous with TV." How the continued pairing will evolve, however, brought on a variety of predictions. For Subject E, it was all about the numbers. As television and social continue to hone their mutually beneficial creative and financial relationship, he argued, social media platforms will refine their ability to measure user activity in meaningful ways. Videology—a video advertising company—and its new partnership with Nielsen is a recent example of the advertising world's desire to hone its targeting ability in the digital arena.[35] For the slew of social media platforms jockeying for the eyes and ears of the world, metrics will be the key to a solid financial foothold in the future. When a social platform used by a network can win over prospective corporate partners with hard data, it will strengthen its status in the world of television. This challenging future for social channels could lead to more financially viable platforms phasing out weaker competitors and shrinking the currently saturated social media marketplace.

Another subject focused on social's growing user base. In a modern television environment, Subject B believed the audience voice could be "deluded or watered down." While this may limit the voice of the individual participant, the interviewee believed the power of fan communities could increase. These digital groups, he argued, would accelerate the fragmentation of the media audience, leading to more niche groups that media professionals would have to address. The reality show *Duck Dynasty* was referenced as an example of a digital fan community's power. After a cast member on the show was suspended by A&E network for controversial comments, some fans of the show took to the digital space to air their grievances on the network's decision. The suspension was lifted shortly after it was handed out, with fan pressure being cited as a possible reason. Subject K spoke on the incredible value of the growing fan voice aided by social media, summing it up with, "Let the villagers help!" Increased viewer feedback, she added, could turn into more of a collaborative relationship between media maker and consumer, with the audience even taking the role of co-creator in certain media endeavors.

From a technological perspective, Subject B described the increasingly shorter "life cycles" of new communication tools. The current leading platforms like Facebook and Twitter would more than likely change, he argued, leaving television with different digital partnerships. Subject D spoke of social's popularity on portable devices like smartphones and tablets, even calling mobile phones the future technological "center of our lives." Younger media consumers, in particular, have grown the market for branded content apps provided by networks such as PBS and Cartoon Network. DreamWorks

Animation is even developing a branded tablet for the new multiplatform savvy generation.[36] As integration between social media and television progressed, the media professional stated, TV would continue to move to alternative screens as well. The "seamless experience" offered by a media product that blurred the concept of a first or second screen would change the very definition of television. TV—or whatever it may be called in the future—would be a multifaceted media source that, as Subject K described, could "attack from all angles with material." At this juncture, screens would no longer be numbered and audience engagement would become the primary concern of the media professional.

Subject J was not ready to crown social media as the true television revolution. Stating that his job included the need to "see past the buzz, clutter, and snake oil," the digital executive viewed social as a valuable tool, but stuck with a perpetually limited user base. Content would always be king, he reasoned, and creating a multitasking friendly media experience could easily move from aiding the conversation to becoming a distraction. Social media is "*a* thing," he reasoned, but "not *the* thing." Another television professional drew a clear line between taped and live programming. For him, episodic television and other pre-produced content was already reaching its limit with social integration. Live television, however, still had a great deal of potential when it came to its relationship with social media. Predictive gaming was viewed as one future area of growth for live TV and social platforms. In fact, the concept of "game-ifying" content or introducing loyalty programs was described as a critical tool for engaging a growing number of younger multitasking consumers. As Subject A acknowledged when watching programming with his kids, "I don't watch linear TV." For him, being part of the new digital media lifestyle did not just mean changing the way one used television, it was also about changing the very idea of what one thought television was.

On the topic of media creation, many subjects saw social platforms as the destination for increasing amounts of original content. One interviewee believed the growing presence of social in television meant media creators would rely less on user content and contribute more of their own material to digital channels. In fact, the continued melding of social and TV has led to networks founding their brand identity on social media. Sean Comb's recently launched Revolt music channel and HLN's network rebranding as a social media news network show the extent to which social media has become a part of television's makeup.[37] Subject E took this concept one step further, saying a dominant social platform like Twitter could be the "next Netflix." Instead of continuing to integrate with television networks, the media executive introduced an alternative future where a social media platform became its own content creator and distributor. While media professionals may not a share a unified view of TV and social media's future partnership, their excitement and desire to be a part of the evolving media relationship was clear. Moreover, the social media producers of television were ready to take responsibility for helping the marriage between social and television reach its full potential. Even with the demands of a fractured audience, prospective sponsors and a diverse group of work colleagues—the TV professional was primed for media's uncertain digital future.

CONCLUSION

This study's look at the relationship between social media and TV paints a picture of two entities that grow more and more intertwined with each passing day. In the digital era, the ongoing collaboration between the two media forms is critical to survival of both entities. While research respondents shared a level of uneasiness with television's future, no one was ready to say goodbye to one of the world's most dominant modern media devices. In fact, while media scholars and television critics have foretold of a fractured world of content where the glowing box cannot keep up, the hard data continues to tell a much different story. A recent study by Nielsen comparing the broadcast and digital spaces of media content reinforces the glaring gulf between the two worlds. In the United States, there are approximately 283 million monthly television viewers compared to 155 million online viewers. Even more convincing is the amount of content being consumed in each space: an average of 146 hours a month for TV versus approximately 12.5 hours a month split between mobile and Web.[38] For individuals interviewed in this study, network television seems to be in a fairly stable place, which leads to the question: If the television industry is thriving, why should we care about its relationship to social media?

As always, a statistic can only tell you a part of the story. One major reason traditional television continues to thrive is that its digital counterparts lack content equivalency. In other words, digital platforms cannot yet offer an audience all of the great content they desire in a timely fashion. However, the rising popularity of "over the top" devices such as Roku and Apple TV, in addition to digital distribution platforms like Hulu and Netflix, is altering the media machine at a rapid rate. Faced with the changing media climate of emerging digital platforms and "cord cutting" consumers, networks are pushing more content to digital platforms than ever before. While the glaring discrepancy between the television and online content does not point to a digital revolution on the horizon, TV has a profound head start in both infrastructure and viewer loyalty over emerging digital channels. Most importantly, the money has not caught up to the online content space. In fact, more glaring than the user statistics of TV and online media is the financial size of each industry: $74 billion for television and $3.5 billion for digital.[39]

Here is where TV's integration with social media finds its significance: Due to its innate communicative openness and technological flexibility, social is the necessary bridge between traditional television and a growing digital space.

As a natural companion to various types of TV programming, social media is a language creative professionals feel comfortable speaking. It is a language that gets them closer to their viewers, while allowing them to expand their creative visions in an industry with otherwise limited space and resources. For now, the relationship between social and TV works because both need each other—social media has found some solid financial footing and media industry relevance, while TV has established a novel avenue for content and a new way to reach a growing digital consumer base. As the digital generation narrows the gap between TV and digital's media market shares, there may come a day where the synergetic relationship between television and social media has run its course. Until then, TV professionals will continue to utilize the incomparable power and reach of social media.

Know your audience and meet their needs—it was the mantra preached by the study's participants. For them, social media was the platform used to push their user-centric strategy to new heights. In all industries, in fact, it can be argued that understanding and serving one's customer is the primary goal. When any business encounters great change that threatens its very survival, they face the same choices detailed by Wilzig and Avigdor's media life cycle model: adaptation, convergence, or obsolescence.[40] For the study's subjects, adaptation in the form of social media integration was being embraced. In fact, the dynamic pairing of social and TV could be taken one step further and viewed as an example of media convergence. Much like the debate over the importance of television's "second screen," the reality of integration versus convergence could be viewed as a semantic debate with little practical reward.

Like the study's participants, skilled professionals in a variety of fields face a world where technology continues to move at an accelerated pace— what was relevant today is often irrelevant the next. The modern worker is being called upon to be an apt student, an individual willing to learn new skills and adapt at a frenzied pace. The television professionals of today provide some remarkable insight with their open attitude toward their own emerging digital world. In their candid narratives of social's gradual implementation into the evolving world of television, adaptability and open mindedness are characteristics that override any other valued work trait such as creativity, work ethic, or individuality. It is this fresh and optimistic perspective that transcends the television industry and provides a sound blueprint for any industry to follow when facing its own great unknown.

NOTES

1. Andreas M. Kaplan and Michael Haenlein, "Users of the World, Unite! The Challenges and Opportunities of Social Media," *Business Horizons*, 53 (2010): 59–68.

2. Sam Lehman-Wilzig and Nava Cohen-Avigdor, "The Natural Life Cycle of New Media Evolution: Inter-Media Struggle for Survival in the Internet Age," *New Media Society*, 6 (2004): 707–30.

3. Lehman-Wilzig and Avigdor, "The Natural Life Cycle of New Media Evolution," 712.

4. Mark Milian, "YouTube Exec: We're Heading for 'Third Wave' of TV," *CNN*, February 1, 2012. http://whatsnext.blogs.cnn.com/2012/02/01/youtube-third-wave/.

5. Gillian Doyle, "From Television to Multi-Platform Less from More or More for Less?" *Convergence: The International Journal of Research into New Media Technologies*, 16 (2010): 1–19; Teg Grenager, "Revolution in a Box," *AdWeek*, October 25, 2010. http://www.adweek.com/news/advertising-branding/revolution-box-103609.

6. Mark A Urista, Dong Qingwen, and Kenneth Day, "Explaining Why Young Adults Use MySpace and Facebook Through Uses and Gratifications Theory," *Human Communication*, 12 (2009): 215–29.

7. Vindu Goel and Brian Stelter, "Social Networks in a Battle for the Second Screen," *New York Times*, October 2, 2013. http://www.nytimes.com/2013/10/03/technology/social-networks-in-a-battle-for-the-second-screen.html?pagewanted=all&_r=0.

8. Susan Jacobson, "Does Audience Participation on Facebook Influence the News Agenda? A Case Study of The Rachel Maddow Show," *Journal of Broadcasting and Electronic Media*, 57 (2013): 338–55.

9. Jacobson, "Does Audience Participation on Facebook Influence the News Agenda?" 350.

10. Paige Albiniak, "Daytime Gets More Than Just One Life to Live," *Broadcasting and Cable*, May 13, 2013. http://www.broadcastingcable.com/news/news-articles/daytime-gets-more-one-life-live/114459.

11. Mike Shields, "SyFy, Facebook Team Up on Social Gaming," *MediaWeek*, October 5, 2009. http://www.adweek.com/news/technology/syfy-facebook-team-social-gaming-113656.

12. Wayne Friedman, "Younger Demos Dislike Social Media Displays on TV," *Media Daily News*, October 24, 2013. http://www.mediapost.com/publications/article/211986/younger-demos-dislike-social-media-displays-on-tv.html.

13. Jeff Baumgartner, "Shazam Gets Interactive with Jimmy Kimmel Live," *Multichannel News*, October 1, 2013. http://multichannel.com/news/content/shazam-gets-interactive-jimmy-kimmel-live/357386.

14. Kaplan and Haenlein, "Users of the World, Unite!" 64.

15. Goel and Stelter, "Social Networks in a Battle for the Second Screen," 1.

16. Todd Spangler, "Facebook Is Sharing TV Chatter with Networks, but What Does It Really Mean?" *Variety*, September 30, 2013. http://variety.com/2013/digital/news/facebook-is-sharing-tv-chatter-with-networks-but-what-does-it-really-mean-1200681820/.

17. Michael O'Connell, "Nielsen and Twitter Partner for Social Media TV Ratings," *Hollywood Reporter*, December 17, 2012. http://www.hollywoodreporter.com/live-feed/nielsen-twitter-partner-social-media-403360.

18. R.Thomas Umstead, "Next TV: Nat Geo Building Interactivity within Original Content," *Broadcasting and Cable*, March 21, 2013. http://www.broadcastingcable.com/news/technology/next-tv-nat-geo-building-interactivity-within-original-content/49849?rssid=20065.

19. Liz Giuffre, "The Coming of the Second Screen," *Metro Magazine*, Winter, 2012. http://connection.ebscohost.com/c/articles/79451059/coming-second-screen.

20. Gary Vaynerchuk, *The Thank You Economy*. New York, NY: Harper Business, 2011.

21. Mark Walsh, "Most Online Time Spent on Smartphones, Tablets," *Online Media Daily*, September 5, 2013. http://www.mediapost.com/publications/article/208569/most-online-time-spent-on-smartphones-tablets.html.

22. Brian Stelter, "Same Time, Same Channel? TV Woos Kids Who Can't Wait," *New York Times*, November 10, 2013. http://www.nytimes.com/2013/11/11/business/media/same-time-same-channel-tv-woos-kids-who-cant-wait.html.

23. Rory Jones, "Facebook Users in the Mideast and North Africa: 28 Million Daily," *Wall Street Journal*, October 1, 2013. http://blogs.wsj.com/middleeast/2013/10/01/facebook-users-in-the-mideast-and-north-africa-28-million-daily/.

24. Barney G. Glaser and Anselm L. Strauss, *The Discovery of Grounded Theory: Strategies for Qualitative Research*. Piscataway, NJ: Transaction Publishers, 2009.

25. Glaser and Strauss, *The Discovery of Grounded Theory*, 79.

26. Giuffre, "The Coming of the Second Screen," 142.

27. Samuel Ebersole and Robert Woods, "Motivations for Viewing Reality Television: A Uses and Gratifications Analysis," *Southwestern Mass Communication Journal*, 23 (2007): 23–42.

28. Vaynerchuk, *The Thank You Economy*.

29. George Winslow, "Discovery to Add 140 Characters to Many Series," *Broadcasting and Cable*, May 13, 2013. http://www.broadcastingcable.com/news/technology/discovery-add-140-characters-many-series/49983.

30. Hope Jensen Schau, Albert M. Muniz Jr. and Eric J. Arnould, "How Brand Community Practices Create Value," *Journal of Marketing*, 73 (2009): 30–51.

31. Ebersole and Woods, "Motivations for Viewing Reality Television," 15.

32. Malcolm Gladwell, *The Tipping Point: How Little Things Can Make a Big Difference*. New York: Hachette Digital, Inc., 2006.

33. Eric Berry, "Instagram-Omnicom Signals the Future of Digital Advertising," *AdWeek*, March 14, 2014. http://www.adweek.com/news/advertising-branding/instagram-omnicom-deal-signals-future-digital-advertising-156308.

34. Sam Thielman, "Viacom, Tumblr Team Up to Offer Co-Branded Campaigns," *AdWeek*, March 20, 2014. http://www.adweek.com/news/television/viacom-tumblr-team-offer-co-branded-campaigns-156431.

35. Garett Sloane, "Videology Taps Nielsen to Boost Marketing Across TV and Digital," *AdWeek*, March 3, 2014. http://www.adweek.com/videowatch/videology-taps-nielsen-boost-marketing-across-tv-and-digital-156062.

36. Molly Wood, "Turning a Tablet into a Child's Interactive TV," *New York Times*, March 19, 2014. http://www.nytimes.com/2014/03/20/technology/personaltech/from-toytalk-a-tablet-that-lets-children-control-their-tv-viewing.html.

37. Jeanine Poggi, "Crime-centric HLN Repositions Itself as Hub for Social Media and Millennials," *Advertising Age*, February 10, 2014. http://adage.com/article/media/hln-rebrands-hub-social-media/291598/.

38. Sam Thielman, "You Won't Believe How Big TV Still Is," *AdWeek*, March 2, 2014. http://www.adweek.com/news/technology/you-wont-believe-how-big-tv-still-156039.

39. Thielman, "You Won't Believe How Big TV Still Is," 1.

40. Lehman-Wilzig and Avigdor, "The Natural Life Cycle of New Media Evolution," 712.

REFERENCES

Albiniak, Paige. "Daytime Gets More Than Just One Life to Live." *Broadcasting and Cable*, May 13, 2013. http://www.broadcastingcable.com/news/news-articles/daytime-gets-more-one-life-live/114459.

Baumgartner, Jeff. "Shazam Gets Interactive with Jimmy Kimmel Live." *Multichannel News*, October 1, 2013. http://multichannel.com/news/content/shazam-gets-interactive-jimmy-kimmel-live/357386.

Berry, Eric. "Instagram-Omnicom Signals the Future of Digital Advertising." *AdWeek*, March 14, 2014. http://www.adweek.com/news/advertising-branding/instagram-omnicom-deal-signals-future-digital-advertising-156308.

Doyle, Gillian. "From Television to Multi-Platform Less from More or More for Less?" *Convergence: The International Journal of Research into New Media Technologies* 16, no. 4 (2010): 1–19.

Ebersole, Samuel, and Robert Woods. "Motivations for Viewing Reality Television: A Uses and Gratifications Analysis." *Southwestern Mass Communication Journal* 23, no. 1 (2007): 23–24.

Friedman, Wayne. "Younger Demos Dislike Social Media Displays on TV." *Media Daily News*, October 24, 2013. http://www.mediapost.com/publications/article/211986/younger-demos-dislike-social-media-displays-on-tv.html.

Giuffre, Liz. "The Coming of the Second Screen." *Metro Magazine*, Winter, 2012. http://connection.ebscohost.com/c/articles/79451059/coming-second-screen.

Gladwell, Malcolm. *The Tipping Point: How Little Things Can Make a Big Difference*. Hachette Digital, Inc., 2006.

Glaser, Barney G., and Anselm L. Strauss. *The Discovery of Grounded Theory: Strategies for Qualitative Research*. Piscataway, NJ: Transaction Publishers, 2009.

Goel, Vindu, and Brian Stelter. "Social Networks in a Battle for the Second Screen." *New York Times*, October 2, 2013. http://www.nytimes.com/2013/10/03/technology/social-networks-in-a-battle-for-the-second-screen.html?pagewanted=all&_r=0.

Grenager, Teg. "Revolution in a Box." *AdWeek*, October 25, 2010. http://www.adweek.com/news/advertising-branding/revolution-box-103609.

Jacobson, Susan. "Does Audience Participation on Facebook Influence the News Agenda? A Case Study of The Rachel Maddow Show." *Journal of Broadcasting and Electronic Media* 57, no. 3 (2013): 338–55.

Jones, Rory. "Facebook Users in the Mideast and North Africa: 28 million daily." *Wall Street Journal*, October 1, 2013. http://blogs.wsj.com/middleeast/2013/10/01/facebook-users-in-the-mideast-and-north-africa-28-million-daily/.

Kaplan, Andreas M., and Michael Haenlein. "Users of the World, Unite! The Challenges and Opportunities of Social Media." *Business Horizons* 53, no. 1 (2010): 59–68.

Lehman-Wilzig, Sam, and Nava Cohen-Avigdor. "The Natural Life Cycle of New Media Evolution: Inter-Media Struggle for Survival in the Internet Age." *New Media Society* 6, no. 6 (2004): 707–30.

Milian, Mark. "YouTube Exec: We're Heading for 'Third Wave' of TV." *CNN*, February 1, 2012. http://whatsnext.blogs.cnn.com/2012/02/01/youtube-third-wave/.

O'Connell, Michael. "Nielsen and Twitter Partner for Social Media TV Ratings." *Hollywood Reporter*, December 17, 2012. http://www.hollywoodreporter.com/live-feed/nielsen-twitter-partner-social-media-403360.

Poggi, Jeanine. "Crime-centric HLN Repositions Itself as Hub for Social Media and Millennials." *Advertising Age*, February 10, 2014. http://adage.com/article/media/hln-rebrands-hub-social-media/291598/.

Schau, Hope Jensen, Albert M. Muniz Jr., and Eric J. Arnould. "How Brand Community Practices Create Value." *Journal of Marketing* 73, no. 5 (2009): 30–51.

Shields, Mike. "SyFy, Facebook Team Up on Social Gaming." *MediaWeek*, October 5, 2009. http://www.adweek.com/news/technology/syfy-facebook-team-social-gaming-113656.

Sloane, Garett. "Videology Taps Nielsen to Boost Marketing Across TV and Digital." *AdWeek*, March 3, 2014. http://www.adweek.com/videowatch/videology-taps-nielsen-boost-marketing-across-tv-and-digital-156062.

Spangler, Todd. "Facebook Is Sharing TV Chatter with Networks, but What Does It Really Mean?." *Variety*, September 30, 2013. http://variety.com/2013/digital/news/facebook-is-sharing-tv-chatter-with-networks-but-what-does-it-really-mean-1200681820/.

Stelter, Brian. "Same Time, Same Channel? TV Woos Kids Who Can't Wait." *New York Times*, November 10, 2013. http://www.nytimes.com/2013/11/11/business/media/same-time-same-channel-tv-woos-kids-who-cant-wait.html.

Thielman, Sam. "You Won't Believe How Big TV Still Is." *AdWeek*, March 2, 2014. http://www.adweek.com/news/technology/you-wont-believe-how-big-tv-still-156039.

———. "Viacom, Tumblr Team Up to Offer Co-Branded Campaigns." *AdWeek*, March 20, 2014. http://www.adweek.com/news/television/viacom-tumblr-team-offer-co-branded-campaigns-156431.

Umstead, R.Thomas. "Next TV: Nat Geo Building Interactivity within Original Content." *Broadcasting and Cable*, March 21, 2013. http://www.broadcastingcable.com/news/

technology/next-tv-nat-geo-building-interactivity-within-original-content/49849?rssid=
   20065.
Urista, Mark A., Qingwen, Dong, and Kenneth Day. "Explaining Why Young Adults Use
   MySpace and Facebook Through Uses and Gratifications Theory." *Human Communication*
   12, no. 2 (2009): 215–29.
Vaynerchuk, Gary. *The Thank You Economy.* New York, NY: Harper Business, 2011.
Walsh, Mark. "Most Online Time Spent on Smartphones, Tablets." *Online Media Daily*, Sep-
   tember 5, 2013. http://www.mediapost.com/publications/article/208569/most-online-time-
   spent-on-smartphones-tablets.html.
Winslow, George. "Discovery to Add 140 Characters to Many Series." *Broadcasting and
   Cable*, May 13, 2013. http://www.broadcastingcable.com/news/technology/discovery-add-
   140-characters-many-series/49983.
Wood, Molly. "Turning a Tablet into a Child's Interactive TV." *New York Times,* March 19,
   2014. http://www.nytimes.com/2014/03/20/technology/personaltech/from-toytalk-a-tablet-
   that-lets-children-control-their-tv-viewing.html.

Chapter Two

Spoiler Alert

*Understanding Television Enjoyment in
the Social Media Era*

Benjamin Brojakowski

Stories have been used to entertain people since formal languages were first developed. Stories have a way of helping people pass the time, forcing them to think critically, and even teaching them about the world, their culture, and their history. Historically, people have experienced storytelling by gathering with others, watching dramatic performances, and eventually reading. Of course modern individuals still experience stories in these ways, but one primary storyteller has emerged in American homes: the television.[1] When the television started to become popular in America, it was seen as more than a black-and-white screen in a heavy, wooden frame. Manufacturers sold the device as a new member of the family to be loved and taken care of.[2] Now, instead of being treated like part of the family, televisions outnumber family members in most households.[3] Even consumers that don't buy television sets can access programming just by looking at a smartphone or logging into Netflix. It is hard to fathom that, in less than a century, one product has become the key storyteller of our society. So how did the television become such a significant factor in our culture? The answer is simply: because people enjoy it.

Yes, people enjoy television. By expanding on a thriving film industry, television brought all the enjoyment of a movie theater into a household. Decades later, people still enjoy television but few understand what enjoyment really is. It is important to understand how enjoyment is defined, how it is measured, and why it causes so much pleasure and satisfaction. Researchers have sought these answers and theorized about experimental findings but the results are often subjective.

These questions about television enjoyment are consolidated into one famous quote from the TV sitcom *Parks & Recreation*. This interaction between two characters on the show is a comedic (and accurate) description of how television enjoyment can be lessened in the modern media environment. In the scene, Morris (played by Joe Mande) blames Donna (portrayed by Retta) for ruining the potential enjoyment he would feel when watching a film because Donna's graphic tweets spoiled the film's plot and content. Even though the film was twenty-five years old, he felt a warning was necessary for viewers that have not seen it.[4] Donna replies with the popular line, "That movie's twenty-five years old, Morris. And if you don't like how I tweet, then don't follow me."[5] These feelings are not uncommon in the current era of hyperserialized, prestige television and point to the social media user's own control of the content they are exposed to. Although enjoyment theories are popular in media research, few researchers have explored the idea of social media users' impact on enjoyment.

The primary objective of this research is to explore and synthesize numerous theories relating to television use and enjoyment. Additionally, changes in television enjoyment are described by examining the evolution of television as a linear, push environment to an unending, viewer-generated pull environment. A second goal of this research is to consider social media tactics that viewers and television networks often use to enhance viewer enjoyment. This is investigated by explaining the use of hashtags, live tweets, and interactive experiences through digital and mobile applications. Finally, options for social media users to utilize their accounts without spoiling content of their favorite programs are explored.

THEORETICAL FRAMEWORK

Before exploring media enjoyment theories, it is important to understand the paratexts of television. Gray defines paratexts as naturally occurring content that exists in the spaces between texts, audiences, and industry. By connecting text, audience, and industry, paratexts create meaning for viewers and can reach an audience before, during, or after the viewing process.[6] One aspect of entryway paratext is speculative consumption. When viewers experience speculative consumption, they measure known information (such as promos, advertisements, reviews, and peer recommendations) about a film or program and determine what pleasures or benefits they will receive from watching it.[7] Viewers then have the opportunity to choose media they believe will provide the most enjoyment.

Uses and Gratifications

After choosing to view a particular program or film it is important to consider what benefits individuals will experience. One popular media theory used to study television and the Internet is the uses and gratifications perspective. Papacharissi and Rubin define uses and gratifications as a psychological communication theory that "assumes people communicate or use media to gratify needs or wants. It focuses on motives for media use, factors that influence motives, and outcomes from media related behavior."[8] In sum, the uses and gratifications approach focuses on why people choose to use a particular media instead of the media content.

Rubin explains five assumptions that current scholars employ when studying uses and gratifications. First, people are active participants in choosing media. Their reasons for choosing a particular media are goal-directed, purposive, and motivated. Second, the media does not use people. People actually use media to satisfy their needs and desires or seek information. Third, social and psychological factors mediate people's communication behavior. Past experiences, paratexts, and interpersonal interactions impact the ways people are drawn to and use media. Fourth, media competes with other forms of communication for selection, attention and use to satisfy people's needs. Last, people are usually more influential than media in this process because individual initiative mediates how media is used. Through this process, media may impact individual characteristics or societal structures and the ways people rely on different communication channels.[9]

Some of the uses and gratifications satisfied by television include the need for information, social utility, arousal, and escape, but the most common use of television is entertainment.[10] Sherry notes that many people view entertainment as enjoyment and that "entertainment subsumes a wide variety of activities, both mediated and unmediated (e.g., playing a game or a musical instrument), that serve both intrinsic enjoyment and enlightenment."[11] Sherry also writes that entertainment gratifications differ from gratifications related to social utility or learning. Television entertainment gratifications can be both relaxing and arousing.[12] This is supported by a recent finding that 76 percent of television viewers watch multiple episodes of one TV series as a way of relaxation.[13] Since television is used so often for enjoyment, it is important to understand how enjoyment is defined.

Enjoyment

Enjoyment has been broadly defined as a pleasurable affective response to a stimulus.[14] Media enjoyment, therefore, is a pleasing response to stimuli such as television, films, books, and music. In 2002, researchers expanded the definition to include cognitive factors as an influential component of enjoyment. Therefore, media enjoyment should include a viewer's identification with and sympathy toward fictional characters (affective response) as well as how they assess the actions of characters and the themes inherent to the messages of the narrative (cognitive responses).[15] Not all researchers view enjoyment in the same manner, though. Zillman defines enjoyment by applying disposition theory.[16] This research notes that enjoyment is a combination of affective character disposition and gratifications sought from witnessing the justification of character actions.[17]

Transportation into a Narrative World

Researchers note that one crucial element of enjoyment is that it whisks viewers away from reality and places them into a story world. Green, Brock, and Kaufman write that this transportation into a narrative world occurs when attention, imagery, and feelings merge to cause a viewer to "transport" into the television world.[18] It is also important to note that this transportation is not meant to signify a mental disorder or a break from reality. Instead, transportation into a narrative world is a desirable state that viewers seek out through media. It has been operationalized as a phenomenon that occurs when a viewer is emotionally or cognitively engaged in media.[19]

Transportation theory has been used to explain the persuasive effects of narratives and contributes to the conceptual understanding of enjoyment. By specifying mechanisms that cause enjoyment, scholars have a clearer idea of how viewers respond to media. The first mechanism is the phenomenological experience of enjoyment through immersion in a narrative world. Enjoyment through favorable outcomes of media exposure is the second mechanism. The third mechanism is a circumstance that enhances or lessens enjoyment.[20] An example of a circumstance that reduces media enjoyment may be an unexpected phone call or email during a program. Conversely, communal viewing of a program may enhance the enjoyment.

Although a pleasurable story or a song can transport individuals the focus of this research is specifically on television. This distinction is important to make because television provides viewers with visual imagery and audible cues instead of causing an individual to construct mental images of characters. The combination of imagery, audible clues, and knowledge about the experiences of other characters help viewers connect with characters to increase enjoyment.

Parasocial Relationships

When viewers are transported into the narrative world they often form relationships with the characters in the programs. These parasocial relationships are "symbolic, one-sided quasi-interactions between a viewer and a media figure."[21] Research suggests that these relationships are hedonically driven and linked to enjoyment.[22] This phenomenon has been studied for decades and occurs often among viewers of all ages; however, it is most common among adolescents. In a 2010 study, Theran, Newberg, and Gleason surveyed 107 teenage girls and found that 94 percent of them had engaged in parasocial relationships to some degree. They found that the bond between media figure and adolescent was mild but did exist in an overwhelming number of participants.[23]

In 2012, Eyal and Dailey expanded this research to observe parasocial relationships among an adult population. Their study of college students shows that parasocial interactions operate similarly to real-life friendships. Although the parasocial relationships have less emotional satisfaction than interpersonal relationships, they are significant in shaping viewers' attitudes, behaviors, and beliefs.[24]

Despite being mild, parasocial relationships are viewed as "a healthy alternative or supplement to normal personal interaction."[25] One reason that these relationships are so attractive to viewers is that they offer positive interactions with very limited risk of rejection.[26] These relationships also have a direct connection to media enjoyment. Literature regarding parasocial relationships and media enjoyment explain that both phenomena must be viewed as multidimensional constructs with several components of involvement (primarily affective, cognitive, and behavioral) that contribute to the amount of pleasure one receives from media content.[27]

Flow Theory

Flow theory is another example of how viewers use television for enjoyment. This theory, similar in many ways to transportation to a narrative world, was initially developed by Csikszentmihalyi to explain the enjoyment felt by everyday experiences. Early studies of flow observed the enjoyment artists and musicians experienced by immersing themselves in their work until they were "lost" in it and the world around them seemed not to exist.[28] Activities most likely to induce flow are characterized by four factors. First, an activity must have tangible goals with clear rules. This means that individuals should choose a task that is possible to complete. Second, it must be possible for individuals to adjust opportunities for action to reach their capabilities. Third, the activities must provide a progress report so the individual can determine how successful the project is. Finally, the activity should be the only endeavor a person is working on and distractions should be kept to a minimum.[29]

Eventually flow theory was expanded to include daily activities such as watching television or playing video games.[30] For example, Jane McGonigal observed flow in video games such as *Tetris*.[31] McGonigal found that, when playing similar games, users are on the edge of their own skill level. The feeling of working at peak ability is so engaging that both quitting and winning are unsatisfying options. Instead, the player must continue to play as a way of maintaining this feeling.[32] In the case of television, researchers note that people have agency to choose what programs they watch as a way of solving perceived problems. People burdened by basic needs, individual differences, and societal factors use television as a way of seeking gratifications in their own ways.[33]

Common Characteristics

Although these theories were developed at different times and for different reasons, they share several common characteristics related to television enjoyment. First, viewers must be actively engaged in the program. Viewers must make the viewing experience a high priority and not become distracted by other activities. Second, the program must be cognitively appropriate. For example, most adults do not enjoy cartoons as much as children. Since adults understand cartoons are not real, they are less likely to be transported into a narrative world or achieve flow by observing these programs. Conversely, small children are less likely to find enjoyment in news programs or shows with adult themes. Third, viewers must make an emotional connection with some character or thematic element of the program. Without any emotional connection, viewers are unlikely to maintain interest or experience pleasure during the program.

Social Media

All of these theories of media use and enjoyment apply to social networking as well. Social media Web sites like Facebook and Twitter provide gratifications for users such as news-gathering[34] and social interaction.[35] Following journalists and news corporations on Twitter, using Facebook to make plans with others and interacting with celebrities through a myriad of digital platforms, are ways that users may achieve these gratifications.

One significant change from television enjoyment is the parasocial relationships social media users experience because the frequency and intensity of these relationships can be amplified by social media platforms. Although many television viewers form relationships with television characters, social media gives users an opportunity to interact with the actors behind their television friends. Users that follow a celebrity on Twitter can learn more intimate details about their personal lives and character development than what is seen or heard on television. Despite the increased likelihood of interaction, the relationship is typically one-sided. Except for occasional Twitter responses and Facebook "likes," the media figure is largely unaware of an individual's attempts at contact.

Roberts addresses the parasocial relationships of teens on Twitter. Although having these relationships with media figures through television has been viewed as a normal practice, social media has opened a new avenue for viewers to interact with celebrities. Roberts explains that teens are more likely to view celebrities as potential friends.[36] The current generation of teens uses multiple Twitter accounts to tweet thousands of messages per day as a way of seeking positive affirmations or follows from their favorite celebrities. Interacting with celebrities, or being followed by one on Twitter, may increase enjoyment if a user feels a personal connection to a media figure.[37]

Finally, social networking is related to flow theory due to the ease in which users can immerse themselves in profile pages. For example, Instagram users can easily view photos of friends, family members, celebrities, athletes, and even strangers.[38] They are encouraged to follow these people to stay up-to-date on their digital lives. Also, a Twitter user can find messages, videos, pictures, and hyperlinks to more than 270 million active users.[39] This immersion into the Twitterverse contains many of the same characteristics of Csikszentmihalyi's observations of flow during daily activities.

SAME BAT TIME. SAME BAT CHANNEL: HOW TIME-SHIFTING TECHNOLOGY REDUCED THE NEED FOR APPOINTMENT-VIEWING

Although transportation into a narrative world and parasocial interactions aren't new to modern viewers, the way viewers enjoy television today is much different. Until recently, researchers had studied television as a linear process with a pre-determined schedule that controls when and what viewers will watch. Simons describes this as a "push environment" because networks had the power to push any material it wanted on the viewers.[40] This environment existed for decades and made appointment-viewing necessary for people to watch and stay interested in programming. The appointment-viewing was such a significant part of the television environment that viewers were often reminded when the next episode would air. The original Batman program popularized this practice by reminding viewers to "Tune in next week. Same bat time. Same bat channel."

The push environment began to change with the introduction of the new technology. The videocassette recorder (VCR) allowed consumers to record television programs with the intent of viewing the episodes at a later time or to save them for multiple viewings. This practice, along with early elements of social media interaction and binge-watching, accelerated with the introduction of the Internet. Reports indicate that fans of the television show *The X-Files* utilized UseNet to meet and interact with other fans. These fans shared recorded episodes by exchanging previously recorded VHS tapes to others that they met in chat rooms.[41] *The X-Files* fans built one of the first social networks and used time-shifting to transition into a "pull environment" of television. The pull-environment allows viewers to create and produce their own viewing habits.[42]

Around the same time, more advanced time-shifting technologies were developed.[43] These technologies and services represent the first level of time shifting: technology.[44] New examples of time-shifting technology include TiVo and other forms of DVRs, such as the DISH Hopper and DirecTV Genie. These technologies may also include VCRs and DVD recording devices or streaming services like Netflix or Hulu.

The second level of time-shifting is release date.[45] This means that viewers can utilize the Internet and satellite systems to find international material before it is released in their home country. Networks like the British Broadcasting Corporation (BBC) have become popular in the United States and viewers now have access to these programs without waiting weeks or months to view programs such as *Sherlock* and *Downton Abbey*.

The third level of time shifting is rhythm.[46] This occurs when users do not watch television at the regularly scheduled times. Instead of watching at the original time of broadcast, viewers have the ability to choose when they will watch a program and how many episodes of the program they will watch. Modern technology has made this practice more popular in recent years, but this is when viewers usually experience spoilers, which will be described in more detail later.

THE INTRODUCTION OF HYPERSERIALIZED TELEVISION

The X-Files sharers did more than interact with other fans and utilize technology in new ways. They also helped usher in the era of hyperserialized programming. Hyperserialized programs are defined as scripted television series with story arcs that cover multiple episodes or seasons.[47] In other words, instead of creating several individual episodes, television series are created as a sequence of programs with links that connect each episode together in a meaningful way. Producers and writers develop shows by imagining an entire series and consider factors such as how many seasons they need to tell their story and how they can transform viewers into binge-watchers. Romano describes high production value, adult themes, anti-heroic lead characters, and, most importantly, gripping, mysterious endings as common characteristics of hyperserialized programs.[48] They are crafted for the DVR-generation so viewers will feel compelled to watch multiple episodes of the program in one sitting or seek additional content after the broadcast.

The introduction of time-shifting technology and online VHS-sharing coincided with the introduction of one of the first well-known hyperserialized dramas, *The Sopranos*.[49] *The Sopranos* was a historic series for multiple reasons but it became known for continuity between episodes. HBO executives virtually dared viewers to miss an episode because one Sunday night away from the television meant the viewer would miss content vital to the popular series. Viewers with time-shifting technology would need to avoid blogs, friends, co-workers, and others that might ruin their enjoyment of the episode before they were able to see it.

This type of programming has increased with series such as *Lost*, *Breaking Bad*, *Dexter*, *The Walking Dead*, and *Game of Thrones*. These shows are known for high ratings, suspenseful conclusions, and loyal fans that interact through social media. They are also popular for being binge-watched by users with DVRs, Netflix, Hulu, and box sets. Research shows that viewers binge-watch these programs because they produce a chemical brain reaction similar to that of addictive drugs or hypnosis.[50] Kubey and Csikszentmihalyi explain that binge-watching is common due to a viewer's brain activity switching from the left side to the right side. This releases endorphins, which relaxes viewers, and encourages them to continue watching so they can maintain this relaxed feeling.[51] The enjoyment created by binge-watching allows networks to build loyal audiences that purchase merchandise and box sets or attend conference panels. The biological response to these programs, the communal aspect of watching with others online, and the new style of storytelling make combine for a more enjoyable viewing experience.

HISTORY OF THE ONLINE FAN

So far, the term "fan" has been used to describe television audiences with favorable attitudes toward individual programs, but the term often invokes many different images of fans that need further explanation. Some people believe fans are nothing more than regular viewers of certain television programs. Others consider fans to be viewers that own merchandise, know trivia, and interact with other viewers of the show. Even more people view television fandom as an extreme, dangerous, or strange practice.[52] Fans are seen as crazed viewers that establish strong bonds with television characters, write fan fiction, join online fan communities, and engage in cosplay with other fans at large conferences and events. Even though this behavior is not usually harmful, these fans are often ostracized and assigned derogatory titles. Examples of these titles can be "Trekkies" (Star Trek fans) or fanboys/fangirls (general fans obsessed with particular programs or characters).

Although the extreme fan label is still the predominant stereotype, the general attitude toward these viewers has softened during this Golden Age of

Television. The high production value and mature themes of series like *The Sopranos*, *Lost*, and *The Leftovers* has transcended "nerd culture" and begun to normalize fandom. Now, instead of being viewed as outsiders for their fandom, many viewers of hyperserialized programs are accepted by others at conferences and conventions. For example, the famed Hall H at the annual San Diego Comic Con is a 6,500-seat auditorium [53] reserved for many of the programs described in this article. So many fans flock to Hall H to see the casts of programs such as *Breaking Bad* and *Game of Thrones* that the Comic Con organizers need to ban people from camping outside. [54]

Before fans make interpersonal connections at conventions, many of their relationships begin online. Online fans are unique because they are a specific group with the means and motivation to find, create, and share information with others. [55] Online fans also have the unique ability to choose their level of interaction with others. Some "lurk" by only reading content, others contribute to conversations, some write fan fiction, and others create lasting friendships. Costello and Moore used qualitative interview methodology to examine these levels of interaction. They found that some of the most involved fans celebrate major life events together and even create informal conventions and gatherings. [56]

SOCIAL MEDIA AND THE VIRTUAL LOUNGEROOM

Online fandom has evolved with the growth of social media and new time-shifting technology. First, time-shifting technology introduced a more convenient viewing experience for both casual viewers and online fans. Although this technology reduces the need for appointment-viewing, it does not reduce the desire to watch live. On a related note, as social media grows, so do the benefits of viewing programs at the regularly scheduled times. One benefit is the communal experience of watching popular programs with others. Although other viewers may not be in the same room or watching the same television set, they may live tweet their thoughts and reactions during the programs as a way of interacting. Communal live tweeting has become very popular and has drawn comparisons to an excitable crowd at a movie theater. [57] In his review of Syfy's B-movie blockbuster *Sharknado*, Sepinwall writes that "I doubt I would have watched more than a few minutes of this schlocktacular if my Twitter feed hadn't been clogged with people watching and cracking jokes right along with me." [58] This phenomenon of communal tweeting is part of the "virtual loungeroom." [59] Harrington, Highfield, and Bruns, define the virtual loungeroom as "an online space where an audience can commune and centrally share the television experience." [60] Although *Sharknado* and hyperserialized programs fall on opposite ends of the television spectrum, Sepinwall's comment emphasizes the total enjoyment experienced by viewers.

Audience Engagement

Some popular celebrities that take advantage of the virtual loungeroom are *Shameless* stars Emmy Rossum and Shanola Hampton. Rossum and Hampton frequently live tweet episodes with behind-the-scenes information as a way of engaging fans. Additionally, producers and the network use the official *Shameless* Web site as a tool for increasing fan interaction. The Web site offers links to all of the show's social media profiles, each actor's Twitter account, and a scrolling conversation box for viewers to engage with other fans.[61]

Another example comes from AMC's *The Walking Dead*. The show Web site uses a forum pattern instead of the conversation box.[62] This allows visitors to choose or create specific discussions with others that are interested in the same topic. Producers encourage Twitter interaction by creating unique hashtags for each episode. Viewers also use the "Dead Yourself" app so they can add zombie features to their selfies and share them with a community of other fans.[63] Finally, *The Walking Dead* is followed by a talk show titled *The Talking Dead* that encourages viewers to call-in or tweet at the panel of guests, usually including a character from the show.[64]

Split Screen Viewing

A third interactive example of television and social media is the use of split screen viewing. Split screen viewing occurs when a viewer uses a computer or mobile device to access exclusive content to be watched simultaneously with the program or during commercials. A 2012 Pew Research Internet Project study indicates that 52 percent of adult smartphone users access their phones while watching television.[65] Instead of being distracted by their phones, these people are more likely to use this technology to stay occupied during commercial breaks, verify information they heard on the program, or interact with other viewers.[66] Additionally, 23 percent of all mobile users are estimated to use their phones to text message other viewers of the program.[67] This smartphone usage differs greatly from viewers that lose interest after being distracted by a phone call or email because the phone is being voluntarily used to stay engaged with the program.

These added benefits serve as paratexts to the actual television show and expand the idea of television evolving from the linear environment of past decades. Social media content, such as tweets, Facebook statuses, Instagram posts, and webisodes turn television into a seemingly endless experience. Along with time-shifting devices, the start and finish times of television programs have been completely removed. This overflow content exists forever in multiple digital spaces and allows users to create enjoyable viewing experiences long after a program, or even an entire series, has ended.

Spoilers

All of these interactive technological advancements are designed to make television more enjoyable for viewers by inviting them into virtual lounge-rooms, but it greatly increases the odds of creating spoilers and reduced enjoyment for fans. Spoilers are defined as "information about the plot of a motion picture or TV program that can spoil a viewer's sense of surprise or suspense."[68] They occur when information about a series or episode leaks, most often through social media, to viewers that have not seen it yet. When spoilers reach an individual, it is often creates diminished enjoyment and lower viewership. If the problem becomes widespread, it is possible for networks to receive lower ratings and less advertising revenue.

Spoilers have been viewed comically in the "Monday Night Football" episode of the sitcom *How I Met Your Mother* when the main characters try to live in New York City for twenty-four hours without learning the result of the Super Bowl.[69] Spoilers were also used as a plot device during an episode of *Scrubs* when a primary character warned all of his coworkers not to ruin the result of a basketball game that he recorded for a later viewing.[70]

Spoilers are more than a plot device for sitcoms, though. Many networks and producers take them very seriously. On a May 2014 episode of his late night show, Conan O'Brien warned *Game of Thrones* fans to mute the television or change the channel because guest, George R. R. Martin, would reveal spoilers about the current season.[71] Entertainment Web site HitFix also disabled the comment section of their Web site for *Game of Thrones* reviews after too many commenters spoiled content for others. Even executives worry about spoiled content of their programs. Executive producer and creator of *Mad Men*, Matthew Weiner, has gained a reputation for his extreme spoiler-phobia.[72]

Twitter users often describe spoilers as a first-world problem, and they would certainly receive support for their opinion; however, online interactions have become much stronger in recent years. Research indicates that online impression-development is nearly equal to face-to-face meetings,[73] users find emotional support in online interactions,[74] and people are more open to expressing true thoughts and feelings because they will not experience direct rejection.[75] When television viewers find a supportive community with similar interests, spoilers present a larger problem than just reduced enjoyment with a program. Viewers may feel betrayed by their online friends and they may stop watching a show if their enjoyment is ruined by a cast member or network.

CATCH 22: HOW DO USERS MAXIMIZE SOCIAL MEDIA AND TELEVISION ENJOYMENT

The recent developments in television and social media have created a unique problem for viewers. On one hand, viewers appear to experience more enjoyment when they can interact with a digital community and stay engaged with programs. On the other hand, users do not want to diminish their television enjoyment if they're unable to view episodes live. As networks urge interactive viewing and social media Web sites encourage increased activity, maximizing enjoyment becomes a more difficult task for viewers. As the social media and television landscapes continue to converge and overlap users must explore options to maintain enjoyment.

Third Party Application

One option for viewers is a third-party application. Apps like Echofon allow users to mute keywords, hashtags, and some of the accounts they follow. By muting accounts and hashtags, Twitter users reduce the risk of spoiled content. This option is especially helpful when programs designate specific hashtags for individual episodes.[76] The official Twitter app for iPhone has also experimented with similar functions in 2014.

Other apps, such as HootSuite, have the ability to manage multiple social media accounts. Users with this application can keep open multiple live streams of their social networking profiles on one page.[77] HootSuite users do not have the option to mute others or hashtags but they can close accounts that are most likely to include spoilers. With this tool, viewers may stay connected to many social media profiles without accessing profiles most likely to contain spoilers.

Third-party applications are not a panacea for spoiled television though. Since users can tweet different hashtags, send direct messages, or create multiple Facebook statuses, social media users never know what they need to mute or which contacts they need to temporarily block. These applications are not guaranteed to eliminate spoilers but they are user-friendly and inexpensive.

Social Media Boycott

Another option is to avoid using social media altogether if the issue of spoilers becomes a significant problem. As previously written, this may be difficult for some users as social media has become a significant source of social interaction, news-gathering,[78] and media enjoyment.[79] Wright explains that college students use social media, particularly Facebook, as a way of creating a network of social support. This is a significant finding because college students make up the demographic that is most likely to use social media while watching television.[80]

Enjoy

This chapter has focused solely on spoilers as a negative characteristic of modern television enjoyment, but this is untrue. Viewers that are unaffected by, or intentionally seek, spoilers can find them on numerous Internet and social media sites. Web sites such as Wetpaint.com provide spoilers, news, and rumors for dozens of TV shows.[81] Bloggers, like Reality Steve, use multiple platforms to provide spoilers and rumors for audiences.[82] In addition to a Web site, Reality Steve boasts nearly 82,000 Twitter followers[83] and over 37,000 fans on his Facebook page.[84] Even individual shows have spoiler-specific Web sites and social media accounts. For example, *Game of Thrones* fans can find the latest news and spoilers about the show on WinterIsComing.net (WIC).[85] WIC also maintains a strong social media presence with over 60,000 Twitter followers[86] and 70,000 Facebook fans.[87]

Costello and Moore used interviews to explain why so many online fans seek out spoilers. Through their interviews, one can assert that "insider information" on these spoiler Web sites is attractive to many fans. However, one of the interview subjects noted that learning information could be "good and bad."[88] Also, fans of many programs may become so immersed in the material that denying themselves the chance to learn more about the show can cause more mental discomfort than waiting for the next episode.

CONCLUSION

In 2013, it was estimated that digital media had finally surpassed television as the dominant form of media in Americans' lives.[89] The study notes that the average U.S. adult uses the Internet and mobile technology for over five hours each day and watches television for more than four hours.[90] By creating digital content and using social media, television networks have found new ways to reach viewers. With the addition of time-shifting technology, television audiences have constant access to material that would not have existed even a decade ago.

Currently, television is so intertwined with social media that it has become difficult for younger demographics to remember the experience of watching television on only one screen. Although this chapter focuses on hyperserialized programs, the interaction between social media and television is not limited to this genre. It is common for sports fans to observe multiple game scores on their smartphone apps or update fantasy teams from their tablets. Additionally, the previously mentioned Pew research study acknowledged the popularity of using cell phones during reality programs to vote on contestants or participate in polls.[91]

This chapter has been an examination of television enjoyment and how social media users can maximize their satisfaction in the current media environment. Much like celebrities that adopt a second career in social media, viewers must adopt a second strategy for television viewing. As they navigate the new world of social media and television convergence it is important to understand the theoretical roots of media enjoyment and ways to create the best experience.

NOTES

1. EMarketer. "Digital Set to Surpass TV."
2. Gray, Jonathan. "From Spoilers to Spinoffs: A Theory of Paratexts." In *Show Sold Separately: Promos, Spoilers, and Other Media Paratexts.*
3. Nielsen. "U.S. Homes Add Even More TV Sets in 2010."
4. *Parks & Recreation.* "Halloween Surprise." Oct. 25, 2012.
5. Ibid.
6. Jonathan Gray, "From Spoilers to Spinoffs."
7. Ibid.
8. Papacharissi, Zizi, and Alan M. Rubin. "Predictors of Internet Use."
9. Rubin, Alan M., "Uses-And-Gratifications."
10. Sherry, J. L.. "Flow And Media Enjoyment."
11. Ibid.
12. Ibid.
13. "Netflix Declares Binge Watching Is the New Normal" (2013).
14. Green, Melanie C., Timothy C. Brock, and Geoff F. Green. "Understanding Media Enjoyment."
15. Ibid.
16. Ibid.
17. Sherry, J. L. "Flow and Media Enjoyment."
18. Green, Melanie C., Timothy C. Brock, and Geoff F. Green. "Understanding Media Enjoyment."
19. Ibid.
20. Ibid.
21. Theran, Sally A., Emily M. Newberg, and Tracy R. Gleason. "Adolescent Girls' Parasocial Interactions with Media Figures."
22. Tsay, Mina, and Brianna Mary Bodine. "Exploring Parasocial Interaction in College Students."
23. Theran, Sally A., Emily M. Newberg, and Tracy R. Gleason. "Adolescent Girls' Parasocial Interactions with Media Figures."
24. Eyal, Keren, and René M. Dailey. "Examining Relational Maintenance in Parasocial Relationships."
25. Theran, Sally A., Emily M. Newberg, and Tracy R. Gleason. "Adolescent Girls' Parasocial Interactions with Media Figures."
26. Ibid.
27. Tsay, Mina, and Brianna Mary Bodine. "Exploring Parasocial Interaction in College Students."
28. Sherry, J. L. "Flow And Media Enjoyment."
29. Ibid.
30. Ibid.

31. McGonigal, Jane. *Reality Is Broken: Why Games Make Us Better and How They Can Change the World* (New York: Penguin Press, 2011).

32. Ibid.

33. Ibid.

34. Hermida, Alfred, Fred Fletcher, Darryl Korell, and Donna Logan. "SHARE, LIKE, RECOMMEND."

35. boyd, Danah, and Nicole Ellison. "Social Network Sites."

36. Roberts, Kayleigh. "The Psychology of Begging to Be Followed on Twitter."

37. Ibid.

38. Instagram Help Center.

39. Twitter. "About." 2014.

40. Simons, Nele. "Watching TV Fiction."

41. Bianco, Julia. "Is Binge Watching As Bad As They Say?"

42. Simons, Nele. "Watching TV fiction."

43. TIVO. History.

44. Simons, Nele. "Watching TV fiction."

45. Ibid.

46. Simons, Nele. "Watching TV fiction."

47. Romano, Andrew. "Why You're Addicted to TV."

48. Ibid.

49. Ibid.

50. Ibid.

51. Ibid.

52. Costello, V., and B. Moore. "Cultural Outlaws: An Examination of Audience Activity and Online Television Fandom." *Television & New Media.*

53. Kim, Tony B. "2013 Comic-Con Tip #4: Hall H Camping."

54. Costello, V., and B. Moore. "Cultural Outlaws: An Examination of Audience Activity and Online Television Fandom." *Television & New Media.*

55. Ibid.

56. "Hall H/Plaza Park Lines." Comic-Con 2014 International: San Diego.

57. Sepinwall, Alan. "Review: Syfy's 'Sharknado.'"

58. Ibid.

59. Harrington, Stephen. "Tweeting about the Telly."

60. Ibid.

61. Showtime. *Shameless.*

62. AMC TV. *The Walking Dead.*

63. AMC TV. *Dead Yourself.*

64. AMC TV. *Talking Dead.*

65. Smith, Aaron. "The Rise."

66. Ibid.

67. Ibid.

68. Merriam-Webster. "spoiler."

69. *How I Met Your Mother.* "Monday Night Football." 2007.

70. *Scrubs.* "My Hypocritical Oath." 2005.

71. Conan O'Brien. May 2014.

72. Sepinwall, Alan. "Mad Men Review."

73. Walther, Joseph. "Impression Development in Computer-Mediated Interaction." 1993.

74. Wright, Kevin B. "Emotional Support and Perceived Stress." 2012.

75. Suler, John. "The Online Disinhibition Effect." 2004.

76. Echofon. Echofon Blog. 2011.

77. HootSuite. "We Are More Than a Social Media Company."

78. Hermida, Alfred, Fred Fletcher, Darryl Korell, and Donna Logan. "SHARE, LIKE, RECOMMEND."

79. boyd, Danah, and Nicole Ellison. "Social Network Sites."

80. Gray, Jonathan. "From Spoilers to Spinoffs."

81. "Wetpaint—Celebrity Gossip, Entertainment News, and Hot TV." Accessed 2014.

82. www.RealitySteve.com.
83. www.facebook.com/RealitySteve.
84. www.twitter.com/RealitySteve.
85. www.wetpaint.com.
86. www.twitter.com/WiCnet.
87. www.facebook.com/WiCnet.
88. Costello, V., and B. Moore. "Cultural Outlaws: An Examination Of Audience Activity and Online Television Fandom."
89. EMarketer. "Digital Set to Surpass TV."
90. Ibid.
91. Smith, Aaron. "The Rise."

REFERENCES

AMC TV. "Dead Yourself." www.amctv.com. http://www.amctv.com/shows/the-walking-dead/dead-yourself.
————. "Talking Dead." www.amctv.com. http://www.amctv.com/shows/talking-dead.
————. "The Walking Dead." www.amctv.com. http://www.amctv.com/shows/the-walking-dead.
Baym, Nancy K.. "The Perils and Pleasures of Tweeting with Fans." In *Twitter and Society*. New York: Peter Lang, 2014.
Bianco, Julia. "Is Binge Watching as Bad as They Say?" *Huffington Post*. http://www.huffingtonpost.com/uloop/is-binge-watching-as-bad-_b_4612996.html (accessed July 13, 2014).
boyd, Danah, and Nicole Ellison. "Social Network Sites: Definition, History, and Scholarship." *Journal of Computer-Mediated Communication* 13 (2007): 210–30.
Costello, V., and B. Moore. "Cultural Outlaws: An Examination of Audience Activity and Online Television Fandom." *Television & New Media* 8, no. 2 (2007): 124–43.
Echofon. "Echofon Blog: Major New Echofon Version for iPad and iPhone." *Echofon Blog: Major New Echofon Version for iPad and iPhone*. http://blog.echofon.com/2011/02/major-new-echofon-version-for-ipad-and.html.
EMarketer. "Digital Set to Surpass TV in Time Spent with US Media." *eMarketer*. http://www.emarketer.com/Article/Digital-Set-Surpass-TV-Time-Spent-with-US-Media/1010096 (accessed January 1, 2014).
Eyal, Keren, and René M. Dailey. "Examining Relational Maintenance in Parasocial Relationships." *Mass Communication and Society* 15 (2012): 758–81.
Gray, Jonathan. "From Spoilers to Spinoffs: A Theory of Paratexts." In *Show Sold Separately: Promos, Spoilers, and Other Media Paratexts*. New York: New York University Press, 2010.
Green, Melanie C., Timothy C. Brock, and Geoff F. Green. "Understanding Media Enjoyment: The Role of Transportation into Narrative Worlds." *Communication Theory* 14 (2004): 311–27.
Harrington, Stephen. "Tweeting about the Telly: Live TV, Audiences, and Social Media." In *Twitter and Society*. New York: Peter Lang, 2014.
HBO. "HBO: Girls Socialize." HBO. http://www.hbo.com/#/girls.
Hermida, Alfred, Fred Fletcher, Darryl Korell, and Donna Logan. "SHARE, LIKE, RECOMMEND: Decoding the Social Media News Consumer." *Journalism Studies* 13 (2012): 815–24.
Hollywood Reporter. "TV Ratings: 'Downton Abbey' Hooks Finale Best 8.5 Million Viewers." *Hollywood Reporter*. http://www.hollywoodreporter.com/live-feed/tv-ratings-downton-abbey-hooks-682845.
HootSuite. "We Are More Than a Social Media Company." About Hootsuite. https://hootsuite.com/about.
"Instagram Help Center." Exploring Photos and Videos. https://help.instagram.com/441951049195380 (accessed July 2014).

Kim, Tony B. "2013 Comic-Con Tip #4: Hall H Camping." *Crazy 4 ComicCon*. April 25, 2013 (accessed September 3, 2014).

McGonigal, Jane. *Reality Is Broken: Why Games Make Us Better and How They Can Change the World*. New York: Penguin Press, 2011.

Merriam-Webster. "spoiler." Merriam-Webster. http://www.merriam-webster.com/dictionary/spoiler (accessed May 27, 2014).

"My Slanted, Sophomoric, and Skewed View of Reality Television." *Reality Steve*. 2014. http://realitysteve.com.

"Netflix Declares Binge Watching Is the New Normal." (2013) http://www.prnewswire.com/news-releases/netflix-declares-binge-watching-is-the-new-normal-235713431.html (accessed June 5, 2014).

Nielsen. "U.S. Homes Add Even More TV Sets in 2010." http://www.nielsen.com/us/en/insights/news/2010/u-s-homes-add-even-more-tv-sets-in-2010.html.

Papacharissi, Zizi, and Andrew L. Mendelson. "An Exploratory Study of Reality Appeal: Uses and Gratifications of Reality TV Shows." *Journal of Broadcasting & Electronic Media*: 51 (2007) 355–70.

Papacharissi, Zizi, and Alan M. Rubin. "Predictors of Internet Use." *Journal of Broadcasting & Electronic Media*: 44 (2000) 175–96.

PBS. "'Sherlock' to Premiere Jan. 19 on PBS, Air Back-to-Back with 'Downton.'" *BBC America*. http://www.bbcamerica.com/anglophenia/2013/10/sherlock-premiere-jan-19-pbs-air-back-back-downton/.

Roberts, Kayleigh. "The Psychology of Begging to Be Followed on Twitter." *Atlantic*, February 26, 2014.

Romano, Andrew. "Why You're Addicted to TV." *Newsweek*, May 15, 2013.

Rubin, Alan M. "Uses-And-Gratifications." In *Media Effects: Advances in Theory and Research*. New York: Routledge, 2009.

Sepinwall, Alan. "'Mad Men' Creator Matthew Weiner Previews Final Season." HitFix. http://www.hitfix.com/whats-alan-watching/mad-men-creator-matthew-weiner-previews-final-season/3.

———. "Review: Syfy's 'Sharknado' Hits the Schlocky Motherlode." (2013). *HitFix*. http://www.hitfix.com/whats-alan-watching/review-syfys-sharknado-hits-the-schlocky-motherlode (accessed 2014).

Serjeant, Jill. "The Office to End Run on U.S. TV in 2013." *Reuters*. http://www.reuters.com/article/2012/08/21/entertainment-us-theoffice-idUSBRE87K0XR20120821.

Sherry, J. L. "Flow and Media Enjoyment." *Communication Theory* 14 (2004): 328–47.

Showtime. "Talk with Other fans about Shameless on Showtime." SHO.com. http://www.sho.com/sho/shameless/socialize.

Simons, Nele. "Watching TV Fiction in the Age of Digitization: A Study into the Viewing Practices of Engaged TV Fiction Viewers." *International Journal of Digital Television* 4 (2013): 177–91.

Smith, Aaron. "The Rise of the 'Connected Viewer.'" Pew Research Centers Internet American Life Project RSS. (2012) http://www.pewinternet.org/2012/07/17/the-rise-of-the-connected-viewer/ (accessed June 18 2014).

Suler, John. "The Online Disinhibition Effect." *CyberPsychology Behavior* 7 (2004): 321–26.

Theran, Sally A., Emily M. Newberg, and Tracy R Gleason. "Adolescent Girls' Parasocial Interactions with Media Figures." *Journal of Genetic Psychology* 171 (2010): 270–77.

Tivo. "History." TiVo. https://www3.tivo.com/jobs/questions/history-of-tivo/index.html (accessed August 2, 2014).

Tsay, Mina, and Brianna Mary Bodine. "Exploring Parasocial Interaction in College Students as a Multidimensional Construct: Do Personality, Interpersonal Need, and Television Motive Predict Their Relationships with Media Characters?" *Psychology of Popular Media Culture* 1 (2012): 185–200.

Twitter. "About Twitter, Inc. About." Twitter About. https://about.twitter.com/company (accessed August 2, 2014).

Walther, Joseph B. "Impression Development in Computer-Mediated Interaction." *Western Journal of Communication* 57: 381–98.

"Wetpaint—Celebrity Gossip, Entertainment News, and Hot TV." Accessed September 3, 2014.

Wright, Kevin B. "Emotional Support and Perceived Stress Among College Students Using Facebook.com: An Exploration of the Relationship Between Source Perceptions and Emotional Support." *Communication Research Reports* 29: 175–84.

Chapter Three

Rhetorical Strengths and Limitations of Interactivity for Activism in the Stewart and Colbert Universe

Christopher A. Medjesky

"I'm mad as hell, and I'm not going to take it anymore!" Who among us has not wanted to open their window and shout that at the top of their lungs? Seriously, who? Because we're looking for those people. We're looking for the people who think shouting is annoying, counterproductive, and terrible for your throat; who feel that the loudest voices shouldn't be the only ones that get heard; and who believe that the only time it's appropriate to draw a Hitler mustache on someone is when that person is actually Hitler. Or Charlie Chaplin in certain roles.
—Rally to Restore Sanity Web site [1]

On the anniversary of Martin Luther King Jr.'s "I Have a Dream" speech, Glenn Beck held a Tea Party–driven "Restoring Honor" rally on the same steps of the Lincoln Memorial where King delivered his famous address. Although Beck's rally was said not to be politically related, the presence of Beck and his guest Sarah Palin inevitably marked the event as political and a criticism of the state of the current government. [2] Although the *Washington Post* described the mood of Beck's rally as "peaceful and calm," Jon Stewart proposed the rally was part of a larger rhetorical move that encouraged the "us versus them" mentality of politics, an approach that ultimately leads toward the normalization of incommensurability. [3] The Rally to Restore Sanity, designed to encourage open and respectful discourse on politics and society, was proposed by Stewart on *The Daily Show* as a response to Beck's rally and a call to action for the audience.

Almost immediately after announcing the planned rally on *The Daily Show*, Stephen Colbert used *The Colbert Report* to suggest an alternative rally be planned. Urged by users of the online sites Reddit and Facebook as well a group calling themselves Restoring Truthiness, Colbert sensed his Colbert Nation wanted him to respond with his own rally, what would become the March to Keep Fear Alive.[4] In the next months, both Stewart and Colbert planned their respective rallies, announcing any updates on their programs. Shortly before the rallies, however, Colbert admitted he had been unable to secure a location for his rally. Stewart stepped in and offered to share his space on the Washington Mall. The result was a merger of the two rallies into the Rally to Restore Sanity and/or Fear, a rhetorical move on the part of the Stewart and Colbert universe that encouraged the audience to become active in politics by prompting rational discussion for a healthy American democracy.[5]

The quotation that opens this chapter demonstrates interactivity on two levels. First, it is interactive in the sense that someone must actively go to the text and read it. Typically, this sort of interactivity would not truly be regarded as interactive by media scholars. Those like Sheizaf Rafaeli, for example, argue true interactivity is part of a three-step process of message transaction.[6] On the surface, there appears to be only be two steps. The user sends an information request to the World Wide Web for the Rally to Restore Sanity and/or Fear's Web site, and the returned information is supplied to the user. True interactivity would suggest a third step is necessary, and in this quotation, this third step comes in the form of the call to action. The Stewart/Colbert universe tells the reader they are looking for a particular type of person. As it continues, the Web site asks, "Are you one of those people? Excellent. Then we'd like you to join us in Washington, DC, on October 30—a date of no significance whatsoever—at *The Daily Show*'s 'Rally to Restore Sanity.'"[7] Sally McMillan describes various forms of interactivity as typically occurring in new media such as user-to-system, user-to-user, and user-to-text. The Rally to Restore Sanity and/or Fear, however, suggests another form of interactivity, one in which the text encourages the audience to go out and interact with the world and the people within it.[8]

Media scholarship has tended to focus on the way interactivity places the audience into action through options the media provide. However, this example from the Rally Web site shows that interactivity in the media can function rhetorically to encourage interactivity outside of the media, an important use of interactivity for rhetorical purposes Barbara Warnick calls the development of offline peer-to-peer networks.[9]

This chapter explores the rhetoric of interactivity through electronic and social media as it is used in the Stewart/Colbert universe. Interactivity provides the opportunity for the constituted audience to go beyond simply consuming rhetorical messages and, instead, encourages the audience members to become active civic participants. However, there are limitations to using interactivity for activism, as seen in the Rally to Restore Sanity and/or Fear as well as the Stewart/Colbert universe's use of interactivity through capitalism and rhetorical pranking. Despite these limitations, interactivity has the potential to be a powerful rhetorical tool, and examples of interactivity in the Stewart/Colbert universe such as Twitter, mobile applications, merchandising, fan sites, online forums, and moments on the television shows themselves show the ways interactivity functions rhetorically to get the audience members actively engaging with the media and, by extension, each other to encourage and, at times, produce engaged democratic participation.

INTERACTIVE MEDIA, CAPITALISM, AND THE ACTIV(IST?) AUDIENCE

In an interview with Chris Wallace on *Fox News Sunday*, Stewart denied he was an activist and instead was simply a comedian, but then went on to point out his comedic sensibility was "informed by an ideological background, there's no question about that."[10] Perhaps Stewart and company are not activists in their minds, but the Stewart/Colbert universe produces rhetorical messages that have increasingly encouraged the audience to take action in politics and society. Through interactivity, that activation of the audience can take material forms, and the embodied nature of the audience permits messages to spread in ways the producers of the Stewart/Colbert universe alone could not achieve. Often, this is done through comedy in a way that permits the audience members to deny they are true activists just as Stewart did in his interview with Chris Wallace. Even through comedy, though, the inspired actions of the audience members are unmistakably about making a difference and raising awareness, making it less likely they would not be considered activists, let alone desire as an audience to separate itself from that label.

Robert Asen argues what is considered as citizenship should be reconsidered in the current age. Specifically, Asen is concerned with engaged citizenship, the type of active rhetorical engagement interactivity can provide. Asen believes "we may wish to consider how citizenship engagement proceeds generatively, exhibits risk, affirms commitment, expresses creativity, and fosters sociability."[11] Asen's view of engaged citizenship bears many similarities to Nola Heidlebaugh's active artistic judgment.[12] By encouraging citizenship through creative thinking and social activity, Asen argues citizenship can be engaged in a number of different modes that allow democratic participation in ways that otherwise would not be seen as active citizenship. Many uses of interactivity demonstrate such engaged citizenship and make use of this creativity to inspire greater civic engagement.

Perhaps the clearest example of the Stewart/Colbert universe using interactivity through electronic and social media to call the audience to action is the development of the Rally. On September 20, 2010, Stewart announced the launch of the Web site RallytoRestoreSanity.com and the activation of a Twitter account where information about the Rally would be shared.[13] The same night, Colbert warned that going to Stewart's Web site would produce a virus—herpes—and instead the audience should go to KeepFearAlive.com and join the March's Facebook page.[14] As the rally developed, it would gain celebrity support such as through Oprah Winfrey's Twitter endorsement.[15] In fact, interactive media such as social media would play a pivotal role in shaping the preparation for the Rally as well as the development of its ideological message.

Both RallytoRestoreSanity.com and KeepFearAlive.com, which were nearly identical in layout, had links to the official accounts on social media sites such as Facebook and Twitter. Through these sites, the audience members were able to gather and share information about preparations for the Rally to Restore Sanity and/or Fear. Warnick argues this type of information sharing between users through online interactivity can produce a beneficial rhetorical effect. Not only does this sharing of information promote the event within the audience, content posted on these sites "accompanied by encouragement to share it with others extends site-specific news and information to other online audiences."[16] Features such as the "Share" button on Facebook and the use of hashtags on Twitter enabled the audience to use these sites to promote the Rally and its message quickly and easily to those beyond the Stewart/Colbert universe's audience.

To help the audience members stay in touch with the producers as well as each other during the event, the Stewart/Colbert universe created an application (app) for iPhone and Android users. The app featured some information found on the Rally Web sites, such as news updates and FAQs, but also had features that allowed the audience to explore and participate in the rhetorical construction of the Rally. The app, for example, featured a link to Google Maps that allowed users to direct themselves toward the Rally itself. The app also allowed the audience to "check-in" with the social media site Four-Square to let others know not only that they were at the Rally but where they were geographically, helping others physically there to find them.

The Rally app was one of many ways the Stewart/Colbert universe took advantage of interactivity through social media to bring the audience together quickly and have it help in the process of constructing the shared message. Zachary Sniderman praised this use of social media, noting the use of many different available interactive communication options like Facebook, Twitter, FourSquare, and texting to keep the Rally audience constantly in the loop.[17] Writing about political campaigns, Warnick argues that Web sites that provided spaces for the audience to contribute through interactive means aided in the rhetorical constructed of the audience because audiences felt a greater sense of connection to the campaign.[18] The use of social media in the preparation and execution of the Rally took this a step further by not only allowing the audience members to interact with the producers but also by actually responding to and using those audience contributions to shape the overall message. Sniderman argues, "As a way of engaging with their audience, earning new followers and maximizing the reach of social media, the Rally to Restore Sanity and/or Fear might just be setting the model."[19]

Many examples such as the Rally focus on some of the nobler ways the Stewart/Colbert universe puts its audience into action, but it must be remembered the universe remains an entertainment property that is permitted to exist because it generates a profit for a larger media conglomerate. As such, one common way money is generated is through providing merchandise for audiences to buy. Comedy Central offers merchandise for both television programs through its Web site.[20] *The Daily Show*'s supply of merchandise is limited, consisting of things such as a mug, copies of *America (The Book)*, and a few Rally to Restore Sanity bumper stickers and shirts.[21] *The Colbert Report* store is much larger, likely as a parodic response to Bill O'Reilly's entire online storefront that would rival some department stores.[22]

Comparing these shows to O'Reilly's store, however, suggests much of the merchandise exists alongside a money making endeavor to spread the rhetoric of the Stewart/Colbert universe in ways the television programs could not do alone. O'Reilly's store features numerous articles of clothing, for example, that are updated constantly. In fact, a member of O'Reilly's audience has the opportunity to become a walking billboard for the show due to the availability of hats, jackets, shirts, travel mugs, and exercise clothes, not to mention other items like doormats, key chains, and playing cards. The merchandise available for the Stewart/Colbert universe is significantly limited in comparison, and the products that would be seen as a public display of identification with the universe are limited. This is an important point, as it suggests the merchandise like shirts for the Rally were intended to help establish the offline peer-to-peer networks through interactivity discussed by Warnick.[23] The clothing becomes part of a performance of identification with the universe and a means to put the audience into action for the purposes of allowing the audience members to hail each other as clearly part of the universe. Moreover, Asen argues that, in neoliberal America, such consumerism may be one mode in which citizens are able to perform civic engagement in a creative and sociable manner.[24]

Take, for example, when Colbert announced on the August 20, 2007 episode that WristStrong bracelets would be available to purchase. He added that money from every sale of the bracelets would be given to the Yellow Ribbon Fund, an organization that helps injured service members when they return home from war.[25] Tying the bracelet to a charity with indirect links to a government endeavor, the Stewart/Colbert universe creates an opportunity for action. This link between the bracelet and the charity is noticeably important for some of the audience. For example, the product reviewer "A. Fox

'FuCH'" noted three years after the bracelet became available, "I [sic] bought two just so ten dollars would go to the yellow ribbon fund."[26] Importantly, the money does not go to the war effort directly, but itself serves as a rhetorical maneuver by supporting the people who have been sent to war and are victims of it. Again, the ideology of the Stewart/Colbert universe controls the rhetoric of the bracelet by inscribing upon it a specific statement that the audience members choose to purchase the bracelet support soldiers without necessarily supporting the wars in Afghanistan and Iraq, wars frequently criticized throughout the Stewart/Colbert universe. In this case, money becomes speech for the audience, speech that is permitted by the interactive opportunities created by the producers.

The WristStrong bracelet is one of many examples in which Colbert used interactivity to encourage the audience to make donations to charities that support the ideological cause of the universe. Colbert used his show to encourage the audience for an entire week to support the Japanese relief effort following the 2011 tsunami by donating to the Red Cross through texting.[27] In 2007, Colbert introduced his Ben & Jerry's ice cream flavor, Americone Dream, and encouraged his audience to consume as much as possible because the proceeds went to charities such as the Yellow Ribbon Fund.[28] Colbert has also used the power of the audience to support the 2010 U.S. Olympic speed skating team financially, leading almost 10,000 audience members into donating over $300,000 to the team.[29] Even the Rally was used as a platform to raise charitable donations for DonorsChoose.org and the Trust for the National Mall.[30]

One of the most significant events to activate the audience by Colbert came through an exchange with Jimmy Fallon in 2011. Fallon, then host of the NBC show *Late Night with Jimmy Fallon*, had previously been feuding with Colbert over their rival ice cream flavors in an attempt to raise the most money for charity. Having made up on *The Colbert Report* in early March of 2011, Colbert and Fallon became quick on-air friends. In late March, Colbert auctioned off a self-portrait for $26,000 and donated the proceeds to Donors-Choose.org to help art programs in schools. Colbert also praised Fallon for agreeing to match the donation personally.[31] The next Monday, however, Fallon announced on his program that he never made such a promise and was immediately confronted by Colbert, who argued he has every right to give away Fallon's money since they are "best friends." Colbert stormed off in disgust, and Fallon told his audience members they needed to fix the problem by raising the money. If they could do it by that Friday, Fallon promised Colbert would return on Friday to sing Rebecca Black's *Friday* on his show.[32] The next day on *The Colbert Report*, Colbert expressed his anger with Fallon, and told his audience he put a donation button on the ColbertNation.com Web site so large they could easily avoid it.[33] Three hours after the episode aired, donations surpassed the $26,000 mark.[34] As the week continued, donations continued to pour in and Colbert and Fallon together sang the song at the end of the week. The result was an extended intertextual exchange that activated the audience to become a part of the solution for strained art programs throughout the country. By combining the audiences of both the Stewart/Colbert universe and Fallon's fans, as of April 2012, money from the stunt resulted in donors raising over $100,000 to help nearly 55,000 art students.[35]

These examples show the way the Stewart/Colbert universe can activate the audience through interactivity to send a message about the ideological values of the universe. Charitable donations have become a form of commodity activism audience members can use to speak back to the Stewart/Colbert universe and shape the message on their own. This form of participation serves as one way the audience can perform the sort of engaged citizenship through capitalism of which Asen speaks.[36] For example, audience members used the Reddit Web site and created the ColbertRally.com Web site to encourage Colbert to hold a rally, but this was not their only form of interactive speech to gain the attention of the universe. The Reddit posts and ColbertRally.com also encouraged supporters to donate money to one of Colbert's favorite charities, DonorsChoose.org. The audience spoke loudly, donating over $100,000 in just a few days, and breaking the servers for DonorsChoose.org in the process.[37] While it is quite likely that the wheels were already in motion for what would eventually become the Rally to Restore Sanity and/or Fear, the actions of these members of the audience demonstrate the power of interactivity in allowing the audience to signal to the producers what it desires. If there had been doubts about the appeal of a rally, such a quick and large donation uses interactivity and money to send a rhetorical message, one that, done through the charity, acknowledges the corresponding ideological perspective of the Stewart/Colbert universe.

It is important to note the act of donating to charities is not necessarily a substantial form of civic engagement for the audience. Although it does represent one form of the creative and generative form of engaged citizenship of which Asen speaks, determining how the impact of that engagement may be impossible. Asen uses the example of someone choosing a local coffee shop over Starbucks. While some may make the choice out of protest to fair trade laws, others may make the choice because one is closer than the other. As such, it becomes difficult to suggest purchasing from a local coffee is necessarily an act of civic engagement. Similarly, making other forms of consumer purchases such as charitable donations do not necessarily equate to active civic engagement. Asen argues such determinations would need to be based on contextualizing the rhetorical act. The Stewart/Colbert universe's rhetoric contextualizes such a determination. However, as Asen argues, the choice to perform engaged citizenship is often individualized, making it difficult to suggest the option for charitable donations or any form of citizenship mentioned here was a purposefully performed act of citizenship on the part of the audience members.[38] What is important, then, is that interactivity, at least through capitalism, provides opportunities for modes of engaged citizenship and expression of judgment.

PRANKING THROUGH INTERACTIVE MEDIA

Of course, those opportunities for active engagement extend beyond capitalism. Another rhetorical strategy employed throughout the Stewart/Colbert universe to activate the audience through interactivity is pranking. Pranking as a rhetorical strategy provides an opportunity for audiences to interject themselves into the normal flow of mediated discourse. A rhetorical prank, according to Christine Harold, is not a stunt to provide opposition to a viewpoint, but instead a "stylistic exaggeration" that can "render a qualitative change by turning and doubling a material or text."[39] Pranks work through appropriating mediated texts and adding a turn of the text, then re-presenting that text back to the audience in a way that disrupts the ability for the media to appear as a seamless and infallible machine. In one example of pranking, Harold talks about the Barbie Liberation Organization, who switched the voice boxes of talking Barbie and G.I. Joe dolls and returned the dolls to stores with a note hidden in the package urging angry families to call their TV news stations to complain.[40] Harold, discussing cultural jamming of which she considers pranking to be a part of, argues that:

> "[J]amming" as a metaphor does not have to be interpreted only as a damming or stopping of corporate media. More interestingly, it can be a strategy that artfully proliferates other media and messages that challenge the ability of corporate messages to make meaning in predictable ways—to jam with rather than against.[41]

Harold argues that pranks are typically associated with arguments about the meaning of the prank, "but such arguments are translations of pranks. They do not account for the power of the prank itself. One might even argue that such translations dilute the rhetorical power of pranks to confuse and provoke."[42]

Amber Day, however, disagrees with Harold, noting, "While I agree that the cleverness of a prank can obviate the need for direct didacticism, I would argue that if it is to have any political or rhetorical effect, a prank must imply some moderately clear critique . . . or risk being nothing more than an exercise in art-student narcissism."[43] Noted rhetorician and prankster Kembrew McLeod appears to agree with Day, arguing, "A good prank uses deception to speak truth to power, or at least crack jokes that expose fissures in power's façade, often using the news to do the work."[44] For example, one of

McLeod's most notable pranks involved him copyrighting the phrase "Freedom of Expression," then sending out cease and desist letters to users of the phrase in order to get the media talking about the way copyright laws have been manipulated to expand corporate power over creativity.[45] McLeod proposes the formula of "Pranks = Satire + Performance Art × Media," a formula that highlights the combination of irony, intertextuality, and interactivity that permits such pranks to have the rhetorical force spoke of by Day and McLeod.[46]

The Stewart/Colbert universe has used the rhetorical strategy of pranking to help encourage an active audience in a way that allows the audience to at least participate in a criticism that inserts itself into other parts of media and culture. Colbert's book *I Am America (And So Can You!)* provides one such example. Since the character of Colbert is so self-assured, it would seem inevitable to him that the book would win numerous awards. With no sense in delaying the inevitable, Colbert awarded his book the first recipient of the Stephen T. Colbert Award for the Literary Excellence. At the end of the book, Colbert tells the audience, "You need to take the lessons of this book and apply them to the community."[47] Colbert's examples in the book stress verbal communication with others as a means to spread his message, but included between this page and the next is a sheet of stickers with the following instructions:

> Heroes, by buying and reading this book, you've proven that you get it—and are therefore now members of the nominating committee for the Stephen T. Colbert Award for the Literary Excellence. Use the medallions below to nominate any book that you feels embodies the values of the Colbert Nation.

The book includes twelve stickers for the reader to affix on to other books that would presumably match Colbert's values. *I Am America*, thus, sets up the potential for the audience to engage in an elaborate prank by placing upon other books the fake award sticker.

There is little evidence as to how common it was for the audience to act upon this potential prank. A brief discussion among readers on the Web site GoodReads.com brainstorms possible recipients of Colbert's award, in particular the works of Ann Coulter.[48] Many of the customer reviews of the book make mention of the award stickers, but none discuss putting the stickers into use.[49] There was also one brief exchange about using the award stickers on the Colbert Nation Forum page.[50] Although Colbert has used the book to perform minor pranks such as placing the book in the stack of books supposedly referenced by America's founding fathers at the National Constitution Center, he never makes reference on *The Colbert Report* to the audience using the stickers as described by the book.[51] Therefore, despite the potential for a prank with rhetorical influence, it becomes difficult to determine whether this lack of discussion about the award means the prank occurred at all or whether it was ineffective or not.

Like the Stephen T. Colbert Award for the Literary Excellence, Colbert's creation of the "Farewell to Postage" postage represents another opportunity for the audience to participate in the process of pranking. On the September 14, 2011 episode of *The Colbert Report*, Stephen Colbert ran a story on the seemingly imminent downfall of the U.S. Postal Service due primarily to budgetary issues. Colbert argues there is only one reason to lament loss of mail service: he has not been on a stamp yet. In true Colbert fashion, he provides the solution by offering his viewers an actual U.S. postage stamp available for purchase that features his picture. By offering a stamp for his audience to purchase, Stephen Colbert provides a means to actually save the institution his character is proposing the American public let die.[52] As part of the received letter, the stamps are literally a way to stick a commentary on the general message that disrupts the normal flow of communication. Again, it is difficult to determine what impact the stamps may have, although, unlike the Stephen T. Colbert Award stickers, there is greater reason to believe that stamps are being used since there are only two reasons to purchase the stamps: usage or collection.

Both the Stephen T. Colbert Award and the "Farewell to Postage" postage represent opportunities presented by the Stewart/Colbert universe for the audience to participate in the rhetorical act of pranking. There are, however, examples where pranks have been conducted involving the audience to influence the process of meaning-making. As I mentioned earlier, social media like Twitter is one space in which people can share and extend a rhetorical message to indicate inclusion in the audience. Aware of this, Jon Stewart used his audience through *The Daily Show*'s Twitter account to influence then presidential candidate Jon Huntsman's open town hall meeting. On November 29, 2011, Huntsman's camp sent the tweet, "I'm taking your questions on Twitter at 4:45 ET TODAY. Use hashtag #Q4Jon. Excited to do it!" *The Daily Show* producers, recognizing the potential for a successful prank, tweeted to their audience, "@MadMen_AMC fans: tweet Jon Hamm your questions by 4:15 Eastern! Hashtag #Q4Jon." In the world of Twitter, hashtags are a means to organize like-minded topics and create a trend that draws the attention of those looking for tweets related to the same topics. Just as the switching of Barbie and G.I. Joe voice boxes described by Harold represents one prank that forced consumers to consider the rhetorical significance of not only the prank but the original messages as well, Stewart and company's switching of the hashtag from a political gathering to fans of the television show *Mad Men* shows how interactivity can be used with social media to impact the meaning-making process.

There are some important limitations to the prank, however. Perhaps unfortunately for Stewart, Huntsman's camp was able to catch on and the quick-thinking Huntsman played along with the prank, answering tweets such as "Jon, in ten years do you see yourself following George Clooney's or Tom Selleck's career path?" with "I definitely want Clooney's career path. But I think that Q was for Jon Hamm. Funny prank @TheDailyShow! #Q4Jon."[53] The Huntsman camp's technological and cultural savvy perhaps diffused the prank before Huntsman could accidentally reveal information that may have impacted his political narrative, making it less of a disruption of meaning-making and more of a humorous linkage between the campaign and the Stewart/Colbert universe.

This issue draws attention to need for those who use interactivity to be one step ahead of the object of the prank. This prank, however, also reinforces the illusion of agency for the audience. In this prank, the audience is not as much in on the prank as it is a medium to institute the prank. *The Daily Show* producers take advantage of their audience's collective ear through Twitter to encourage them to become the active mechanism in the prank. In this situation, though, the audience members are unaware they are part of the prank, and instead truly believe they are talking to Jon Hamm and asking him questions about *Mad Men*. While this shows an important way the producers may use interactivity to produce rhetorical activity, this form of activity reduces the role of the audience. Interactivity does not guarantee an increased role of the audience members as individual contributors to the rhetorical process. Given the overall desire to turn inactive people into active and critically minded citizens, it is counter-productive to trick them into participation without them being aware of their role in the attempted construction of a rhetorical message.

Although there are limitations to these pranks, there have been successful pranks performed in by the Stewart/Colbert universe and the audience. One of the most successful pranks in the history of the Stewart/Colbert universe came in early 2009. NASA announced they were campaigning to let citizens vote for the name of what was then known as Node 3 of the International Space Station. Although NASA suggested a few names to vote on, they permitted write in votes. On March 3, 2009, Colbert, in his words, "mobilized" the audience and told audience members to go to the NASA site and enter in "Colbert" as the name for the station.[54] By the next night, the choice for "Colbert" topped the list of options.[55] On March 10, NASA representative William Gerstenmaier hinted NASA may not honor the results of the vote if "Colbert" won.[56] When the final votes were counted, "Colbert" had won handily. The vote was created by NASA to represent the spirit of cooperation, but Gerstenmaier's prediction had apparently come true with NASA implying they were going to go ahead with the second place choice, "Serenity."[57]

The prank, initialized by Colbert and put into action by the audience, may have started out as what seemed like a joke but turned into a rhetorical opportunity to expose the deception of the vote. Colbert was quick to point to the hypocrisy of ignoring the results of a vote that was inspired by the notion of cooperation when the outcome was brought on by his audience coming together.[58] Democratic Congressman Chaka Fattah took it a step further, releasing a press statement that said, "NASA decided to hold an election to name its new room at the international space station and the clear winner is Stephen Colbert. The people have spoken, and Stephen Colbert won it fair and square—even if his campaign was a bit over the top. . . . We insist on democracy in orbit."[59] NASA eventually settled on an another write-in name, "Tranquility," and opted to name a treadmill sent into space the Combined Operational Load Bearing External Resistance Treadmill or C.O.L.B.E.R.T.[60] Colbert and his audience may not have achieved the goal of naming the node after Colbert, but the prank heightened awareness of the ways government agencies, even for seemingly inconsequential decisions, can strip the democratic process of its meaning. The result was a discussion with greater rhetorical significance than if the node had simply been named "Colbert." The prank became a moment for the Stewart/Colbert universe to use interactivity not only to interject the universe into another situation but to also use that opportunity to change the discussion from a celebration of space achievement to one of the hidden limits of democracy in America.

CONCLUSION

This chapter has used a variety of examples throughout the Stewart/Colbert universe to argue for how interactivity provides modes of participation that allows for the type of active engaged citizenship Asen proposes for today's democracy.[61] The Rally to Restore Sanity and/or Fear shows how involving the audience through social media allows the audience to play an important role in the construction of the message. Although constantly guided by the producers' hands, the audience embraced the interactivity of the event and produced what one writer called a potential "defining moment of [a] generation."[62] As Zachary Sniderman argues, "the Rally to Restore Sanity and/or Fear might just be setting the model" for how to use social media for activism, and the Rally highlights the way such activism can be shaped by and shape the audience.[63]

Interestingly, the Stewart/Colbert universe has appeared to embrace the concept of using capitalism for speech. As Asen argues, in neoliberal America, such forms of activism may be powerful forms of speech.[64] Through direct calls for donations, the Stewart/Colbert universe at varying times was able to support the Olympics, teachers, the National Mall, and many other organizations. However, it is the manner in which the universe was able to integrate the use of capitalist activism with their more direct forms of activism such as through the Rally or through playful forms of activism such as pranking that suggests the power of this form of interactivity as part of the construction of an overarching ideological message.

Separated from the capitalist connection to charities, pranks in the Stewart/Colbert universe put the audience into action in an effort to shape the process of message construction. The Stephen T. Colbert Award for the Literary Excellent and the "Farewell to Postage" postage show how the producers can construct the mechanism for a prank, but rely on the audience to accomplish the task. In these examples, the prank's critique is written into the instructions for the audience that, in turn, confine the ways the objects of the prank may interpret the commentary. Of course, without follow-up, it is unclear how impactful these pranks were in execution. Colbert's NASA prank, however, was less clear in its critique. Although it served early on as a means to highlight the absurdity of the naming process, the fluidity of the prank allowed for the conversation to turn more toward a discussion of democracy after NASA's decision. Colbert's NASA prank suggests there is benefit in letting commentary emerge from an interactive prank as opposed to being inscribed upon it from the beginning.

These examples may not make the Stewart/Colbert universe or its audience activists in a strict sense. In fact, as I have argued, there may be limitations to how powerful some of these uses of interactivity as rhetoric can be in encouraging the audience to act. However, the use of interactivity points to ways the entire Stewart/Colbert universe can be put into action to produce material changes and shape discourse beyond the confines of the universe itself. For now, though, it is important to further consider how interactive media such as social media are a valuable rhetorical tools for bringing the audience into the process of producing and shaping meaning-making well beyond the confines of the television texts.

NOTES

1. The Web site remains titled as the Rally to Restore Sanity. However, as the rally progressed in planning and merged with Colbert's March to Keep Fear Alive, the content merged to become the Rally to Restore Sanity and/or Fear. "Rally to Restore Sanity," RallytoRestoreSanity.com, http://www.rallytorestoresanity.com/.

2. Amy Gardner, Krissah Thompson, and Philip Rucker, "Beck, Palin tell thousands to 'restore America,'" *Washington Post*, August 29, 2010, http://www.washingtonpost.com/wp-dyn/content/article/2010/08/28/AR2010082801106.html.

3. Gardner, Thompson, and Rucker, "Beck, Palin," para. 10; "I Have a Scheme," *The Daily Show*, August 26, 2010, http://www.thedailyshow.com/watch/thu-august-26-2010/i-have-a-scheme.

4. "Geese Witherspoon," *The Colbert Report*, September 7, 2010, http://www.colbertnation.com/the-colbert-report-videos/352238/september-07-2010/geese-witherspoon; "What Is Restoring Truthiness?" ColbertRally.com, http://colbertrally.com/what-is-restoring-truthiness.

5. Throughout this chapter, I refer to the set of paratexts associated with *The Daily Show with Jon Stewart* and *The Colbert Report*, including all their electronic paratexts, as the Stewart/Colbert universe. This is done not only for simplicity but also to reinforce a underlying assumption in this chapter, which is that the interactivity found throughout the texts links them all as well as the audience together in the process of textual construction and meaning-making.

6. Sheizaf Rafaeli, "Interactivity: From New Media to Communication," in *Advancing Communication Science: Merging Mass and Interpersonal Processes*, eds. R. P. Hawkins, J. M. Wiemann, and S. Pingree (Beverly Hills, CA: Sage, 1988), 110–34.

7. "Rally to Restore Sanity," RallytoRestoreSanity.com, http://www.rallytorestoresanity.com/.

8. Sally J. McMillan, "Exploring Models of Interactivity from Multiple Research Traditions: Users, Documents and Systems. In *The Handbook of New Media*, eds. Leah A. Lievrouw and Sonia Livingston (Thousand Oaks, CA: Sage, 2002), 163–82.

9. Barbara Warnick, *Rhetoric Online: Persuasion and Politics on the World Wide Web* (New York: Peter Lang, 2007), 88.

10. *Fox News Sunday with Chris Wallace*, Fox News, June 20, 2011.

11. Robert Asen, "A Discourse Theory of Citizenship," *Quarterly Journal of Speech* 90 (May 2004): 189–211.

12. Nola J. Heidlebaugh, *Judgment, Rhetoric, and the Problem of Incommensurability* (Columbia, SC: The University of South Carolina Press, 2001).

13. "Ride to the Rally to Restore Sanity," *The Daily Show*, September 20, 2010, http://www.thedailyshow.com/watch/mon-september-20-2010/ride-to-the-rally-to-restore-sanity.

14. "March to Keep Fear Alive Media Coverage," *The Colbert Report*, September 20, 2010, http://www.colbertnation.com/the-colbert-report-videos/359631/september-20-2010/march-to-keep-fear-alive-media-coverage.

15. "Rally to Restore Sanity and/or Fear Announcement," *The Daily Show*, October 14, 2010. http://www.thedailyshow.com/watch/thu-october-14-2010/rally-to-restore-sanity-and-or-fear-announcement.

16. Warnick, *Rhetoric Online*, 88.

17. Zachary Sniderman, "How the 'Rally to Restore Sanity and/or Fear' Nailed Social Media," *Mashable*.com, October 27, 2010, http://mashable.com/2010/10/27/stewart-colbert-rally-social-media.

18. Warnick, *Rhetoric Online*, 89.

19. Sniderman, "Nailed Social Media," para. 14.

20. Comedy Central is the host network for both *The Daily Show with Jon Stewart* and *The Colbert Report*. Comedy Central is a property of Viacom.

21. "*The Daily Show* merchandise," Shop.ComedyCentral.com, http://shop.comedycentral.com/category/41229863681/1/The-Daily-Show.htm.

22. To see just how much O'Reilly merchandise exists, see http://www.billoreilly.com/store.

23. Warnick, *Rhetoric Online*, 88.

24. Asen, "A Discourse Theory of Citizenship," 206–7.

25. After injuring his wrist in July of 2007, Colbert went on a crusade to tackle what he called the media's endorsement of wrist violence in America. As a response, on the August 8, 2007 episode of *The Colbert Report* he introduced the WristStrong bracelet. The WristStrong bracelet is a silicone bracelet much like Lance Armstrong's LiveStrong bracelet, which was intended to raise money and awareness for cancer. "Wrist Watch—Fighting Back," *The Colbert Report*, August 8, 2007, http://www.colbertnation.com/the-colbert-report-videos/91153/august-08-2007/wrist-watch---fighting-back/; "WristStrong Bracelets," *The Colbert Report*, August 20, 2007, http://www.colbertnation.com/the-colbert-report-videos/91799/august-20-2007/wriststrong-bracelets.

26. "Customer Reviews. The Colbert Report WRISTSTRONG Bracelet," Amazon.com.

27. "Crisis in the Middle Everywhere—Japan & Libya," *The Colbert Report*, March 21, 2011, http://www.colbertnation.com/the-colbert-report-videos/378276/march-21-2011/crisis-in-the-middle-everywhere---japan---libya; "Californians Respond to Japanese Disaster," *The Colbert Report*, March 22, 2011, http://www.colbertnation.com/the-colbert-report-videos/

378439/march-22-2011/californians-respond-to-japanese-disaster; "The Word—Over-Reactor," *The Colbert Report*, March 23, 2011, http://www.colbertnation.com/the-colbert-report-videos/378624/march-23-2011/the-word---over-reactor; "Bears & Balls – Misery Edition," *The Colbert Report*, March 24, 2011, http://www.colbertnation.com/the-colbert-report-videos/378802/march-24-2011/bears---balls---misery-edition.

28. Colbert has kept Americone Dream in the minds of his audience through several segments in order to raise the charitable donations. The flavor remains a strong seller for Ben & Jerry's. See "Ben and Jerry—Introducing Americone Dream," *The Colbert Report*, March 5, 2007, http://www.colbertnation.com/the-colbert-report-videos/83216/march-05-2007/ben-and-jerry---introducing-americone-dream; "Stephen Colbert Charity Ice Cream at Ben & Jerry's Shops NOW thru Feb. 2012," Yellow Ribbon Fund, August 11, 2011, http://www.yellowribbonfund.org/?p=2514.

29. Howard Berkes, "U.S. Speedskating Finds Savior in Stephen Colbert," National Public Radio, January 19, 2010, http://www.npr.org/templates/story/story.php?storyId=122731145.

30. "Langur Monkey Security," *The Colbert Report*, October 5, 2010, http://www.colbertnation.com/the-colbert-report-videos/361086/october-05-2010/langur-monkey-security; "Rally to Restore Sanity—FAQ," RallytoRestoreSanity.com, http://www.rallytorestoresanity.com/faq/, para. 1.

31. "Raging Art-On—Art 3," *The Colbert Report*, March 23, 2011, http://www.colbertnation.com/the-colbert-report-videos/378623/march-23-2011/raging-art-on---art-3.

32. Rebecca Black's *Friday* was a then popular Internet video that featured a thirteen-year-old Black singing a song about her Fridays. *Late Night with Jimmy Fallon*, NBC, March 28, 2011.

33. "Jimmy Fallon Promises a Performance by Stephen," *The Colbert Report*, March 29, 2011, http://www.colbertnation.com/the-colbert-report-videos/379251/march-29-2011/jimmy-fallon-promises-a-performance-by-stephen.

34. "Stephen Practices Rebecca Black's 'Friday,'" *The Colbert Report*, March 30, 2011, http://www.colbertnation.com/the-colbert-report-videos/379370/march-30-2011/stephen-practices-rebecca-black-s--friday-.

35. "The Jimmy Fallon/Stephen Colbert Project," DonorsChoose.org, http://www.donorschoose.org/donors/viewChallenge.html?id=157284.

36. Asen, "A Discourse Theory of Citizenship," 206–7.

37. Megan Friedman, "Reddit Campaign for Colbert Rally Breaks Donation Record," *Time*, September 14, 2010, http://newsfeed.time.com/2010/09/14/reddit-campaign-for-colbert-rally-breaks-charity-records/#comment-14227.

38. Asen, "A Discourse Theory of Citizenship," 206–7.

39. Christine Harold, *OurSpace: Resisting the Corporate Control of Culture* (Minneapolis: The University of Minnesota Press, 2007), 78.

40. Harold, *OurSpace*, 79–81.

41. Harold, *OurSpace*, xxvi.

42. Harold, *OurSpace*, 106.

43. Amber Day, *Satire and Dissent: Interventions in Contemporary Political Debate* (Bloomington: Indiana University Press, 2011), 174.

44. Kembrew McLeod, "Everything Is Connected," *Quarterly Journal of Speech* 96 (November 2010): 421–26.

45. Kembrew McLoed, "My Freedom of Expression® Trademark," Kembrew.com, http://kembrew.com/prank/my-trademark-of-freedom-of-expression/.

46. Kembrew McLeod, "Everything Is Connected," 424.

47. Stephen Colbert, Richard Dahm, Paul Dinello, and Allison Silverman, eds., *I Am America (And So Can You!)* (New York: Grand Central Publishing, 2007), 215.

48. "The Stephen T. Colbert Award Sticker," GoodReads.com, http://www.goodreads.com/topic/show/64611-the-stephen-t-colbert-award-sticker.

49. "Customer Reviews: *I Am America (And So Can You!)*," Amazon.com, http://www.amazon.com/Am-America-And-Can-You/product-reviews/0446582182/ref=cm_cr_dp_all_summary?ie=UTF8&showViewpoints=1&sortBy=bySubmissionDateDescending.

50. "How do I submit my book to get the Stephen T. Colbert award for literary excellence," Colbert Nation Forum, May 21, 2010, http://forums.colbertnation.com/?page=ThreadView& thread_id=141.
51. "National Constitution Center," *The Colbert Report*, April 16, 2008, http://www.colbertnation.com/the-colbert-report-videos/166020/april-16-2008/national-constitution-center.
52. "Return to Sender," *The Colbert Report*, September 14, 2011, http://www.colbertnation.com/ the-colbert-report-videos/396672/september-14-2011/return-to-sender.
53. Aly Semigran, "Jon Stewart and 'The Daily Show' Twitter prank Jon Huntsman, who, for the record, is not Jon Hamm," *Entertainment Weekly*, November 30, 2011, http://popwatch.ew.com/ 2011/11/30/jon-stewart-and-the-daily-show-mad-men-twitter-prank-jon-huntsman/.
54. "Space Module: Colbert—Name NASA's Node 3 After Stephen," *The Colbert Report*, March 3, 2009, http://www.colbertnation.com/the-colbert-report-videos/220492/march-03-2009/space-module--colbert---name-nasa-s-node-3-after-stephen.
55. "Space Module: Colbert—Scientiology's New Galactic Overlord," *The Colbert Report*, March 3, 2009, http://www.colbertnation.com/the-colbert-report-videos/220648/march-04-2009/space-module--colbert---scientology-s-new-galactic-overlord.
56. "Space Module: Colbert—William Gerstenmaier," *The Colbert Report*, March 10, 2009, http://www.colbertnation.com/the-colbert-report-videos/221174/march-10-2009/space-module--colbert---william-gerstenmaier.
57. "Space Module: Colbert—Democracy in Orbit," *The Colbert Report*, March 30, 2009, http://www.colbertnation.com/the-colbert-report-videos/223137/march-30-2009/space-module--colbert---democracy-in-orbit.
58. "Space Module: Colbert—Democracy in Orbit," *The Colbert Report*.
59. "NASA Appropriator Says: Stephen Colbert Is Out of this World," Congressman Chaka Fattah, 2nd District of Pennsylvania, March 26, 2009, http://fattah.house.gov/press/nasa-appropriator-says-stephen-colbert-is-out-of-this-world/.
60. "New Space Station Module Name Honors Apollo 11 Anniversary," NASA, http://www.nasa.gov/externalflash/name_ISS/index.html.
61. Asen, "A Discourse Theory of Citizenship."
62. Zak Kinnaird, "Stewart and Colbert Gathering May Be Defining Moment of Generation," *Daily Athenaeum*, September 19, 2010, http://www.thedaonline.com/article_832f1c08-7e2b-5bcb-9769-e80a2b006b70.html, para. 13.
63. Sniderman, "Nailed Social Media," para. 14.
64. Asen, "A Discourse Theory of Citizenship," 206–7.

REFERENCES

Asen, Robert. "A Discourse Theory of Citizenship." *Quarterly Journal of Speech* 90 (May 2004): 189–211.
Berkes, Howard. "U.S. Speedskating Finds Savior in Stephen Colbert," *National Public Radio*, January 19, 2010. http://www.npr.org/templates/story/story.php?storyId=122731145.
The Colbert Report. "Bears & Balls—Misery Edition." *The Colbert Report* video, 5:38. March 24, 2011. http://www.colbertnation.com/the-colbert-report-videos/378802/march-24-2011/bears---balls---misery-edition.
———. "Ben and Jerry—Introducing Americone Dream." *The Colbert Report* video, 6:00. March 5, 2007. http://www.colbertnation.com/the-colbert-report-videos/83216/march-05-2007/ben-and-jerry---introducing-americone-dream.
———. "Californians Respond to Japanese Disaster." *The Colbert Report* video, 2:42. March 22, 2011. http://www.colbertnation.com/the-colbert-report-videos/378439/march-22-2011/californians-respond-to-japanese-disaster.
———. "Crisis in the Middle Everywhere—Japan & Libya." *The Colbert Report* video, 5:33. March 21, 2011. http://www.colbertnation.com/the-colbert-report-videos/378276/march-21-2011/crisis-in-the-middle-everywhere---japan---libya.

———. "Geese Witherspoon." *The Colbert Report* video. 7:32. September 7, 2010. http://www.colbertnation.com/the-colbert-report-videos/352238/september-07-2010/geese-witherspoon.

———. "Jimmy Fallon Promises a Performance by Stephen." *The Colbert Report* video. 3:44. March 29, 2011. http://www.colbertnation.com/the-colbert-report-videos/379251/march-29-2011/jimmy-fallon-promises-a-performance-by-stephen.

———. "Langur Monkey Security." *The Colbert Report* video. 2:45. October 5, 2010. http://www.colbertnation.com/the-colbert-report-videos/361086/october-05-2010/langur-monkey-security.

———. "March to Keep Fear Alive Media Coverage." *The Colbert Report* video. 3:35. September 20, 2010. http://www.colbertnation.com/the-colbert-report-videos/359631/september-20-2010/march-to-keep-fear-alive-media-coverage.

———. "National Constitution Center." *The Colbert Report* video. 5:33. April 16, 2008. http://www.colbertnation.com/the-colbert-report-videos/166020/april-16-2008/national-constitution-center.

———. "Raging Art-On—Art 3." *The Colbert Report* video. 7:38. March 23, 2011. http://www.colbertnation.com/the-colbert-report-videos/378623/march-23-2011/raging-art-on---art-3.

———. "Return to Sender." *The Colbert Report* video. 3:42. September 14, 2011. http://www.colbertnation.com/ the-colbert-report-videos/396672/september-14-2011/return-to-sender.

———. "Space Module: Colbert—Democracy in Orbit." *The Colbert Report* video. 3:18. March 30, 2009. http://www.colbertnation.com/the-colbert-report-videos/223137/march-30-2009/space-module--colbert---democracy-in-orbit.

———. "Space Module: Colbert—Name NASA's Node 3 After Stephen." *The Colbert Report* video. 2:27. March 3, 2009. http://www.colbertnation.com/the-colbert-report-videos/220492/march-03-2009/space-module--colbert---name-nasa-s-node-3-after-stephen.

———. "Space Module: Colbert—Scientology's New Galactic Overlord." *The Colbert Report* video. 3:53. March 3, 2009. http://www.colbertnation.com/the-colbert-report-videos/220648/march-04-2009/space-module--colbert---scientology-s-new-galactic-overlord.

———. "Space Module: Colbert—William Gerstenmaier." *The Colbert Report* video. 6:46. March 10, 2009. http://www.colbertnation.com/the-colbert-report-videos/221174/march-10-2009/space-module--colbert---william-gerstenmaier.

———. "Stephen Practices Rebecca Black's 'Friday.'" *The Colbert Report* video. 3:24. March 30, 2011. http://www.colbertnation.com/the-colbert-report-videos/379370/march-30-2011/stephen-practices-rebecca-black-s--friday-.

———. "The Word—Over-Reactor." *The Colbert Report* video. 5:26. March 23, 2011. http://www.colbertnation.com/the-colbert-report-videos/378624/march-23-2011/the-word---over-reactor.

———. "WristStrong Bracelets." *The Colbert Report* video. 0:44. August 20, 2007. http://www.colbertnation.com/the-colbert-report-videos/91799/august-20-2007/wriststrong-bracelets.

———. "Wrist Watch—Fighting Back." *The Colbert Report* video. 4:48. August 8, 2007. http://www.colbertnation.com/the-colbert-report-videos/91153/august-08-2007/wrist-watch---fighting-back/.

Colbert, Stephen, Richard Dahm, Paul Dinello, and Allison Silverman, eds. *I Am America (And So Can You!)*. New York: Grand Central Publishing, 2007.

"Customer Reviews: I Am America (And So Can You!)." Amazon. http://www.amazon.com/Am-America-And-Can-You/product-reviews/0446582182/ref=cm_cr_dp_all_summary?ie=UTF8&showViewpoints=1&sortBy=bySubmissionDateDescending.

"Customer Reviews: The Colbert Report WRISTSTRONG Bracelet." Amazon. http://www.amazon.com/The-Colbert-Report-WRISTSTRONG-Bracelet/product-reviews/B0038QUYJ0/ref=cm_cr_pr_btm_link_1?ie=UTF8&showViewpoints=0.

"*The Daily Show* merchandise." Shop.ComedyCentral, http://shop.comedycentral.com/category/41229863681/1/The-Daily-Show.htm.

The Daily Show with Jon Stewart. "I Have a Scheme." *The Daily Show with Jon Stewart* video, 10:19. August 26, 2010. http://www.thedailyshow.com/watch/thu-august-26-2010/i-have-a-scheme.

———. "Rally to Restore Sanity and/or Fear Announcement." *The Daily Show with Jon Stewart* video, 10:29. October 14, 2010. http://www.thedailyshow.com/watch/thu-october-14-2010/rally-to-restore-sanity-and-or-fear-announcement.

———. "Ride to the Rally to Restore Sanity." *The Daily Show with Jon Stewart* video, 2:49. September 20, 2010. http://www.thedailyshow.com/watch/mon-september-20-2010/ride-to-the-rally-to-restore-sanity.

Day, Amber. *Satire and Dissent: Interventions in Contemporary Political Debate.* Bloomington: Indiana University Press, 2011.

Fox News Sunday with Chris Wallace. Fox News. June 20, 2011.

Friedman, Megan. "Reddit Campaign for Colbert Rally Breaks Donation Record." *Time,* September 14, 2010. http://newsfeed.time.com/2010/09/14/reddit-campaign-for-colbert-rally-breaks-charity-records/#comment-14227.

Gardner, Amy, Krissah Thompson, and Philip Rucker. "Beck, Palin Tell Thousands to 'Restore America.'" *Washington Post,* August 29, 2010. http://www.washingtonpost.com/wp-dyn/content/article/2010/08/28/AR2010082801106.html.

Harold, Christine. *OurSpace: Resisting the Corporate Control of Culture.* Minneapolis: The University of Minnesota Press, 2007.

Heidlebaugh, Nola J., *Judgment, Rhetoric, and the Problem of Incommensurability.* Columbia: The University of South Carolina Press, 2001.

"How Do I Submit My Book to Get the Stephen T. Colbert Award for Literary Excellence?" Colbert Nation Forum. May 21, 2010. http://forums.colbertnation.com/?page=ThreadView&thread_id=141.

"The Jimmy Fallon/Stephen Colbert Project." DonorsChoose. http://www.donorschoose.org/donors/viewChallenge.html?id=157284.

Kinnaird, Zak. "Stewart and Colbert Gathering May Be Defining Moment of Generation." *Daily Athenaeum,* September 19, 2010. http://www.thedaonline.com/article_832f1c08-7e2b-5bcb-9769-e80a2b006b70.html, para. 13.

Late Night with Jimmy Fallon. NBC. March 28, 2011.

McLeod, Kembrew. "Everything Is Connected." *Quarterly Journal of Speech* 96 (November 2010): 421–26.

———. "My Freedom of Expression® Trademark." *Kembrew.* http://kembrew.com/prank/my-trademark-of-freedom-of-expression/.

McMillan, Sally J. "Exploring Models of Interactivity from Multiple Research Traditions: Users, Documents and Systems." In *The Handbook of New Media,* eds. Leah A. Lievrouw and Sonia Livingston, 163–82. Thousand Oaks, CA: Sage, 2002.

"NASA Appropriator Says: Stephen Colbert Is Out of this World." Congressman Chaka Fattah Congressional Site. March 26, 2009. http://fattah.house.gov/press/nasa-appropriator-says-stephen-colbert-is-out-of-this-world/.

"New Space Station Module Name Honors Apollo 11 Anniversary." NASA. http://www.nasa.gov/externalflash/name_ISS/index.html.

Rafaeli, Sheizaf. "Interactivity: From New Media to Communication." In *Advancing Communication Science: Merging Mass and Interpersonal Processes,* eds. R. P. Hawkins, J. M. Wiemann, and S. Pingree, 110–34. Beverly Hills, CA: Sage, 1988.

"Rally to Restore Sanity." Rally to Restore Sanity. http://www.rallytorestoresanity.com/.

"Rally to Restore Sanity—FAQ." Rally to Restore Sanity. http://www.rallytorestoresanity.com/faq/, para. 1.

Semigran, Aly. "Jon Stewart and 'The Daily Show' Twitter prank Jon Huntsman, who, for the record, is not Jon Hamm." *Entertainment Weekly,* November 30, 2011. http://popwatch.ew.com/2011/11/30/jon-stewart-and-the-daily-show-mad-men-twitter-prank-jon-huntsman/.

Sniderman, Zachary. "How the 'Rally to Restore Sanity and/or Fear' Nailed Social Media." *Mashable.* October 27, 2010. http://mashable.com/2010/10/27/stewart-colbert-rally-social-media.

"Stephen Colbert Charity Ice Cream at Ben & Jerry's Shops NOW thru Feb. 2012." Yellow Ribbon Fund. August 11, 2011. http://www.yellowribbonfund.org/?p=2514.

"The Stephen T. Colbert Award Sticker." *Good Reads*. http://www.goodreads.com/topic/show/64611-the-stephen-t-colbert-award-sticker.

Warnick, Barbara. *Rhetoric Online: Persuasion and Politics on the World Wide Web*. New York: Peter Lang, 2007.

"What Is Restoring Truthiness?" Colbert Rally, http://colbertrally.com/what-is-restoring-truthiness.

Chapter Four

Fandom Communication in a Mediated Age

The Use of Twitter and Blogs for Dissent Practices among National Basketball Association Fans

Corey Jay Liberman, Michael Plugh, and Brian Geltzeiler

On July 11, 2014, LeBron James, who is arguably (and statistically) one of the greatest players to ever play the game of professional basketball, decided that he would rejoin the Cleveland Cavaliers, after having been a member of the Miami Heat for four years. When news of this decision was released, there was a myriad of different (and publicly communicated) reactions by fans of both the Cavaliers and the Heat on blogs and Twitter feeds. Most were positively framed, such as "We will miss you in Miami";[1] "Thanks for the memories";[2] "You are an amazing player and man";[3] "We wish you all the best";[4] "Good luck in Cleveland";[5] "Good choice and congrats";[6] and "He's coming home."[7] Yet, other fans took to social media outlets for their opportunity to communicate disappointment, hatred, dismay, and regret for James's decision, which was manifested in such postings as "LeBron just made the worst decision of his life";[8] "We don't want you";[9] "Burn your LeBron James Heat jersey";[10] "Screw LeBron James for abandoning a team who brought him his two championships";[11] "I hope you never win a ring again";[12] "You have to question a person's integrity when you are willing to play for someone who called you a coward and betrayer";[13] and "Good job, LeBron, you went back to a city that hates you."[14] Despite the fact that these two series of comments are differentially framed, they do share one thing in common: they are emblematic of the opportunity provided by social media outlets to fans for communicating to, with, and about sports athletes which has, in many ways, reformed and reconceptualized the very notion of fandom interaction.

This chapter is about the use of social media for purposes of dissent (or disagreement) and why, how, and with what effect(s) fans use particular blogs and Twitter feeds to produce such communication regarding players, coaches, management, and even owners within the National Basketball Association. In so doing, the authors will first present a detailed discussion regarding dissent communication and why it is of particular concern from a fandom perspective. Next, the authors will present the results from a dissent communication content analysis regarding fans' use of two major NBA blog outlets (*ESPN*'s "True Hoop Network" and *Yahoo*'s "Ball Don't Lie"), as well as fandom tweets, to explain and describe the antecedents (what created the drive for one to communicate dissent through blogging and tweeting), processes (through what communicative practices did bloggers and tweeters dissent), and effects (what were the results of such dissent) of communicating dissent via blogging and Twitter. The chapter concludes with an overall discussion of the importance of studying the communication of dissent, from a mediated perspective, and how this comes to inform the fandom literature.

WHAT IS DISSENT?

Dissent is a term that has appeared in the organizational communication literature for the past three decades and, although there are slight variations embedded in its different definitions, Jeffrey Kassing most aptly defines it as "[both] feeling apart from one's organization . . . and expressing disagreement or contradictory opinions about one's organization."[15] Based on this definition, it is possible for employees to feel psychologically discontent with certain workplace issues (i.e., dealing with fellow coworkers, organizational practices, organizational policies), yet not overtly manifest this discontent through communication. As Jeffrey Kassing argued in a future article,[16] three salient predictors of one's willingness to communicate dissent are individual factors (certain employees have personal traits more conducive for dissent communication), relational factors (certain employees have cultivated relationships with others so that the process of dissent communication is psychologically easier), and organizational factors (certain organizations are more fostering of dissent practices, and are more open to change, as compared to others). Jeffrey Kassing further argued that dissent can perhaps best be defined as "a communicative reaction to . . . dissatisfaction."[17] One might very well ask why choose an organizational communication definition for a study on fandom communication and its link to television? Using the paradigmatic logic undergirding Daniel Katz and Robert Kahn's[18] General Systems Theory, fans are as important to the National Basketball Association (NBA) as the players themselves. From a synergistic, holistic perspective, then, without fans, there would be no players; without players, there would be no teams; without teams, there would be no coaches; without coaches, there would be no owners; without owners, there would be no franchises; without franchises, there would be no league. As such, an organizational communication definition of dissent is employed here because, logistically and theoretically speaking, fans are part of the NBA organization about which this chapter speaks.

It is important to note, at this point, that this chapter is also framed from a constituency-based approach to dissent communication. That is, the nature of the dissent is predicated on the source of the message. For example, the owner of the New York Knicks is going to dissent about something quite different as compared to the coach, physical trainer, team doctor, or player of the same team. Given this logic, fans provide an interesting forum for dissent.

According to Jeffrey Kassing and Todd Armstrong's[19] study, there seems to be nine major categories of what they call dissent-triggering events (things about which most individuals will publicly dissent within the organizational confines): (1) employee treatment (employees dissent because they are treated unfairly or unequally), (2) organizational change (employees dissent because of alterations that have changed the organization's culture and its ways of performing work), (3) decision-making (employees dissent because leaders seem to make decisions using strategies that are not always most beneficial), (4) inefficiency (employees dissent because work seems to take much longer than it should), (5) roles/responsibilities (employees dissent because of their own, or others', work-related identities, (6) resources (employees dissent because either there are no, or few, means for getting work done or the few means available are used improperly), (7) ethics (employees dissent because behavior is emblematic of unethical), (8) performance evaluation (employees dissent because they disagree about their own, or others', behavioral assessment), and (9) preventing harm (employees dissent because the organization does little to help protect its coworkers).

When it comes to dissent through mediated communication regarding NBA players, fans are likely to be discontented about such things as poor play on the court, badly managed behavior(s) off the court, relational disputes with both teammates and rivals, communicative dealings with the media, contract negotiations, and poor sportsmanship. These categories, although different, are aligned with the thought processes that produced the aforementioned dissent-triggering events forwarded by Jeffrey Kassing and Todd Armstrong.[20] This is not to say that fans have not had things about which to complain regarding players since 1946 (the inaugural year of the National Basketball Association). Nor is this to say that fans have not had an opportunity to voice their concerns in the past: fans could scream at players during the game or voice their opinions as a function of "call-in" talk shows or have their testimonials published as part of a sportscaster's daily column. The mass utilization of media has, however, changed three very important variables associated with fandom communication as it relates to the National Basketball Association.

The first is immediacy. That is, fans from all over the world can instantaneously post messages that can be accessed, read, and responded to within a moment's time. This certainly has both its advantages and disadvantages. On the one hand, it allows for more of a quasi-synchronous form of interaction. According to Surinder Kahai and Randolph Cooper,[21] media that provide such synchrony both increases social presence (the psychological feeling that two co-interactants are, in fact, sharing a social space) and provides increased cue accessibility (most importantly, according to the authors, emotional elicitation). On the other hand, however, if one's blog or tweet goes unnoticed or does not receive a response, communication, based on its original definition, does not, in actuality, occur. Thus, one might be forced to reconceptualize a seemingly fundamental question: if a dissent message is posted and does not receive a response (and there is no way to determine if others have read it), has communication occurred? Using traditional communication theory, although interaction might not necessarily occur, action certainly did.[22] The variable of immediacy, in this case, does not necessitate a co-interactant for communication to take place. That is, feedback and immediacy are not one and the same. Rather, the mere posting of a dissent message is indicative of information dissemination and media provide the opportunity for this to be immediate.

The second, which does require both notice of, and reaction to, a dissent post, is response. That is, rarely do dissenters communicate disagreement or discontent without retort being among the overarching goals. For example, Theodore Avtgis, Candice Thomas-Maddox, Elycia Taylor, and Brian Patterson,[23] in their study linking one's propensity to communicate dissent based on organizational burnout, discuss three types of dissent practices used by employees. The goal of the first type, articulated dissent, which is dissent communicated directly to one's boss, is organizational change. Assume, for example, that an employee is disgruntled about the way that her superior provides negative feedback about workplace practices. This employee is going to increase her chances of this boss implementing a change if he is the direct recipient of the discontent. The goal of the second type, latent dissent, which is dissent communicated to coworkers who are equally perturbed but have no opportunity for rectification, is social support. Psychological feelings of social solidarity do, in fact, alleviate many of the stressors created by the dissent itself. The goal of the third type, displaced dissent, which is dissent communicated to individuals not connected, even remotely, to the organization, is to, as Theodore Avtgis, Candice Thomas-Maddox, Elycia Taylor, and Brian Patterson claim, "serve as a 'catharsis,'" ultimately providing an emotional release to a neutral, outside party or constituent.[24] In sum, the goal of articulated dissent is change, the goal of latent dissent is social solidarity, and the goal of displaced dissent is emotional liberation.

What, then, is the goal of fans communicating dissent about players? One very interesting goal is a direct response from the player himself. This has happened on many isolated occasions, mainly in an effort, on behalf of the player, to save face and to repair a potentially damaged image. For example, in 2011, the National Basketball Association witnessed its fourth historical lockout, which resulted in a delayed start to the season (by a total of fifty-five days) and a reduced season schedule for teams (by a total of sixteen games). Clearly, fans were none too happy and many took to Twitter and blogs to communicate their dissent, most of who blamed the players for the lockout because of their greed and desire for increased salaries. Tony Allen, the shooting guard for the Memphis Grizzlies, finally got so annoyed with fandom dissent that he publicly responded with a Twitter post on October 11, 2011: "If I see one more person on my timeline thinkin' the players want more money . . . I'm gone go crazy."[25] The myriad of comments directed to Allen resulted in his communicative response to the dissent itself: something that could not have happened before the advent of social media. Another goal might be, very similar to latent dissent, social solidarity. That is, how many other fans agree with my dissent regarding a particular player? It is quite likely that solidarity among fans existed thirty years ago, but, without social media providing us with testimonial proof, there was no way to empirically gauge this. Yet another goal might be merely to give the public access to one's viewpoint not for purposes of social solidarity, but rather for purposes of argumentation. Recently, there was a blog posting by Josh Benjamin, an NBA correspondent, entitled *The 50 Most Controversial Figures in NBA History*. Alongside this column, there was an opportunity to post about each (and all) of these fifty NBA players. As one might assume, some fans enjoy and thrive on controversy, while others believe that it impinges on an otherwise great professional sport. After a quick read of the posts, it became readily apparent that the majority of messages were dissent-laden in an effort to create a particular emotion in the recipient, which would ultimately lead to refutation in the form of argumentation. In the end, regardless of the impetus for one's dissent, response is another variable associated with fandom communication that has changed as a result of mediated communication.

Third, online communication has provided fans of the NBA to socially construct a sense of community, creating social capital, which Robert Putnam[26] popularized within the field of sociology, and which has, subsequently, become a popular way of framing network studies for communication scholars.[27] Predicated on the principle of homophily, wherein similar individuals are more likely than dissimilar individuals to congregate, fans who are not collocated with one another have, with the introduction of blogs and Twitter, been afforded with a uniting tool conditioned by dissenting. Does one necessarily know, for certain, whether there is a positive correlation between following one's dissent tweets and the extent to which she agrees with the opinions forwarded? Certainly not. Does one necessarily know, for certain, whether there is a positive correlation between reading one's blog and the extent to which she agrees with ideas? Absolutely not. However, one can assume that by following such tweets, and by reading such blogs, these fans feel a connection with the dissenters and, from a psychological perspective, this might very well be (and likely is) enough to construct a sense of shared community; the one about which Robert Putnam speaks.[28]

Since this edited volume has embedded chapters that all link fandom culture(s), social media, and television programming, an important question that needs to be answered at this point, before moving on to our analysis of dissent practices among NBA fans, is the relationship among these three variables. From our perspective, television viewing both affects, and is affected by, fandom dissent. Beginning with the former perspective, that viewing televised NBA games affects fandom dissent, the argument is quite clear and rational: that the major impetus for dissenting about players, coaches, managers, and owners, whether it be about their sportsmanship or performance or intra-team relationships or inter-team feuds or salary disputes or poor decision-making practices, much of this information is ascertained through televised programming. This is not to say that everything about which fans dissent is presented within the games themselves. This is certainly not the case. For example, a salary dispute or player disagreement with upper management is not something to which fans will be privy during a game's broadcast. However, during interviews and sports talk shows, fans are likely to be exposed to such information, ultimately providing what we call rational dissent: disagreement that has, on the surface level, reason to create disagreement. Thus, to say that television viewing affects fandom dissent is another way of saying, in essence, that television viewing creates the very dissent that fans ultimately communicate via social media outlets.

The second perspective, however, is that television viewing is affected by dissent itself. When fans become dissenters themselves, or read the dissent posts of other dissenters, it certainly comes to affect and shape, for better or worse, the lenses through which they view the behavioral habits of players. Assume, for example, that, a fan reads a blog post about a poor on-the-court behavioral decision enacted by Los Angeles Clipper Blake Griffin, considered one of the best players in the NBA this season. From a communicative/ behavioral perspective, this fan has several choices. She could ignore the posting and decide not to respond. She could respond by agreeing. She could respond by disagreeing and retaliating. However, by having been exposed to this original dissent post, the lens through which she subsequently views the behavior(s) of Blake Griffin is, nonetheless, altered. If she agrees with the posting, her positively valenced emotions regarding Blake Griffin might be tainted. If she disagrees with the posting, her positively valenced emotions regarding Blake Griffin might be strengthened. If she neither agrees nor disagrees with the posting, although her emotions might not be altered, her cognizance of his on-the-court behaviors will be heightened and she will be more reactive to any infraction. Regardless of agreement, disagreement, or neutrality regarding the dissent posting, one thing is certain: dissent communication, while it certainly is affected by television viewing, also has a hugely salient effect on television viewing as well.

CONTENT ANALYSIS OF BLOGS AND TWEETS

In order to determine the strategies used to dissent (articulated, latent, or displaced) and the effects of such dissent by NBA fans using social media outlets, the authors chose to conduct a qualitative content analysis. Although a detailed discussion of the genesis of content analysis extends well beyond this chapter, it can be defined as an empirical methodology that provides the researcher with a tool to systematically determine the extent to which there exist certain themes within particularly specified texts. According to Ole Holsti, content analysis entails "making inferences by systematically and objectively identifying special characteristics of messages."[29] Despite the fact that many content analysis studies in the communication field are inductive in nature, inasmuch as researchers begin with a basic research question and then let the data (in this case, texts) determine the emergence of categories, other studies are deductive in nature, where researchers have predetermined categories, into which they attempt to place [potentially isolated] textual examples. Since the authors knew that they were interested in looking, specifically, for the three aforementioned types of dissent (articulated, latent, and displaced), and the extent to which these are manifested in communication using mediated forms of technology, this research methodology could aptly be framed as deductive, rather than inductive, despite its qualitative approach. In addition, this content analysis is predicated on what Klaus Krippendorff calls a "categorical distinction," wherein "units [are defined] by their membership in a class or category—by their having something in common."[30] For purposes of this study, the texts had dissent form as a commonality: dissent that was articulated (from fan to target), dissent that was latent (from fan to fan), or dissent that was displaced (from fan to non-fan).

As Bruce Berg[31] argues, one of the most important decisions that a content analyst must make prior to methodological engagement is to determine the unit(s) of analysis. In brief, a unit of analysis for content analysis purposes is, in essence, the text(s) that will be analyzed. For purposes of this analysis, all dissent messages were posted to one of two blogs (*ESPN*'s "True Hoop Network" and *Yahoo*'s "Ball Don't Lie") or appeared on Twitter. The rationalization behind choosing these two blogs was threefold. First, these two blogs are regularly mentioned by sports reporters, writers, and analysts as the blogs that attract not only the most NBA fans, but also the most intellectual NBA fans (regarding their knowledge of the sport, players, coaches, management, and ownership). Second, these two blogs are, based on both fan testimonials and trade reports, two of the largest electronically based communities that have created a network of NBA fans, and mediated dissent communication has clearly become an important effect of such networked membership. Finally, these two blogs enable public access and, as such, subscription was not a prerequisite for content analysis.

RESULTS

Blogs and blog networks initially spawned a community of NBA writers who were afforded an opportunity to have their work read widely. Originally, the communication response(s) from readers was limited to the comments section at the bottom of the page and it was extremely rare for the bloggers (and even rarer for the commenters) to have any kind of interaction with the individuals about whom they were writing and discussing. However, those walls have since been torn down by the introduction, and mass utilization, of Twitter. With this technological innovation, fans can communicate directly with media, players, coaches, executives, and even owners. Some of the messages communicated via Twitter are what one might call meaningless. There is, though, still an abundance of communication, whether direct or indirect, that can cause enough dissent to spur action. In fact, one tweet (of no more than 140 characters in length) can be incredibly powerful. In today's mediated age, the great majority of tweets provide direct links to blogs and vice versa. In fact, Twitter has largely replaced the comments section of blog sites. These comments sections still exist, but the "mentions" function on Twitter allows the target of the interaction to be aware that he (or she) is the initial impetus for the post. In the end, both blogs and Twitter have become fruitful avenues for dissent communication among NBA fans.

Articulated dissent is the first category of messages that will be discussed. As mentioned earlier in this chapter, articulated dissent is disagreement or potential protest shared by an individual to one who is in a position to make a change. Although articulated dissent is not always directed at the source of such dissent, it was found, in this study, that this was, in fact, the case. On December 31, 2013, Carmelo Anthony, a small forward for the New York Knicks, communicated a "seemingly innocent" tweet about a fan giveaway of a few pairs of his newest line of basketball sneakers. A Twitter follower took issue with the tweet and responded, publicly, to Anthony, by delivering an expletive-filled tweet that said "How about if you win a ring . . . you fucking kill me man . . . rooting for you all the fucking time and you always disappoint me."[32] In essence, the message here was that this fan would much rather have Carmelo Anthony win a championship than give away a pair of sneakers. Most players, by and large, do a great job of avoiding these types of hateful messages, as they are oftentimes (especially with technology) bombarded with them constantly. In this particular instance, however, Anthony let the send button get the best of him and responded to the message by telling his obscene Twitter follower that "[I] didn't ask for your glazed donut face ass to root for me anyway."[33] Anthony later admitted that he should never have responded and the direct result of his response to the fan's articulated dissent was that Anthony lead the Knicks to their most impressive three-game stint of a very difficult season, on the much-dreaded three-game roach trip in Texas.

In June of 2012, a second example surfaced, when a fan sent a tweet to Amare Stoudemire, who, at the time, was the center for the New York Knicks, strongly suggesting that Stoudemire return next season in much better athletic shape, tweeting that he "Better come back a lot stronger and quicker to make up for this past season."[34] Stoudemire took particular exception to the comment and sent the follower a direct, profane message, while referring to the poster using a particular homosexual slur: "Fuck you . . . I don't have to do anything, fag."[35] Stoudemire quickly realized the error of his ways and apologized, electronically, to the follower ten hours later, with another direct message: "I apologize for what I said earlier. . . . I just got off the plane and had time to think about it . . . sorry bro . . . no excuses . . . won't happen again."[36] In the world of Twitter, a direct message is a private communication episode that a user can send to one of his or her followers, which is designed to be private. However, in today's information age, and the increasing ability for smartphone users to capture a copy of a message (or a portion of it) and publicize it, the term "private," as it relates to the Internet, might now be but an antiquity. Unfortunately for Stoudemire, his apology was too little too late, as the follower had taken a screenshot of the tweet and sent it to the sports scandal Web site known as *Deadspin*, which provides stories of infamous players, coaches, owners, and teams who engage in poorly manifested behaviors. The net result, in the end, was public embarrassment for Stoudemire, and his team, and [potentially] the league at large.

Arguably one of the most significant moments in the last four years of the NBA occurred on July 8, 2010, when LeBron James, who, at the time, was a free agent for the first time in his professional basketball career, announced, after meeting with six different teams, was ready to announce the team that he would join. In a one-hour *ESPN* special entitled "*The Decision*," James announced that he was going to play basketball for the Miami Heat, which was the start of the third example of articulated dissent. In so doing, he made what many basketball analysts and critics consider to be a critical public relations error when he, in an egomaniacal nature, announced, on public television, that he was "taking [his] talents to South Beach."[37] From the time that James announced his decision, using this verbiage, the public backlash that surfaced was extreme. In addition, the backlash was so swift and harsh that it, according to many, changed how James approached the sport of basketball, his celebrity status, and his life in general. This backlash came through multiple mediated formats, but perhaps most strongly through fan postings on blogs and Twitter. One particular tweet gained extreme media coverage, which was the message that said "Hey LeBron James . . . you're the cockiest player to never win a title."[38] Although this did not lead to a direct response by LeBron James, it was certainly communicated through a medium conducive for a discursive rebuttal.

In James's first year with the Miami Heat, the team made the finals and James played much of the series portraying an appearance that seemed to communicate lack of focus and mental engagement. He later admitted that he would stay up late at night, after games, reading all of his direct mentions in blogs and on Twitter. It is possible, in the end, that this form of articulated dissent was having an adverse effect on his performance on the court. In mid-January of 2011, the Cleveland Cavaliers, the team that James left to join the Heat, was among the worst in the entire NBA, with a winning percentage of 21.1 percent. This, of course, was the first season without the "talent" of LeBron James. After the team had lost its twenty-first of twenty-two games in a row, including a humiliating defeat to the Los Angeles Lakers by a score of 112–57, James took to Twitter and posted the following message: "Crazy . . . Karma is a b**** . . . gets you every time . . . it's not good to wish bad on anybody . . . God sees everything."[39] Clearly, this was a communicated retaliation by a player who was publicly humiliated by fans through the use of social media.

A fourth example occurred during the summer of 2013 when Dwight Howard, the center for the Houston Rockets, was engaged in a series of pedestrian Twitter exchanges with fans regarding his decision to join the Houston franchise after a tumultuous, single-year stay with the Los Angeles Lakers; a move that received a lot of media attention. According to an interview with Dwight Howard during the fall of that season, he decided to join the Rockets because he "wanted to win."[40] Marcus Kirk, an avid Twitter user, tweeted, in response to Howard's decision, that "it's the truth . . . bro . . . I know Dantoni sucks big time . . . but you could've gave us one more year."[41] This tweet was implying that despite Dantoni (the Lakers coach at the time) being subpar, Howard's decision to leave the team was both abrupt and poorly executed. In a direct response to Kirk's tweet, Howard attempted to brush away talk of his short, but dramatic, time with the Lakers by reply-ing "let it go . . . no need to dwell on the past."[42] However, this seemingly innocent (and isolated) moment of communication grew into an exacerbated exchange between Howard and another Twitter user by the name of Brittni. Within moments of Howard's tweet, Brittni chimed in by tweeting two emoji faces, laughing so hard that tears were pouring from their eyes, with the comment "u ain't never getting a ring."[43] Howard quipped in reply by indi-cating that "with that face ion think u getting one either . . . lol."[44] Although this can be interpreted, especially from the micro-level of message analysis, as a humoristic exchange, what began as communication about one fan's disappointment about a player's decision to join another team ended in an emotionally fueled, dialogue that is emblematic of the potential that media provide for fandom dissent.

NBA superstar Kevin Durant, the power forward for the Oklahoma City Thunder, has engaged with fans using online media on more than one occa-sion, but, most recently, during an event that was triggered on July 1, 2014. Within the course of a live, mediated "question and answer" session with fans, which surfaced around Durant, his team, and the rest of the National Basketball Association, one user of Twitter, named PEGEE, tweeted, directly to Durant, asking "is your barber blind?"[45] Durant, in a moment of mediated rage, replied by saying "does your barber need binoculars to shape up that skinny ass chin strap goatee?"[46] Over the next ten minutes, PEGEE tweeted, in reference to the brief encounter, three times, saying the following: "KD just killed me . . . I'm outta here . . . lol";[47] "I quit . . . you the real MVP";[48] and "I'm bout to go play 2K an injure this nigga KD on purpose."[49] The final tweet in this series makes reference to the popular NBA-themed video game NBA2K, in which players take control of avatars bearing the likeness of actual NBA players in simulated basketball competition. Searches on Twitter reveal that the account used by PEGEE to engage with Kevin Durant has since been suspended, but the series of Tweets has been captured and pre-sented by multiple online blogs and Web sites. This is another example of the ability afforded by media to communicate dissent directly from a fan to a player and the emotional elicitation that both tweets and responses to tweets can have on both parties.

Nearly two months after the PEGEE incident, Twitter user Markus tweeted at Kevin Durant, saying "if you could brush your hair with any type of brush, which brush would it be?"[50] As in the PEGEE incident, Durant chose to respond to Markus, directly, tweeting "u ain't got nothing else . . . I'm tired of seeing the same jokes."[51] The exchange continued, with Markus replying "people would stop using the same jokes if you brush your hair . . . bruh . . . I'm a wave master . . . I can help you out . . . FRFR."[52] Kevin Durant, showing some restraint, once again replied by saying "bro . . . stop . . . it's not even funny . . . it's weird actually . . . you gonna group chat your boys like 'you see me roast KD.'"[53] After several more exchanges, Markus referenced Durant's basketball abilities, tweeting "Dawg . . . whenever you in NC, we having a shootout . . . winner gets your Twitter account and a pair of KDs."[54] In essence, this is implying that Markus, the basketball fan, is a better player than an NBA superstar. Kevin Durant finally ended the dialogic exchange when he tweeted "not smart, sir . . . that's like me telling The Rock I can bench more than him."[55] Durant's remarks during the exchange are emblematic of the motivation undergirding confrontational communication in a mediated world. Since Twitter is a public forum, and since Kevin Durant is a high profile member of that forum, the expression of dissent may act as a platform for publicity or for building social capital among peers and other fans.

Yet another fan-to-player communication episode transpired on the morning of August 29, 2014, when Jared Dudley, the small forward for the Milwaukee Bucks, innocently tweeted a message to a friend saying "congrats to my man @Ra5ik and his beautiful bride . . . what a special night . . . so happy for you guys . . . god bless you guys."[56] Unexpectedly, this was met by a reply from Master Wongton, a fan of the NBA, which read "yo Jared . . . you're fucking trash bruh . . . that's why they traded your bum ass away . . . lmao,"[57] referencing the recent trade of Dudley from the Phoenix Suns to the Bucks. Dudley retweeted Master Wongton's comment with the addendum "maybe . . . but I'm in the league and u watching me . . . lol."[58] Although in this example the use of "lmao" and "lol" might serve to de-intensify the magnitude and effect(s) of the emotionally laden dissent, it was, in all likelihood, used as a nonverbal, paralinguistic strategy: not to hide either constituent's motivation, but rather to potentially soften the impact of the comments. Either way, this is yet another example of the opportunity for dissent communication as a result of social media.

In early February of 2014, Tike Mouey, a Twitter user, tweeted the following comment to Roy Hibbert, the center for the Indiana Pacers: "[Hibbert] is lucky to be in the NBA because he is not a very educated man based on his tweets."[59] Clearly perturbed by this communication episode, Hibbert replied by saying "haha . . . cuz I got my degree in government at Georgetown . . . I'm so uneducated bro."[60] In a remarkable twist of events, Tike Mouey sent a second message to Hibbert saying "you're my man Roy . . . can I get a follow?"[61] Hibbert's succinct reply, perhaps, offers a pristine model for other NBA players in their interactions with troublesome fans, as he tweeted "nah . . . u get blocked."[62] Witty comebacks, such as this, are a type of currency in social media environments. A testy interaction, on its own, is enough to create a story worthy of blogging. In the end, Hibbert's exchange was featured on several prominent sports Web sites, including Yahoo's well-known blog "Ball Don't Lie."[63]

Vinny D'Amadeo, a fan of the NBA, had a dissent-laden interaction with Damian Lillard, the point guard for the Portland Trailblazer, which was later featured on "Yardbarker," which is part of *Fox Sports Digital*, whose tagline is "real-time rumors, gossip, opinions, and humor from the best sports blogs."[64] On August 22, 2014, D'Amadeo tweeted the following message to Lillard: "we played the same amount of mins today bro,"[65] implying that he spent much of the game on the bench rather than helping his team on the court. At the time of this tweet, Lillard was among several young guards vying for a final spot on the U.S. Men's Basketball Team, training to play in the International Basketball Federation (FIBA) World Championship in Spain. Lillard responded to D'Amadeo by saying "did we make the same amount of money today, too?"[66] This was clearly an example of a fan's dissent cultivating a negative emotional reaction in a player and how social media becomes a communicative platform for the exchange of messages between these two constituencies.

As a last example of articulated dissent, Twitter user Mike Edwards sent a direct message to Eric Bledsoe, the point guard for the Phoenix Suns, which said "you're the worst player in the NBA."[67] Bledsoe, in a moment of temporary rage, chose to strike back at Edwards by replying with a simple "yo mama."[68] This response to the articulated dissent communicated by Mike Edwards was received with much apparent delight, as he went on to tweet more than twelve times about this isolated encounter, including "thank you @EBled2 for the free publicity . . . I couldn't pay for this much attention,"[69] "Yoo . . . I can't believe Eric Bledsoe tweeted at me . . . this is hilarious,"[70] "@EBled2 . . . please don't ever delete your comeback . . . I need the followers,"[71] "man I love it . . . free publicity,"[72] and "I don't think my Twitter has had this many views in its entire existence . . . Eric Bledsoe is the man."[73] Edwards continued his tirade, perhaps trying to bait Bledsoe into replying again, Tweeting the following message: "this dude really just said . . . yo mama tho . . . that's a one year at Kentucky education comeback."[74] Bledsoe never responded. In the end, however, Mike Edwards accomplished his goal: to communicate dissent, openly, and to receive a direct response in return.

Another category of dissent messages, latent dissent, is the second that will be discussed. As mentioned earlier in this chapter, latent dissent is disagreement or potential protest shared by an individual, though not directed to an individual who can make an important and necessary change. Rather, the purpose of this type of dissent is to provide social solidarity and, in a sense, social support. Since much of the communication that occurs in and through the mediated world does created networked communities, latent dissent was found to be a very powerful interaction vehicle through which fans congregate around certain issues surrounding players, coaches, owners, and staff.

At the end of the 2012–13 NBA season (April 12 to be exact), Kobe Bryant, the shooting guard for the Los Angeles Lakers, tore his Achilles tendon while playing a game against the Golden State Warriors. In fact, after he felt (and, according to press releases, heard) the "pop" in his left foot, the star player went back on the court to shoot two free throws: a decision that Bryant regretted afterward. Just over a year later, LeBron James, mentioned earlier in this chapter, suffering from dehydration, was forced to bench himself during an NBA Finals game against the San Antonio Spurs on June 5; a series which the Spurs would lose in four straight games. During this game, which was the first of the four-game series, Jon Schoonover, a Twitter user and a fan of the NBA, tweeted the following message: "Kobe tears Achilles, shoots two free throws, and walks off on his own . . . LeBron cramping . . . taps out . . . SMH."[75] In the social media world, the acronym SMH stands for Shaking My Head. Situations like this are not uncommon when it comes to fandom communication via social media outlets, where they publicly duel about their favorite players, taking the low moments of one popular player to boost the reputation of their own favorite. In response to this tweet, Jon Schoonover received ten different replies from four distinct users, including Just Keep Livin's "#stupidity,"[76] and Gabe Antrobus's "seems pretty obvious who wants it."[77] A comment thread like this never amounts to much more than mild ridicule and "chest bumping," but illustrates one of the ways that fans play the age-old game of "my guy is better than your guy" via social media.

Deron Williams, the point guard for the Brooklyn Nets, was once considered by fans and NBA analysts, one of the top two or three position players in the NBA. His move from the Utah Jazz to the Brooklyn Nets during the 2010–11 season was seen as a huge triumph for the Nets franchise, as they prepared to move from their long-time home in Northern New Jersey to a new, downtown Brooklyn arena. Williams's play, when he joined the Nets team, lagged in comparison to his days in Utah and, eventually, he began to miss significant portions of game time with nagging ankle injuries. Fans of the Brooklyn Nets grew disillusioned with Williams and began to openly mock him in social media, often fixating on his hairstyle as a proxy for his play. In December of 2012, Twitter user The Mister Marcus tweeted the following message: "Deron Williams is not that nice . . . he's out of shape and his hair looks like burnt Ramen Noodles."[78] This single tweet produced three different replies from three distinct Twitter users (and fans of the NBA), including Art X Drug Clothing's "man . . . thank u . . . I got people saying he's the best PG in the league"[79] and CRVE's "that was funny, but let's be real . . . D Williams is a Top 5 PG in the league . . . really Top 2 if u ask me behind CP3."[80] It is clear that this latent dissent on behalf of The Mister Marcus created two different communicative responses: agreement, on one hand, and a mild defense of Williams on the other. Clearly, even one, isolated latent dissent message can create two partitioned fan communities within the online, mediated world in which we live.

On October 29, 2014, Eric Freeman, a writer for *Yahoo*'s online NBA blog "Ball Don't Lie," authored a piece about a recently played game between Kobe Bryant's Los Angeles Lakers and Dwight Howard's Houston Rockets.[81] Howard and Bryant were famously teammates for a single season (2012–13), before Howard shunned the Lakers in free agency to play elsewhere. Reports frequently surfaced during Howard's brief tenure in Los Angeles about the difficult and tumultuous relationship that he shared with Kobe Bryant and it was apparent, during the game, that some bad blood remained. In a late-game encounter between the players, Howard aggressively swung his elbows around Bryant's head, catching him across the chin. The incident was relatively minor as far as on-court aggression is concerned, but the history between the two players fueled a great deal of media coverage and speculation in the following days' news. Eric Freeman's article garnered hundreds of comments, including one from *Yahoo* user Peter, who remarked:

> Seeing Kobe once again jacking a lot of shots just to get his 19 points is depressing to watch . . . 6 of 17 shooting is very poor percentage. His problem is that he's losing and playing the wrong way of basketball. Every single time he catches the ball, two dribbles, turn around, and fade away jumper, and shoot the ball, without even taking a good look at the basket. That's very poor shot selection and at the same time disrupting the entire flow of offense without any kind of ball movement and getting other guys involved. The majority of the people don't like this style of selfish play anymore and would like to see this guy leave the Lakers for good. At 36 years old, he needs to share the ball and play basketball the right way, not selfishly continue to jack up many bad shots and missed them a la Allen Iverson.[82]

This comment received nine replies from six distinct users. Genuine, a *Yahoo* user, agreed with Peter's comment, noting "when the tam falls behind, he seems to forget about the 'team' concept . . . then it becomes 'KoMe.'"[83] RipHamilton, another *Yahoo* user, shared some of Peter's initial concerns, but qualified some of his remarks, saying "Kobe was 6-15 FG (40 percent) instead, which is not efficient . . . I must agree . . . but if you watched the game, you would find out that he was 6-13 FG in the first half and 0-2 FG in the second half . . . but this is just the beginning of a long season . . . with a new roster and new coach . . . thus many adjustments are needed . . . well . . . we will see what happens."[84] In addition to the support of Peter's comment on behalf of Genuine and RipHamilton, full and qualified, latent dissent was expressed by another *Yahoo* user, New to Corona, who communicatively defended Kobe Bryant, saying "poor shot selection . . . maybe you should go coach him . . . I'm sure he'd love that . . . Kobe's won 5 titles, which I'm pretty sure are 5 more than you have . . . but you're the expert, so."[85] This example, again, illustrates the role of mediated latent dissent in producing a partitioned NBA fandom community.

As a fourth example of latent dissent, *Yahoo*'s Eric Freeman wrote another piece for "Ball Don't Lie," dated October 9, 2014, in which he described advice offered by Chris Bosh, the power forward for the Miami Heat, to Kevin Love, a power forward who had recently been traded from the Minnesota Timberwolves to the Cleveland Cavaliers; to play alongside LeBron James.[86] As James's former teammate in Miami, Bosh described the type of personal sacrifice required to succeed with LeBron as the "alpha star," noting and citing his own experiences at a member of the Heat organization. The article received well over one thousand comments, including one from *Yahoo* user Shadow, who wrote:

> Chris Bosh should just close his mouth and thank LeBron James for getting him a couple of rings because Bosh will not get another one. He was clearly the 3rd best player on the team and quite honestly didn't deserve to make anywhere near what Wade and LeBron made. So LeBron gets him two rings and a huge payday and he is now complaining. Can't wait to watch Kevin Love eat Bosh alive when the Cavs pound Miami.[87]

Shadow received three replies to the comment, the first of which shows support for the comment and anticipation for the eventual matchup between Chris Bosh and Kevin Love, posted by user John, which read: "this is extremely accurate . . . love Bosh . . . instaclassic."[88] Another comment, however, from *Yahoo* user ler, produced defense for Bosh, when he said "he was the 3rd best player by design, not default . . . you are clueless if you think Bosh's numbers won't explode this season."[89] Yet again, the partitioning of the NBA fandom community based on communication via social media platforms.

On November 3, 2014, Henry Abbott, a writer for *ESPN*'s online NBA blog "True Hoop Network," posted an article about the Miami Heat's strong start to the 2014–15 season, despite the loss of LeBron James to the Cleveland Cavaliers.[90] The article prompted thirty-eight comments, including one from user Terrance Forehead Jackson, who wrote "watch how they get exposed Tuesday night against the Rockets . . . then everyone will be quiet on the low."[91] The comment was met with six replies from six unique users, including the following: Kyal Rainey's "get exposed how . . . at worst, they will get beat in a shootout . . . the Rockets don't play the best D in the world";[92] Isaac Jean's "it's funny how people keep on saying the Heat not going nowhere";[93] and Jeff Alan Moe's "So . . . 6th best in the West beats the Heat by 20 . . . sounds about right."[94] As this example illustrates, media writers and analysts, themselves, have the opportunity to create a fandom discourse, emblematic of latent dissent, which seems to create a fandom divide.

Kevin Arnovitz, a writer for *ESPN*'s "True Hoop Network," authored an article in May 2013 about Roy Hibbert, the center for the Indiana Pacers, and his questionable defensive play during the Eastern Conference playoff series against the Miami Heat.[95] This posting created much fandom communication, including ssiccone's "love how 20/20 second guessing is always right . . . Hibbert and his 2 cement blocks for feet was going to be able to come swooping in like batman to save the day from the other side of the court . . . second guessing is always right."[96] This single comment prompted a back-and-forth of dissent between ssiccone and geeman217, who initially reacted by replying "fact of the matter is . . . you don't give up a lay-up in that situation . . . if Bosh, Allen, or whomever hits a J . . . so be it . . . but to give up a lay-up was just plain dumb defense and dumb coaching."[97] Much of the dialogue between these two fans was confrontational, illustrating pro-Hibbert and anti-Hibbert mentalities, and the communication episode finally ended when geeman217 made the following comment: "that's only because your limited knowledge of basketball won't allow you to understand that Hibbert could've parked his butt in the lane, while the Pacers switched down on Bosh . . . THEN Bron would've had to kick the ball out or go to the rim HOPING for a foul."[98] The capital letters used in the comment reflect points of emphasis, which, in this case, was driven by an intense emotional response on behalf of geeman217. Based on these three comments, latent dissent produced an obvious difference in player allegiance, ultimately fueled by Kevin Arnovitz: the author who posted the original article to *ESPN's* "True Hoop Network."

The New York Knicks found themselves floundering and mediocre, despite high expectations, during the 2011–12 NBA season. A little-known point guard named Jeremy Lin entered a game, his first as a member of the Knicks organization, on February 4, 2012, in what most sports analysts call a moment of coaching desperation, and started a three-month-long spectacle known within the NBA fandom community as "Linsanity." In short, Lin set the sports world on fire by playing like an All-Star and single-handedly brought life back to the Knicks season. Lin became a free agent at the end of the year and received a high dollar contract offer from the Houston Rockets; an offer that the Knicks elected not to match. Lin, as a result, departed for Texas. Rockets fans initially celebrated their acquisition, but Lin failed to recapture the same level of play during his time in Houston and, as such, fans reacted poorly. On April 28, 2014, Twitter user Donnell Davis tweeted the following: "I will tweet @JLin7 every day to let him know how much he sucks."[99] J.C., another Twitter user and fan of the NBA, replied to Davis' tweet, expressing displeasure, saying "wow . . . you really gonna be that fan."[100] Another Twitter user, CHARGER, also wanted to include James Harden, a shooting guard for the Houston Rockets and who most sports analysts consider the star of the team, into the dissent communication episode, tweeting "yet he don't blame the biggest problem . . . Harden."[101] In response, Davis returned to Twitter in an effort to elaborate on his disdain for Lin by tweeting the following: "doesn't matter . . . you don't dribble the ball around two defenders that late in the game when all you have to do is hold it."[102] Clearly, a built-up sense of frustration had reached a breaking point during a particularly poor period of play, prompting the exchange among fans and the particularly harsh criticism of Jeremy Lin. Such dissent would not have been possible without social media platforms enabling such communication.

As an eighth example, during the 2013 NBA offseason, the New York Knicks executed a trade with the Toronto Raptors to acquire former number-one draft pick Andrea Bargnani: a power forward who originally, from 2002 to 2006, played professional basketball in Italy. Bargnani, unfortunately, never lived up to the lofty expectations set for a number-one draft pick and was frequently injured and kept on the bench as a result. Fan reactions to the trade were clearly mixed. For example, Twitter user Andrew Sutton remarked, on June 30, 2013 that "Bargnani can really help us chase a ring next year . . . don't believe me now then wait till the season starts."[103] Ed Boulanger, another Twitter user, retweeted the message, adding the simple "WUT"[104] to show his disdain for the idea. In short, the acronym "WUT" is a communicative response to what is deemed an absurd, unclear, unbelievable, questionable comment. In other words, Ed Boulanger was questioning the veracity of Andrew Sutton's post. In support of Boulanger's post, Twitter user, and NBA fan, Chris, replied "4.8 career rebounds per game average . . . he's a seven footer."[105] This example illustrates, again, the opportunity, provided by social media platforms, for the communication of latent dissent, ultimately creating a partitioned fandom community.

As LeBron James has matured and his championship credentials have been established, a debate in the NBA community has emerged in which James is oftentimes compared to Michael Jordan, the former star of the Chicago Bulls, who has long been considered the greatest basketball player to have ever played the game. This type of dialogic debate makes for perfect social media fare, as evidenced on October 28, 2014, when Twitter user Joe tweeted the following: "when facing elimination in the playoffs (PPG – RPG – APG – FG%) . . . LeBron . . . 31.9 – 10.1 – 6.4 – 46.6% . . . Jordan...31.3 – 7.9 – 7.0 – 45.8% . . . LBJ>MJ."[106] A dissent back-and-forth ensued among Joe and several other Twitter users, most of who favored Michael Jordan in the debate. For example, Twitter user Giants=8-8 wrote "idiotic, arbitrary stats is the only way you'll ever be able to say LeBron>MJ . . . just give it up already."[107] User Manmeet Singh attempted to steer the conversation away from the contentious debate about the players, tweeting "does it matter who won . . . a team won . . . not an individual player."[108] Manmeet Singh's attempts were to no avail, as a number of other comments were posted in support of Michael Jordan, including a comment from user Irish Ducusin, which read "just face it . . . LeBron will always be behind Jordan."[109] Debates such as these will rage on across social media and these social media networks are perfectly conducive for these types of dissent practices.

As a final example of latent dissent, the Los Angeles Lakers struggled to find a suitable coaching replacement for the legendary Phil Jackson, once he opted to move into retirement following the 2010–11 NBA season. The first head coach hire was Mike Brown, who had a brief and tumultuous tenure with the team, lasting just a shade over one season (and the 2011–12 season was shortened by sixteen games). Mike Brown's successor, Mike D'Antoni, had an equally turbulent run with the Lakers, known for his innovative and dynamic offense, but often denigrated for his lack of focus on the defensive side of the ball. D'Antoni finished the 2012–13 season that Mike Brown began and then resigned at the end of the 2013–14 season, during which the Lakers had a final record of 27–55. Just prior to his resignation in March of 2014, Twitter user Ryan Ward tweeted the following: "at this point, fire D'Antoni and let Kurt Rambis take over for the rest of the season . . . at least you're letting players know change is coming."[110] Kurt Rambis, a former NBA player from 1980 to 1995, who played for the Los Angeles Lakers, the Charlotte Hornets, the Sacramento Kings, and the Phoenix Suns, and former assistant coach of the Lakers and former head coach of the Minnesota Timberwolves, has long been a fan favorite. Twitter user DoughBalls replied to Ryan Ward's tweet with the following: "I'd love to see Rambis take over . . . front office has to do something."[111] However, these two comments were met with dissent when KMC, a Twitter user and NBA fan, posted the following: "Rambis . . . did you just ignore his head coaching job in Minny . . . and he's in charge of this Laker defense."[112] Again, this type of latent dissent produces the elicitation of emotion on one hand and a segregated community of NBA fans on the other.

There is a third type of dissent, known as displaced dissent, which is most aptly defined as disagreement that is communicated neither to an individual who can make a change, nor to a person who can relate to the disagreement in question for purposes of social support. Rather, displaced dissent is oftentimes communicated for cathartic purposes: to merely release emotions to individuals who have no interest (vested or otherwise) in the variable(s) creating the dissent in the first place. Since the two major independent variables predictive of one's desire and willingness to post a dissent message on a blog or Twitter are (1) interest in the National Basketball Association and (2) some impetus for disagreement, the same argument can rationally be made for those who read NBA blogs and NBA tweets: that without having an interest in the NBA, and an impetus for disagreement, there would be no rational reason for one to follow such mediated messages. As such, whereas displaced dissent certainly occurs when it comes to workplace dissent and relational dissent, we found that it is absent within the realm of NBA fandom dissent in the social media world.

DISCUSSION AND CONCLUSION

As nearly all publications that have analyzed, and theorized about, how communication has changed with the mass utilization of mediated technologies, one of the quasi-universal claims (truths being too open to criticism) is that media are less likely to inhibit one's decision to communicate emotionally laden comments to a myriad of different constituents. In fact, in a recent episode of *Real Time with Bill Maher*, one of his guests, Chris Hardwick, began speaking about the ease at which people who watch his show, @ *Midnight*, publicly criticize him using online, social media. His major claim was twofold. First, individuals are more likely to make very harsh, negatively framed, negatively valenced comments about public figures because, in essence, there are no negative ramifications for doing so due to anonymity. That is, social beings can very easily hide behind their statements because of the lack of social presence and the effects (or lack thereof) that accompany this. This is a claim that has been proposed, and supported, since research about online communication first surfaced.

The second, more poignant, argument was that, unfortunately, people are quick to communicate their disagreement with, or disapproval of, public figures with a sense of immediacy which, as Hardwick claimed, provides but a mere "emotional snapshot." In other words, individuals are quick to communicate dissent via online technologies in that moment of hatred, dislike, disapproval, abhorrence, and loathing. This is not to say that such dissent does not occur during face-to-face interaction. It certainly does. Decades of literature on conflict studies indicates that one hugely important independent variable predictive of poorly managed conflict is one's [potentially irrational] emotionally driven dissent framed directly about a particular social agent. However, technologies have provided such dissenters with a much more immediate forum for dissent communication and, in all likelihood, have provided already existing dissenters with new ammunition . . . and nonexistent dissenters with an "easy to use" instruction manual.

The findings of the content analysis indicate that fans do, in fact, communicate dissent regarding NBA issues and that this dissent can be either articulated or latent. One variable is common for dissenters, regardless of whether their communication is articulated (framed directly to the source or framed to an individual in a unique position to make a change) or latent (framed to sources who likely feel the same way to create social cohesion, rather than behavioral change): knowledge about a dissent-triggering event. That is, without fans having information about a wrongdoing committed by a player, coach, owner, or president, there would be no need to dissent in the first place. Each of the examples provided began with knowledge of such a wrongdoing, which was then followed by a series of messages that appeared on blogs and/or Twitter. The choice to communicate articulated dissent or latent dissent, then, is truly affected by one extremely salient query: what is the dissenter's major motivation?

Those who communicate articulated dissent were, by and large, attempting to accomplish two things. First, some who communicate articulated dissent are, in actuality, attempting to receive a direct response from the target of the dissent itself. Before the introduction of social media as a communication platform, this was not a viable option for fans. However, with the introduction of social media, many NBA players have their own Twitter accounts (as evidenced by this chapter), and their own blogs, as a way of connecting with fans (at least psychologically conditioning fans to think that there is a connection). In the foregoing examples of articulated dissent, this goal of player response (or what might be called retaliation) was accomplished, including Carmelo Anthony's "[I] didn't ask for your glazed donut face ass to root for me anyway," Dwight Howard's "let it go . . . no need to dwell on the past," and Jared Dudley's "maybe . . . but I'm in the league and u watching me . . . lol." None of these three responses were suggesting change. Rather, they were, in essence, verbal retaliation against the dissent. However, if the goal of the dissent was a response or reaction from the player, it certainly worked. A second goal, and one which might be more overt and purposive as compared with the first, is to stir controversy. The examples provided regarding Kevin Durant's on-the-court performances, Roy Hibbert's lack of education, and Eric Bledsoe's lack of basketball talent created public controversy by, again, putting these athletes in the public spotlight for wrongdoings (i.e., personal or athletic) committed. Although by not responding, the controversy may, in fact, have been heightened, responding, too, heightened the controversy.

Those fans who communicate latent dissent are also attempting to accomplish two goals. First, the dissenters are creating a forum for social solidarity. That is, individuals with similar cognitive, emotional feelings get a certain sense of shared identity and collective normalcy. When fans notice that others post about LeBron James being a better NBA player than Michael Jordan or Deron Williams's post-trade play on the Brooklyn Nets paling in comparison to his pre-trade play on the Utah Jazz, it creates a networked community. Although this might not be a cognizant effect of latent dissent among NBA fans, it certainly is an effect when one takes a closer look at the way that messages are framed, designed, and structured. Second, and what might seem a necessary byproduct of creating a networked community, latent dissent messages create, as we found in our examples, two partitioned communities.

That is, there is a population of fans who supported Dwight Howard's sentiments regarding his ex-teammate Kobe Bryant, yet there was another population of fans who defended Bryant at all costs; there was a population of fans who agreed that Roy Hibbert's defense play during the 2013 Eastern Conference Playoff, yet there was another population of fans who believe that he played well. It was not the partitioned communities that provided dissenters with the knowledge about how to create/frame the messages. Rather, it was how the messages were created and framed that socially constructed the communities. This, in essence, is a goal of latent dissent: to create communities of individuals, who cannot make any changes, but who can, collectively, offer social support and engage in dialogues based on mutual understandings and perspectives.

Based on this content analysis, it should be clear that dissent is a predominant form of communication among fans using social media outlets. Certainly fans of the NBA who are using such outlets might very well be attending live games, though it is quite likely that many are viewing these games through mediated formats. As such, television viewing/programming can be framed as an antecedent to, and effect of, both articulated and latent dissent practices. In other words, fans are provided with information regarding players, coaches, management, and the like through televised aspects of the NBA, be it live games, highlights, interviews, or talk shows. It is because of the accruement of such information that fans are able to cogently dissent. However, television also becomes an effect of dissent itself. Once fans learned about LeBron James's claim that he will "take his talent to South Beach," they likely began watching him and his team and the entire NBA much more diligently and with a much "stricter" eye. Once fans witnessed the breakout performance of Jeremy Lin, this likely came to impact how they subsequently viewed him as a player and the team with which he was associated. Once fans of the New York Knicks saw the play of Andrea Bargnani fall well short of their expectations, they likely began to question the value of top draft picks.

In the end, this chapter has moved those interested in better understanding fandom communication one step closer to comprehending the role of mediated dissent among NBA fans and the effects of such dissent. While immediacy and anonymity can certainly lead to a myriad of negatively framed effects associated with mediated communication, dissent through blogs and Twitter can, in the end, produce the very effects desired by social agents. Because of social media, an NBA sports world that was once divided has become united, electronically speaking. There will certainly remain divisive rivalries between and among teams and players. After all, rivalry is the independent variable most important for the creation of fandom networks. However, the fact that social media have allowed such divisive, rivalry networks to communicate with (and, to some extent, at) one another has created unity in a sports world that once lacked it. At this point, the beneficial effects that such fandom communication is having certainly outweigh the disadvantageous, and is creating a culture of dissenters still predicated on one thing: a love for the game of basketball.

NOTES

1. Maverick Movies, Twitter post, July 11, 2014, http://twitter.com/MaverickMovies.
2. Alberto Ibarguen, Twitter post, July 11, 2014, http://twitter.com/Ibarguen.
3. Legit, Twitter post, July 11, 2014, http://twitter.com/leanaaabsm.
4. Truth Seeker, Twitter post, July 11, 2014, http://twitter.com/xfilestrustno1.
5. Brian Gonzalez, Twitter post, July 11, 2014, http://twitter.com/bigb_gonzalez.
6. Brandon Ives, Twitter post, July 11, 2014, http://twitter.com/chivez.
7. Vicky Lightfoot, Twitter post, July 11, 2014, http://twitter.com/VickyALightfoot.
8. Justin Seflow, Twitter post, July 11, 2014, http://twitter.com/JustinSef.
9. Homicidal Gaming, Twitter post, July 11, 2014, http://twitter.com/HGCLANS.
10. Mike Antonio, Twitter post, July 11, 2014, http://twitter.com/MrMikeAntonio.
11. Omar Farooq, Twitter post, July 11, 2014, http://twitter.com/heat4life03.
12. SupaJoE, Twitter post, July 11, 2014, http://twitter.com/_supajoe.
13. MC, Twitter post, July 11, 2014, http://twiiter.com/user33131.
14. Baby Sosa, Twitter post, July 11, 2014, http://twitter.com/pvck_cvrlos.
15. Jeffrey W. Kassing, "Articulating, Antagonizing, and Displacing: A Model of Employee Dissent," *Communication Studies*, 48 (1997): 312.
16. Jeffrey W. Kassing, "Consider This: A Comparison of Factors Contributing to Employees' Expressions of Dissent," *Communication Quarterly*, 56 (2008): 342–55.
17. Jeffrey W. Kassing, "Articulating, Antagonizing, and Displacing: A Model of Employee Dssent," *Communication Studies*, 48 (1997): 319.
18. Daniel Katz and Robert Louis Kahn, *The Social Psychology of Organizations* (New York: Wiley, 1966).
19. Jeffrey W. Kassing and Todd A. Armstrong, "Someone's Going to Hear about This: Examining the Association between Dissent-Triggering Events and Employees' Dissent Expression," *Management Communication Quarterly*, 16 (2002): 39–65.
20. Ibid.
21. Surinder S. Kahai and Randoph B. Cooper, "Exploring the Core Concepts of Media Richness Theory: The Impact of Cue Multiplicity and Feedback Immediacy on Decision Quality," *Journal of Management Information Systems*, 20 (2003): 263–99.
22. Uma Narula, *Handbook of Communication: Models, Perspectives, Strategies* (Darya Gang, India: Atlantic Publishers and Distributers, 2006).
23. Theodore A. Avtgis, Candice Thomas-Maddox, Elycia Taylor, and Brian R. Patterson, "The Influence of Employee Burnout Syndrome on the Expression of Organizational Dissent," *Communication Research Reports*, 24 (2007): 97–102.
24. Ibid., 99.
25. Terry Allen, Twitter post, October 11, 2011, http://twitter.com/aa000G9.
26. Robert D. Putnam, *Bowling Alone* (New York: Touchstone, 2001).
27. Marya L. Doerfel, Chih-Hui Lai, and Lisa Chewning, "The Evolutionary Role of Interorganizational Communication: Modeling Social Capital in Disaster Contexts," *Human Communication Research*, 36 (2010): 125–62.
28. Robert D. Putnam, *Bowling Alone* (New York: Touchstone, 2001).
29. Ole Rudolf Holsti, "Content Analysis," in *The Handbook of Social Psychology* (2nd ed.), eds. Gardner Lindzey and Elliot Aronson. (Reading: Addison-Wesley, 1968): 608.
30. Klaus Krippendorff, *Content Analysis: An Introduction to Its Methodology* (2nd ed.) (Thousand Oaks, CA: Sage, 2004): 105.
31. Bruce L. Berg, *Qualitative Research Methods for the Social Scientist* (7th ed.) (Boston: Allyn & Bacon, 2001).

32. Sham, Twitter post, December 31, 2013, http://twitter.com/_kingsleyy5.
33. Carmelo Anthony, Twitter post, December 31, 2013, http://twitter.com/carmeloanthony.
34. Brian Ferrelli, Twitter post, June 23, 2012, http://twitter.com/BrianFerrelli.
35. Amare Stoudemire, Twitter post, June 23, 2012, http://twiiter.com/AmareStoudemire.
36. Amare Stoudemire, Twitter post, June 24, 2012, http://twitter.com/AmareStoudemire.
37. James, LeBron, Interview. ESPN, July 8, 2010.
38. Miami Sucks, Twitter post, July 9, 2010, http://twitter.com/MiamiCheat.
39. LeBron James, Twitter post, January 11, 2011, http://twitter.com/KingJames.
40. Howard, Dwight, Interview. NBC Sports, September 29, 2013.
41. Marcus Kirk, Twitter post, August 7, 2013, http://twitter.com/KinG_WlzrD617.
42. Dwight Howard, Twitter post, August 7, 2013, http://twitter.com/DwightHoward.
43. Brittni, Twitter post, August 7, 2013, http://twitter.com/uHateBre.
44. Dwight Howard, Twitter post, August 7, 2013, http://twitter.com/DwightHoward.
45. PEGEE, Twitter post, July 1, 2014, http://twitter.com/VH1PNUT.
46. Kevin Durant, Twitter post, July 1, 2014, http://twitter.com/KDTrey5.
47. PEGEE, Twitter post, July 1, 2014, http://twitter.com/VH1PNUT.
48. PEGEE, Twitter post, July 1, 2014, http://twitter.com/VH1PNUT.
49. PEGEE, Twitter post, July 1, 2014, http://twitter.com/VH1PNUT.
50. Markus, Twitter post, August 29, 2014, http://twitter.com/TheUncurvable.
51. Kevin Durant, Twitter post, August 29, 2014, http://twitter.com/KDTrey5.
52. Markus, Twitter post, August 29, 2014, http://twitter.com/TheUncurvable.
53. Kevin Durant, Twitter post, August 29, 2014, http://twitter.com/KDTrey5.
54. Markus, Twitter post, August 29, 2014, http://twitter.com/TheUncurvable.
55. Kevin Durant, Twitter post, August 29, 2014, http://twitter.com/KDTrey5.
56. Jared Dudley, Twitter post, August 29, 2014, http://twitter.com/JaredDudley619.
57. Master Wongton, Twitter post, August 29, 2014, http://twitter.com/Ching_Ling666.
58. Jared Dudley, Twitter post, August 29, 2014, http://twitter.com/JaredDudley619.
59. Tike Mouey, Twitter post, February 4, 2014, http://twitter.com/touey_z.
60. Roy Hibbert, Twitter post, February 4, 20014, http://twtter.com/Hoya2aPacer.
61. Tike Mouey, Twitter post, February 4, 2014, http://twitter.com/touey_z.
62. Roy Hibbert, Twitter post, February 4, 20014, http://twitter.com/Hoya2aPacer.
63. Kelly Dwyer, "Roy Hibbert Efficiently Downs an Internet Troll, Talks Shop with Steve Austin, as All of Twitter Applauds," *Ball Don't Lie* (blog), Yahoo.com, February 5, 2014, http://sports.yahoo.com/blogs/nba-ball-dont-lie/roy-hibbert-efficiently-downs-internet-troll-talks-shop-174546642--nba.html.
64. Damian Lillard has hilarious reponse for Twitter troll, *Yardbarker* (blog), November 5, 2014, http://www.yardbarker.com/nba/articles/damian_lillard_has_hilarious_response_to_fan_who_poked_fun_at_him_for_lack_of_team_usa_playing_time/17118998.
65. Vinny D'Amadeo, Twitter post, August 22, 2014, http://twitter.com/Vin_Lee.
66. Damian Lillard, Twitter post, August 22, 2014, http://twitter.com/Dame_Lillard.
67. Mike Edwards, Twitter post, April 15, 2014, http://twitter.com/M_Edwards21.
68. Eric Bledsoe, Twitter post, April 15, 2014, http://twitter.com/EBeld2.
69. Mike Edwards, Twitter post, April 15, 2014, http://twitter.com/M_Edwards21.
70. Mike Edwards, Twitter post, April 15, 2014, http://twitter.com/M_Edwards21.
71. Mike Edwards, Twitter post, April 15, 2014, http://twitter.com/M_Edwards21.
72. Mike Edwards, Twitter post, April 15, 2014, http://twitter.com/M_Edwards21.
73. Mike Edwards, Twitter post, April 15, 2014, http://twitter.com/M_Edwards21.
74. Mike Edwards, Twitter post, April 15, 2014, http://twitter.com/M_Edwards21.
75. Jon Schoonover, Twitter post, June 5, 2014, http://twitter.com/Scoonie_12.
76. Just Keep Livin, Twitter post, June 5, 2014, http://twitter.com/MA_Deuce.
77. Gabe Antrobus, Twitter post, June 5, 2014, http://twitter.com/antrobus15.
78. The Mister Markus, Twitter post, December 27, 2012, http://twitter.com/themistermarcus.
79. Art X Drug Clothing, Twitter post, December 27, 2012, http://twitter.com/ARTXDRUGS.
80. CRVE, Twitter post, December 27, 2012, http://twitter.com/JDgoesHAM.

81. Eric Freeman, "Kobe Bryant and Dwight Howard Tussle Late in Lakers-Rockets Opener," *Ball Don't Lie* (blog), Yahoo.com, October 29, 2014, http://sports.yahoo.com/blogs/nba-ball-dont-lie/kobe-bryant-and-dwight-howard-tussle-late-in-lakers-rockets-opener--video-055849530.html.

82. Peter, October 29, 2014, comment on Eric Freeman, "Kobe Bryant and Dwight Howard Tussle Late in Lakers-Rockets Opener," *Ball Don't Lie* (blog), Yahoo.com, October 29, 2014, http://sports.yahoo.com/blogs/nba-ball-dont-lie/kobe-bryant-and-dwight-howard-tussle-late-in-lakers-rockets-opener--video-055849530.html.

83. Genuine, comment on Freeman, "Kobe Bryant and Dwight Howard."

84. RipHamilton, comment on Freeman, "Kobe Bryant and Dwight Howard."

85. New to Corona, comment on Freeman, "Kobe Bryant and Dwight Howard."

86. Eric Freeman, "Chris Bosh Warns Kevin Love That Playing with LeBron James Can Be Frustrating for Stars," *Ball Don't Lie* (blog), Yahoo.com, October 9, 2014, http://sports.yahoo.com/blogs/nba-ball-dont-lie/chris-bosh-warns-kevin-love-that-playing-with-lebron-james-can-be-frustrating-for-stars-031507786.html.

87. Shadow, October 9, 2014, comment on Eric Freeman, "Chris Bosh Warns Kevin Love That Playing with LeBron James Can Be Frustrating for Stars."

88. John, October 9, 2014, comment on Eric Freeman, "Chris Bosh Warns Kevin Love That Playing with LeBron James Can Be Frustrating for Stars."

89. Ler, October 9, 2014, comment on Eric Freeman, "Chris Bosh Warns Kevin Love That Playing with LeBron James Can Be Frustrating for Stars."

90. Henry Abbott, "Why the Miami Heat Are Rollin,'" *True Hoop* (blog), ESPN.com, November 3, 2014, http://espn.go.com/blog/truehoop/post/_/id/71077/why-the-miami-heat-are-rollin#comments.

91. Terrance Forehead Jackson, November 3, 2014 (4:11pm), comment on Henry Abbott, "Why the Miami Heat Are Rollin,'" *True Hoop* (blog), ESPN.com, November 3, 2014, http://espn.go.com/blog/truehoop/post/_/id/71077/why-the-miami-heat-are-rollin#comments.

92. Kyal Rainey, comment on Abbott, "Why the Miami Heat Are Rollin.'"

93. Isaac Jean, comment on Abbott, "Why the Miami Heat Are Rollin.'"

94. Jeff Alan Moe, comment on Abbott, "Why the Miami Heat Are Rollin.'"

95. Kevin Arnovitz, "A Few Thoughts about Roy Hibbert," *True Hoop* (blog), ESPN.com, May 23, 2013, http://espn.go.com/blog/truehoop/post/_/id/58594/a-few-thoughts-about-roy-hibbert.

96. ssiccone, May 23, 2013 (3:17pm), comment on Kevin Arnovitz, "A Few Thoughts about Roy Hibbert," *True Hoop* (blog), ESPN.com, May 23, 2013, http://espn.go.com/blog/truehoop/post/_/id/58594/a-few-thoughts-about-roy-hibbert.

97. geeman217, comment on Arnovitz, "A Few Thoughts on Roy Hibbert."

98. geeman217, comment on Arnovitz, "A Few Thoughts on Roy Hibbert."

99. Donnell Davis, Twitter post, April 28, 2014, http://twitter.com/temetdonnell.

100. J.C., Twitter post, April 28, 2014, http://twitter.com/jcixiv.

101. CHARGER, Twitter post, April 28, 2014, http://twitter.com/icharger.

102. Donnell Davis, Twitter post, April 28, 2014, http://twitter.com/temetdonnell.

103. Andrew Sutton, Twitter post, June 30, 2013, https://twitter.com/oawesomeandrew.

104. Ed Boulanger, Twitter post, June 30, 2013, https://twitter.com/ed_boulanger.

105. Chris, Twitter post, June 30, 2013, https://twitter.com/bkny248.

106. Joe, Twitter post, October 28, 2014, http://twitter.com/sportstalkjoe.

107. Giants = 8-8, Twitter post, October 28, 2014, http://twitter.com/kingjohn6969.

108. Manmeet Singh, Twitter post, October 28, 2014, http://twitter.com/manmeets1ngh.

109. Irish Ducusin, Twitter post, October 28, 2014, http://twitter.com/iducusin02.

110. Ryan Ward, Twitter post, March 7, 2014, http://twitter.com/lakers_examiner.

111. DoughBalls, Twitter post, March 7, 2014, http://twitter.com/chetanp_32.

112. KMC, Twitter post, March 7, 2014, http://twitter.com/kbookid.

REFERENCES

Avtgis, Theodore, Candice Thomas-Maddox, Elycia Taylor, and Brian R. Patterson. "The Influence of Employee Burnout Syndrome on the Expression of Organizational Dissent." *Communication Research Reports*, 24 (2007): 97–102.

Berg, Bruce. *Qualitative Research Methods for the Social Scientist* (7th ed.). Boston: Allyn & Bacon, 2001.

Holsti, Ole. "Content Analysis." In *The Handbook of Social Psychology* (2nd ed.), eds. Gardner Lindzey and Elliot Aronson, 596–692. Reading: Addison-Wesley, 1968.

Kahai, Surinder, and Randolph B. Cooper, "Exploring the Core Concepts of Media Richness Theory: The Impact of Cue Multiplicity and Feedback Immediacy on Decision Quality," *Journal of Management Information Systems*, 20 (2003): 263–99.

Kassing, Jeffrey. "Consider This: A Comparison of Factors Contributing to Employees' Expressions of Dissent." *Communication Quarterly*, 56 (2008): 342–55.

———. "Articulating, Antagonizing, and Displacing: A Model of Employee Dissent." *Communication Studies*, 48 (1997): 311–32.

Kassing, Jeffrey, and Todd A. Armstrong. "Someone's Going to Hear about This: Examining the Association between Dissent-Triggering Events and Employees' Dissent Expression." *Management Communication Quarterly*, 16 (2002): 39–65.

Katz, D., and Robert L. Kahn. *The Social Psychology of Organizations*. New York: Wiley, 1966.

Krippendorff, Klaus. *Content Analysis: An Introduction to Its Methodology* (2nd ed.). Thousand Oaks, CA: Sage, 2004.

Narula, Uma. *Handbook of Communication: Models, Perspectives, Strategies*. Darya Gang, India: Atlantic Publishers and Distributers, 2006.

Putnam, Robert. *Bowling A lone*. New York: Touchstone, 2001.

Chapter Five

What Types of #SportFans Use Social Media?

The Role of Team Identity Formation and Spectatorship Motivation on Self-Disclosure during a Live Sport Broadcast

Shaughan A. Keaton, Nicholas M. Watanabe, and Brody J. Ruihley

Previous research provides evidence that sport team fandom is marked by behavior such as overt emotional responses to viewed performances;[1] however, previous definitions and models also recognize team fandom as a psychological construct, including notions of involvement (for example, the Psychological Continuum Model)[2] and sport team identification (STI).[3] The manner(s) in which fans come to identify with their favorite teams influences connectedness to the team, which then has direct association with emotional outcomes and psychological effects.[4] In essence, the development of a fan's interest in a team (including but not limited to family socialization, geography, team characteristics, athletic performance, or media) drives how and whether they think about, attend, and react to their favorite team's performance.

Thus, this chapter proposes to examine the convergence of team identity formation and self-disclosure via social media during live sport broadcasts. Essentially, the major factors involve identity formation, spectatorship motives, psychological commitment and investment, self-esteem, and self-disclosure via social media and we will attempt to shed light on the question of what types of sport fans use social media during consumption of sport and whether and what types of disclosures they typically make. Following is a discussion of the major variables of interest including relevant hypotheses.

SOCIAL MEDIA

Research into social media has focused on a number of aspects as to how individuals use and engage with other users and organizations through various platforms.[5] Specifically, the environment of social media creates a dynamic platform in which all users and organizations are able to generate content in an equal manner.[6] While there exist many types of platforms through which to consider social media, Twitter has been one of the most popular forums where interactions take place in regard to sport.[7] It is thus that sport consumers have come to use Twitter as an alternative to traditional forms of one-way media, and therefore began using social media to interact and engage with teams, organizations, athletes and other important figures in sport and sport media.[8]

Notably, research into social media has been divided into two lineages:[9] first is content-based inquiry which specifically examines words and texts created by users of the platform. Notably, Hambrick, Simmons, Greenhalgh, and Greenwell[10] have specifically examined how athletes were able to employ Twitter as a tool through which to connect with fans, and note various categories of how communication occurred between different parties. Second is audience-based inquiry, which considers the uses, characteristics, demographics, and other factors that draw individuals to social media. These researchers have greatly focused on college sport fans, with specific studies considering how social media has been used by fans[11] as well as how fans are attracted to coaches based on team performance and other factors.[12]

Considering the focus of this chapter, the social media platform of Twitter is a reasonable choice in regard to attempting to understand the behavior of fans as they interact in the context of sport. As noted, Twitter is a popular channel through which sporting discussions occur,[13] and allows for the examination of real-time interaction between individual consumers, sport figures, and organizations as sporting events are unfolding. Furthermore, the research to date in the realm of social media has been moving toward identification of factors that determine what will attract consumers to a Twitter feed. Specific research focused on college football and Twitter use has employed large scale surveys[14] to investigate factors that draw individuals to Twitter accounts and have helped to establish the importance of understanding this type of fan behavior.

Notably, a large portion of the research that has examined the determinants of getting a sport fan to interact and follow a sport team's Twitter account have employed data which were not collected around the actual sporting event. Furthermore, while research has widely considered social media use from a communications framework, there has been little research that has attempted to bring the understanding of social media use into the realm of consumer and fan behavior in sport. That is, while research studies have often discussed the importance of managing social media accounts to draw in more fans, there is still a need to develop linkages with consumer behavior studies to understand how consumption of social media may be similar or different to other sport products. This research attempts to further integrate the use of social media communication research with consumer behavior studies by employing a methodology that examines the use of social media through both a communicative and consumer behavior lens.

CONSUMER BEHAVIOR

Consumer behavior has been one of the most widely examined research areas within the realm of sport, with research focused on how fans as consumers are drawn or motivated to attend sporting events, purchase sport products, or exhibit other forms of consumption behavior.[15] The examination of fan consumer behavior has been considered an important part of sport academia and practice, as it is noted that understanding how individuals behave in their consumption patterns can allow organizations to behave in a more strategic manner.[16] For organizations to understand the behavior of individuals, there is need to consider the motivations, antecedents, and outcomes which could potentially influence them to make sport related purchases and consume sport-related media.[17] Thus, there is a need from a managerial and academic standpoint to try and identify behavioral, cognitive, and psychological factors that push individuals to identify with organizations, and thus help motivate them to consume sport.

While such a task would seem simple enough, research has noted that the motivations and identifications displayed by consumers to sport entities can often change based on the type of sport and other factors.[18] Thus, because consumers display such a wide range of behavior within a sport context, it is important that each group be specifically targeted through focused research and understanding.[19] With this targeting in mind, and because social media interaction is such a different product than consumption of sport events or merchandise with regard to function, cost, and ease of access for fans, it is likely that the motivations and identity antecedents to use Twitter and other platforms will be different than other products. Thus, these authors examine sport fans' consumption of sport via social media by analyzing the motivations and identity antecedents that influence them to exhibit certain behaviors.

TEAM IDENTITY FORMATION

Research into STI formation has focused on how individuals develop a connection with specific sport organizations. This process includes examination of how individuals build psychological connections with their teams, the importance of specific social factors such as the team's performance and characteristics,[20] family socialization,[21] peer socialization, geographic location of the individuals and the importance of media.[22] Specifically, it has been noted that the antecedents of team identity can impact the connection and involvement a fan has with a sport entity.[23] Because of the varying differences and magnitudes that may exist with regard to the psychological connections between fans and a team, it is necessary to consider how a sport team identity is developed. In considering STI, it is the case that there is a lack of research which has focused on the connection between identification and social media use, as well as how fans on Twitter may develop this identification with sport organizations.

SPECTATORSHIP MOTIVATION

Another well-researched area in the realm of consumer behavior has been motivations of fans and spectators. A sport consumer can be a fan and a spectator simultaneously, but one can view sports without being a fan, and one can have a psychological commitment to a team, but not watch the games. It has been noted that fans and spectators can exhibit somewhat different motivations;[24] that is, the individuals who are only fans of a team and individuals who are only spectators of sport can have different motivations (although they can coincide). The motivational differences that can exist between fans and spectators are traced to factors such as point of attachment[25] and other commonly considered motives. The most commonly focused upon motives for developing models of fandom and spectatorship of sport teams includes escape, economics, group affiliation, family, entertainment, eustress, aesthetics, and self-esteem.[26]

The previously mentioned motives are all important factors when researching and modeling motivations that draw spectators to events, and thus need to be considered in depth.[27] The motivation of escape has been noted to be important in allowing individuals to attend sporting events as a way to avoid the issues and flow of their lives.[28] Economic motives have also been shown to be important because of fan interest in gaining profits from sporting events through behaviors such as wagering. While economic studies have investigated the relation between wagers on sporting events and team identity, the consumer behavior literature has found little evidence that economic gains are important.[29] In this, the potential is that betting and other economic behavior may not be attributed to fans or spectators of sporting events, but may be traced to consumers who are focused on making money.[30]

Prior literature also shows the importance of social connections for sport fans with regard to group affiliations and family, and is there is linkage between social obligations and attending sporting events. It is specifically noted that family is of high importance to sport consumers who are married, in that people may be motivated to attend sporting events because of their family and others from their social circles who also attending the events.[31] Entertainment is another important motivation for fans to consume sport products, as the conception of having fun is said to be of interest to fans.[32] In this, the entertainment can come in many forms such as watching aggressive contact sports[33] or even in watching a rival sport franchise lose.[34] Likewise, eustress can be considered to be important because of its relation to being excited by events in a sporting event, and thus being entertained. Research has shown that this type of beneficial stress influences increases in positive feelings from the uncertain nature of the sport contest, thus attracting fans to attend events.[35]

Aesthetics is also considered to be important with research because of the artistry, grace, or other intrinsic factors which fans enjoy watching at sporting events.[36] Finally, self-esteem is also an important because of the theorization that it may influence how sport spectators respond to team performances.[37] Specifically, the idea that has been widely examined is whether sport fans will distance themselves from teams because of their on-field failures, and will engage themselves because of team success.[38] Considering the importance of these motivations, it is natural that they are all considered in the past and present research that considers consumer motivations. What is important to note in the case of this research is that little connection has been made between social media use and some of these factors. Thus, it is necessary that consumer motivations be further developed with regard to understanding social media use.

PSYCHOLOGICAL TEAM COMMITMENT AND INVESTMENT

Psychological team commitment research delves into the importance of social-psychological factors, which may build attachment between a sport team and a fan's emotions and interest.[39] Daniel F. Mahony, Robert Madrigal, and Dennis R. Howard developed the *psychological commitment to team* (PCT)[40] scale to help investigate constructs involved with fan loyalty to sport organizations. The development of this scale followed research that noted the importance of consumer loyalty to brands and/or products, and this psychological connection is also linked to the behavior of sport fans.[41] Therefore, it is posited within the research that commitment by a sport consumer manifests via loyalty, and can be an important and enduring relationship. Thus, a consumer who manages to develop loyalty to a team can be considered to be invested in the team's on-field (and off-field) performance because of the attachment which has been created.

SELF-ESTEEM AND SPORT FANDOM

Self-esteem has been linked with the performance of teams on the field and how fans will negotiate wins and losses by the teams they have connected with.[42] This behavior is often subdivided into the concepts of basking in reflected glory (BIRGing) and cutting off reflected failure (CORFing), with fans performing these types of behavior as part of their own identity and connection with a sport team.[43] That is, a fan whose team wins may be found to be wearing merchandise from that team the next day to represent the win as part of the building of their own self-esteem, while a fan whose team loses may avoid wearing their team's merchandise to avoid being associated with failure. The behavior of CORFing is not only the process of avoiding failure, but also to help maintain self-esteem by disassociating with an aversive part of their identity.[44]

It is not the case that BIRGing, CORFing, and other self-esteem related fandom behaviors are independent, but also are related with identity theory.[45] The argument is posited that self-esteem behaviors and identity are all part of the performances of the self, and therefore a fan exhibiting self-esteem behavior may also be displaying parts of their own identity. Research investigating this occurrence has found that there may be linkages between identity and self-esteem behaviors.[46] Thus, in conducting research focused on identification with sport teams, such as STI and motivations studies mentioned earlier, it is important to consider the construct of self-esteem. This position may especially be the case in examining sport based social media, where fans are able to exhibit their team identification behavior in real-time through words and images. It would seem to be important that self-esteem must also be considered in the analysis of social media narratives, as they may reflect the identity and self-esteem behaviors which are associated with sport fandom.

SELF-DISCLOSURE AND SPORT FANDOM

In conclusion, it is evident that there are a large number of factors that can play a role in an individual's identity formation and spectator motivation, and those factors' influences on a fan's inclination to consume a sport product. Likewise, it is also the case that researchers into social media have shown a variety of factors important in the communicative process for individuals.

Thus, there is great need for modeling to be developed to help create a better understanding of how consumers are motivated to interact with teams on social media and to explicate on the development of detailed scales for understanding all the potential factors influencing these behaviors. Accordingly, this study poses two research questions:

RQ1: What types of sport fans self-disclose via social media during live broadcasts?

RQ2: What topics are disclosed by fans who self-disclose via social media during live broadcasts?

METHODOLOGY

Procedures

Participants were recruited for this study a variety of ways, including snowball samples gathered by means of social media (Facebook, Twitter, and email lists), offered for extra credit to students at two large midwestern U.S. universities and one small private southeastern U.S. college, and offered to students as part of their course work at a large southeastern U.S. university.

At the university involving course credit, students were recruited for the study via an online scheduling system. Those choosing this study as an option from a list of departmental projects earned 1.5 percent of their course grade for participating. All data collected were confidential, all students provided informed consent, and the appropriate Institutional Review Board approved all procedures.

All participants were directed to an external and secure universal resource locator (URL) where they were shown a list of instructions. The first step was to follow @fanresearch2014 on Twitter. Before the live sport broadcast they chose to watch, they were instructed to take a preliminary survey inquiring about their team identity formation, their spectatorship motivations, psychological commitment to their teams, and self-esteem. During the live sport broadcast, participants were asked to "use Twitter to provide your thoughts and commentary. The content is not restricted and you may disclose any type of information that you feel is appropriate to the situation." After the live broadcast, they were asked to take another short survey. The participants were assured that their survey answers were anonymous, their Twitter user names would not be used outside of research, and that their participation was voluntary and that they could choose to stop at any point. The participants were restricted to those over eighteen years of age.

Participants

A total of 127 (sixty females, sixty-seven males, one transgender) participants constituted the sample for this study. Participants ranged from eighteen to sixty-one years of age ($M = 20.18$, $SD = 4.59$). The participants self-described as white or Caucasian ($n = 91$; 71.7 percent), black or African American ($n = 23$; 18.1 percent), Asian-American or of Asian descent ($n = 4$; 3.2 percent), Asian Indian ($n = 4$; 3.2 percent), Native American ($n = 1$; 0.8 percent), Hispanic or Latino ($n = 3$; 2.4 percent) or Other ($n = 1$; 0.8 percent). Sixty-seven of the participants chose not to tweet at all; out of the sixty who did, the average number of tweets was 6.13.

Preliminary Analyses

Prior to running primary analyses, data were inspected for violations of multivariate assumptions.[47] All measurement scales were assessed for dimensionality and ability to represent current data. Commonly used fit indexes and comparison thresholds were utilized: The comparative fit index (CFI) above 0.90 and the root mean square error of approximation (RMSEA) below 0.08.[48] Internal consistency was estimated using Cronbach's α. A post hoc power analysis was conducted using G*Power software.[49] With $N = 127$ and alpha set at 0.05, power to detect small Pearson correlation effects ($r = 0.10$) was 0.30, 0.97 for medium ($r = 0.30$), and exceeded 0.99 for large ($r = 0.50$) effects.[50]

Quantitative Measures

To measure consumer behaviors and attitudes in relation to sport fandom in regard to RQ1, scales were utilized that target team identity formation, spectatorship motives, psychological attitudes, and self-esteem. The first three measures in regard to identity formation, spectatorship motives, and psychological attitudes were chosen because they have been shown to exhibit consistent and satisfactory psychometric properties. These particular scales were developed in conjunction after a comprehensive and rigorous analysis of the most frequently used scales in sport consumer and sport psychology literature and contain original items from those scales.[51] They have been shown to replicate in at least six data sets in the United States and Japan,[52] have been significantly correlated with behavioral, cognitive, physiological, and psychological variables,[53] and are a cumulative result of sport scale research and development. Each scale is described with accompanying psychometric evidence of reliability and validity.

Sport Team Identification

The Causation of Sport Team Identification Scale (C-STIS)[54] was administered to study participants. The C-STIS measures team identification antecedents on a seven-point Likert scale and contains twenty-two items across five latent constructs (Media Popularity/Influence, Geography, Family, Athletic Performance, and Team Characteristics). Four factors were retained for this analysis: Media Popularity/Influence (e.g., "I chose my favorite team because they receive a substantial amount of national television coverage"; $n = 4$; $\alpha = 0.92$), Geography (e.g., "I am a fan of this team because it is an important connection between me and my hometown or university"; $n = 5$; $\alpha = 0.86$), Family ("I chose my favorite team because my parents and/or family follow this team"; $n = 5$; $\alpha = 0.92$), Athletic Performance (e.g., "Watching a well-executed athletic performance is something I enjoy"; $n = 4$; $\alpha = 0.77$), and Team Characteristics; e.g., "I chose my favorite team because I like their reputation/image"; $n = 4$; $\alpha = 0.81$). After removing two low-loading items, these data fit the proposed factor structure, $\chi^2(176) = 276.49$, $p < .001$, CFI = 0.93, RMSEA = 0.07, CI90% = 0.05, 0.08.

Sport Spectatorship Motives

The second measurement scale, the Scale of Sport Spectatorship Motives,[55] explains reasons why people choose to become spectators of sporting events. The scale contains eighteen items across four latent constructs: Recreational Value (e.g., "One of my reasons to watch and attend sport games is to seek excitement and stimulation"; $n = 4$; $\alpha = 0.75$), the extent of their Fan Self-Concept (e.g., "Being a sport fan is very important to me"; $n = 5$; $\alpha = 0.93$), Aesthetics (e.g., "One of my reasons to watch and attend sport games is the beauty and grace of the game"; $n = 5$; $\alpha = 0.90$), and the extent to which someone watches sport simply to pass time (Casual Spectatorship: e.g., "One of my reasons to watch and attend sport games is to kill time"; $n = 4$; $\alpha = 0.84$). These data fit the proposed factor structure, $\chi^2(111) = 225.02$, $p < .001$, CFI = 0.91, RMSEA = 0.09, CI90% = 0.07, 0.11.

Psychological Attitudes

The third measurement scale, the Sport Team Psychological Commitment[56] scale, explains psychological benefits, involvement, and commitment. The STPC contains fifteen items across three factors that describe psychological effects of STI on a seven-point Likert scale including self-actualization (e.g., "Being a spectator of my favorite sport helps me to develop and grow as a person"; $n = 5$; $\alpha = 0.91$), commitment (e.g., "I have stopped following a team because I had too many commitments and/or I simply did not have time"; $n = 6$; $\alpha = 0.85$) and investment (e.g., "I continue to be a fan of this team because it would be very stressful for me to openly discontinue my association with this team"; n = 4; α =0 .80). These data fit the proposed factor structure, χ^2 (85) = 148.79, p < 0.001, CFI = 0.91, RMSEA = 0.08, CI90% = 0.06, 0.10.

Self-Esteem

For self-esteem, Rosenberg Self-Esteem Scale (RSES)[57] was utilized. The RSES contains fifteen items across one factor on a seven-point Likert scale (e.g., "On the whole, I am satisfied with myself."; α = 0.87). These data did not initially fit the proposed factor structure, χ^2 (77) = 334.65, p < .001, CFI = 0.48, RMSEA = 0.16, CI90% = 0.15, 0.18. After removing low-loading items, we were left with seven items that together approached usable model fit, χ^2 (14) = 36.5, p < 0.001, CFI = 0.94, RMSEA = 0.11, CI90% = 0.06, 0.10.

QUALITATIVE ANALYSIS

Thematic analysis was utilized to qualitatively examine participant tweets to answer RQ2. Thematic analysis assists researchers in "identifying, analyzing and reporting patters (themes) within data."[58] It also "minimally organizes and describes your data set in (rich) detail" and supports interpretation of topics and data. As part of this thematic analysis, data were coded following guidelines of open, axial, and selective coding.[59] Open coding searches for themes within participant responses participant and segments "them into categories of information."[60] Then, axial coding locates themes shared between multiple responses. Finally, selective coding "takes the central phenomenon and systematically relates it to other categories, validating the relationships and filling in categories that need further refinement and developments."[61] Through these processes, categories to the participant tweets were discovered and examined. Themes were then developed based on similarities and differences within coding categories. One author initially conducted the thematic analysis. Then, the other two authors thoroughly examined results; all authors discussed the findings, cleared any disagreement, and came to full agreement on results.

Results

To contribute toward answering what types of fans tend to self-disclose via live sporting events via social media (RQ1), a number of canonical linear discriminant models were estimated to, in effect, create profiles of two groups of sport fans. In other words, it allows us to compare one group of fans to another, namely those who tweeted versus those who chose not to.

For the set of identity formation factors, one discriminant dimension was significant, $F(5,116) = 3.19$, $p < .01$, $r_{canonical} = 0.35$. By examining the canonical loadings and group centroids, we can observe that on the positive (*did not* self-disclose via social media) end of the continuum, family is a more important antecedent of identity formation, whereas on the negative (*did* self-disclose via social media) end, athletic performance was more important in team selection.

In the next model, a significant root indicated the factors of spectatorship motivation have differential impacts on whether or not a participant self-disclosed via social media. The discriminant dimension was significant, $F(4,121) = 2.98$, $p < 0.02$, $r_{canonical} = 0.30$. By examining the canonical loadings, we can see that on the positive (*did not* self-disclose via social media) end of the continuum, none of the spectatorship motivations were influential on these participants, whereas on the negative (*did* self-disclose via social media) end, a watching sport as an opportunity to make their fan identities visible to other was a strong motivator. Finally, in separate model estimations for the psychological variables and demographics, no significant roots were present, suggesting that there is no differential association between these factors and the likelihood that the sport fan would self-disclose via social media.

To answer RQ2 (concerning what topics are disclosed by fans that self-disclose via social media during live broadcasts), a qualitative thematic analysis was conducted. Results of the thematic analysis of participant tweets categorized the communication into three types: statement ($n = 305$, 84.25% of tweets), conversation ($n = 42$, 11.60%), and photo ($n = 15$, 4.14%). When examining the topic area and subject matter, several themes were discovered. The top five themes revolved around and were titled *analysis* ($n = 95$ codes, 26.24% of tweets), *game update* ($n = 93$, 25.69%), *encouragement* ($n = 305$, 12.71%), *emoticon-only* ($n = 40$, 11.05%), and *celebration/disappointment* ($n = 40$, 11.05%). Other themes consisted of *promotion, trash talk, appreciation, and general comments*.

Summary of Results

To summarize the results, the typical person who chose to self-disclose via Twitter during a live sporting event in these data is one to whom athletic performance is important in selecting a favorite team (i.e., a reason why they chose a particular team with which to identify), and the opportunity to perform that allegiance (i.e., to be seen by others as a fan of a particular team or sport) is a primary motivator for watching sports (making social situations such as tailgating or social media such as Twitter important in this process). The only influential factor associated with those who chose *not* to self-disclose via Twitter was the effect of family on team selection, meaning that these individuals' families likely share the same favorite team and identify with that team. The types of self-disclosures shared by those who did Tweet included statements, conversations, and photos. The major themes among these types included game updates, encouragement, emoticons, and celebration/disappointment.

Discussion

The results of this investigation support the notion that there are, in fact, some differences between sport fans who self-disclose via social media and sport fans who do not. First, sport fans who used Twitter to share their thoughts about sport during a live broadcast were more likely to value athletic performance when choosing a team and valued performing their sport identity to others. It makes sense, then, that these people would be more prone to share their fandom via social media because they want to be seen as fans by others. Athletic performance is also important to them because it is probable that those who like to be seen by others as fans would want to be seen as associated with more skillful teams or athletes. This notion is supported in the literature all the way back to the seminal studies on BIRGing and CORFing.[62] Simply put, those who are affiliated with more successful teams are more likely to want to be seen as being associated with that team (or athlete). Indeed, those who tweeted during the live broadcasts liked to be seen as actively engaged through providing game updates to others. They were also emotionally invested, offered encouragement to the teams, displayed different emotional states through emoticons, and made celebratory or disappointing proclamations concerning the teams' performances (which were found to be important during their STI).

Sport fans that did not use Twitter during the live sport broadcast were only distinguished by the fact that they were more influenced by family when choosing a favorite team. Perhaps this socialization process makes social media an afterthought during spectatorship of games. To these fans, it is about the interaction between significant family members during these episodes. Interestingly, there were no marked differences concerning the psychological variables. Whether or not fans used social media showed no differentiation between level of commitment or investment in a team. There were also no differences in self-esteem or feelings of self-actualization. From a practical standpoint, this research has important implications for the management of social media and in understanding consumer behavior on a digital platform. The current research helps to develop the understanding that fans on social media may be reflecting a level of vicarious achievement and with regard to sport teams they are connected with. It is evident that sport fans displayed more potential for engaging and interacting in conversations around their favorite teams after those teams have had good performances.

Thus, a team managing their media may want to heavily promote themselves and the performance right after a positive outcome; in this manner, they can build more connections with consumers.

Limitations

Within this study there are potential limitations. The first is the sample employed to analyze these data. A larger sample from a wider demographic range would have been ideal. However, because of various restrictions, the sample for this study was mostly limited to university students. The use of students is a widely employed strategy for research projects, especially ones that focus on sport identification and consumption. Thus, while the use of student subjects within this research is not entirely ideal, it is an acceptable research practice. In the case of this investigation, the sample may have had the potential to bias results given that these data were collected at schools that are popular partly because of their sport programs. Variables such as geography, family, and social motivations may have been particularly influenced in this sample.

A second limitation that exists within this study is the focused nature of research within the United States. While the sample is suitable for this study, future research studies could consider a larger sample group from various demographic backgrounds both within the United States and other countries. There is need for further testing of this methodology in various contexts.

Finally, these authors wish to address the fact that we only tested a small number of identification and motivation variables. There are myriad influences on the development of an individual's identity and their motivations to consume sport. These authors chose to explicate on a small cross-section of that process, mainly variables relevant to socialization (i.e., family, community, media, recreation, and spectatorship). Future research should begin to elucidate other relevant areas.

Future Research

In further consideration of the results and implications, future researchers should also continue to investigate several important areas and contexts. This study included sport fans in the United States, and though this specific context is unique and important, future research should consider the vast array of sport fans of individual athletes—especially in other cultures—and how they compare to one another. Clearly there are a large number of contexts through which to consider different types of sport consumerism and how they compare to one another. Future research can help to expand theoretical and empirical understanding of sport fans.

NOTES

1. Walter Gantz and Lawrence Wenner, "Fanship and the Television Sports Viewing Experience," *Sociology of Sport Journal* 12 (1995).

2. Daniel C. Funk and Jeffrey D. James, "The Psychological Continuum Model: A Conceptual Framework for Understanding an Individual's Psychological Connection to Sport," *Sport Management Review* 4 (2001).

3. Daniel L. Wann, "The Causes and Consequences of Sport Team Identification," in *Handbook of Sports and Media*, ed. A. A. Raney and J. Bryant (Mahway, NJ: Erlbaum, 2006); Daniel L. Wann and Nyla R. Branscombe, "Sports Fans: Measuring Degree of Identification with Their Team," *International Journal of Sports Psychology* 24 (1993).

4. Shaughan A. Keaton and Christopher C. Gearhart, "Identity Formation, Identity Strength, and Self-Categorization as Predictors of Affective and Psychological Outcomes: A Model Reflecting Sport Team Fans' Responses to Highlights and Lowlights of a College Football Season," *Communication & Sport* (2013).

5. Michael L. Kasavana, Khaldoon Nusair, and Katherine Teodosic, "Online Social Networking: Redefining the Human Web," *Journal of Hospitality and Tourism Technology* 1, no. 1 (2010); Jo Williams and Susan J. Chinn, "Meeting Relationship-Marketing Goals through Social Media: A Conceptual Model for Sport Marketers," *International Journal of Sport Communication* 3, no. 4 (2010).

6. danah m. boyd and Nicole B. Ellison, "Social Network Sites: Definition, History, and Scholarship," *Journal of Computer-Mediated Communication* 13, no. 1 (2007); Constantino Stavros et al., "Understanding Fan Motivation for Interacting on Social Media," *Sport Management Review* (2013).

7. Ann Pegoraro, "Sport Fandom in the Digital World," *Routledge Handbook of Sport Communication* (2013).

8. Chris Anderson, *The Long Tail: Why the Future of Business Is Selling Less of More* (Hachette Digital, Inc., 2006). David Rowe, *Global Media Sport: Flows, Forms and Futures* (A&C Black, 2012); "Following the Followers Sport Researchers' Labour Lost in the Twittersphere?," *Communication & Sport* 2, no. 2 (2014).

9. Galen Clavio, "Social Media and the College Football Audience," *Journal of Issues in Intercollegiate Athletics* 4 (2011).

10. Marion E. Hambrick et al., "Understanding Professional Athletes' Use of Twitter: A Content Analysis of Athlete Tweets," *International Journal of Sport Communication* 3, no. 4 (2010).

11. Clavio, "Social Media and the College Football Audience."; Galen Clavio and Patrick Walsh, "Dimensions of Social Media Utilization among College Sport Fans," *Communication & Sport* (2013).

12. Jonathan A. Jensen, Shaina M. Ervin, and Stephen W. Dittmore, "Exploring the Factors Affecting Popularity in Social Media: A Case Study of Football Bowl Subdivision Head Coaches," *International Journal of Sport Communication* 7, no. 2 (2014).

13. Pegoraro, "Sport Fandom in the Digital World."

14. Clavio and Walsh, "Dimensions of Social Media Utilization among College Sport Fans."

15. David Shilbury, Shayne Quick, and Hans Westerbeek, *Strategic Sport Marketing* (Sydney, Australia: Allen and Unwin, 1998).

16. Shaughan A. Keaton, Nicholas M. Watanabe, and Christopher C. Gearhart, "A Comparison of College Football and Nascar Consumer Profiles: Identity Formation and Spectatorship Motivation," *Sport Marketing Quarterly* 24, no. 1 (2015); William A. Sutton et al., "Creating and Fostering Fan Identification in Professional Sports," *Sport Marketing Quarterly* 6 (1997).

17. Robert K. Stewart, Aaron C. T. Smith, and Matthew Nicholson, "Sport Consumer Typologies: A Critical Review," *Sport Marketing Quarterly* 12 (2003).

18. Jeffrey D. James and Stephen D. Ross, "Comparing Sport Consumer Motivations across Multiple Sports," *Sport Marketing Quarterly* 13 (2004).

19. Robert K. Stewart, Aaron C. T. Smith, and Matthew Nicholson, "Sport Consumer Typologies: A Critical Review," *Sport Marketing Quarterly* 12 (2003).

20. William A. Sutton et al., "Creating and Fostering Fan Identification in Professional Sports," *Sport Marketing Quarterly* 6 (1997); Wann, "The Causes and Consequences of Sport Team Identification."

21. Daniel L. Wann, Kathleen B. Tucker, and Michael P. Schrader, "An Exploratory Examination of the Factors Influencing the Origination, Continuation, and Cessation of Identification with Sports Teams," *Perceptual and Motor Skills* 82 (1996).

22. Christopher C. Gearhart and Shaughan A. Keaton, "The Influence of Motives for Selecting a Favorite Team on Sport Team Identification and Fan Behavior" (paper presented at the National Communication Association, New Orleans, LA, 2011); Shaughan A. Keaton, "Sport Team Fandom, Arousal, and Communication: A Multimethod Comparison of Sport Team Identification with Psychological, Cognitive, Behavioral, Affective, and Physiological Measures" (dissertation, Louisiana State University, 2013).

23. Keaton and Gearhart, "Identity Formation, Identity Strength, and Self-Categorization as Predictors of Affective and Psychological Outcomes: A Model Reflecting Sport Team Fans' Responses to Highlights and Lowlights of a College Football Season."

24. Galen. T. Trail et al., "Motives and Points of Attachment: Fans Versus Spectators in Intercollegiate Athletics," *Sport Marketing Quarterly* 12 (2003).

25. Ibid.

26. Daniel L. Wann et al., "Motivational Profiles of Sport Fans of Different Sports," *Sport Marketing Quarterly* 17 (2008).

27. Ibid.

28. Lloyd R. Sloan, "The Motives of Sports Fans," in *Sports, Games, and Play: Social and Psychological Viewpoints*, ed. J. H. Goldstein (Hillsdale, NJ: Erlbaum, 1989).

29. Daniel L. Wann, "Preliminary Validation of the Sport Fan Motivation Scale," *Journal of Sports and Social Issues* 19(1995).

30. Wann et al., "Motivational Profiles of Sport Fans of Different Sports."

31. Karen H. Weiller and Catriona T. Higgs, "Fandom in the 40's: The Integrating Functions of All American Girls Professional Baseball League," *Journal of Sport Behavior* 20, no. 2 (1997).

32. Jennings Bryant, "Viewers' Enjoyment of Televised Sports Violence," in *Media, Sports, and Society*, ed. Lawrence A. Wenner (Newbury Park, CA: Sage, 1989).

33. Jennings Bryant, Paul Comisky, and Dolf Zillmann, "The Appeal of Rough-and-Tumble Play in Televised Professional Football," *Communication Quarterly* 29, no. 4 (1981).

34. Barry S. Sapolsky, "The Effect of Spectator Disposition and Suspense on the Enjoyment of Sport Contests," *International Journal of Sport Psychology* (1980).

35. Sloan, "The Motives of Sports Fans."

36. Daniel L. Wann and Christi L. Ensor, "Family Motivation and a More Accurate Classification of Preferences for Aggressive Sports," *Perceptual and Motor Skills* 92, no. 2 (2001).

37. Galen T. Trail, Dean F. Anderson, and Janet S. Fink, "Consumer Satisfaction and Identity Theory: A Model of Sport Spectator Conative Loyalty," *Sport Marketing Quarterly* 14, no. 2 (2005).

38. Ibid.

39. Daniel F. Mahony, Robert Madrigal, and Dennis R. Howard, "Using the Psychological Commitment to Team (Pct) Scale to Segment Sport Consumers Based on Loyalty," *Sport Marketing Quarterly* 9 (2000).

40. Ibid.

41. Garry J. Smith et al., "A Profile of the Deeply Committed Male Sports Fan," *ARENA Review* 5, no. 2 (1981).

42. Trail, Anderson, and Fink, "Consumer Satisfaction and Identity Theory: A Model of Sport Spectator Conative Loyalty."

43. Ibid.

44. Daniel L. Wann and Nyla R. Branscombe, "Die-Hard and Fair-Weather Fans: Effects of Identification on BIRGing and CORFing Tendencies," *Journal of Sports and Social Issues* 14 (1990).

45. Laurie H. Ervin and Sheldon Stryker, *Theorizing the Relationship between Self-Esteem and Identity*, Extending Self-Esteem Theory and Research: Sociological and Psychological Currents (New York, NY: Cambridge University Press, 2001).

46. Robert Madrigal, "Cognitive and Affective Determinants of Fan Satisfaction with Sporting Event Attendance," *Journal of Leisure Research* 27 (1995); Trail, Anderson, and Fink, "Consumer Satisfaction and Identity Theory: A Model of Sport Spectator Conative Loyalty."

47. Barbara G. Tabachnick and Linda S. Fidell, *Using Multivariate Statistics*, 5th ed. (Boston: Pearson Education, 2007).

48. Barbara M. Byrne, *Structural Equation Modeling with Amos*, 2nd ed. (Ottawa, Ontario, Canada: Routledge, 2010); Rex B. Kline, *Principles and Practice of Structural Equation Modeling*, 2nd ed. (New York: Guilford, 2005); Tenko Raykov and George A. Marcoulides, *A First Course in Structural Equation Modeling*, 2nd ed. (Mahwah, NJ: Erlbaum, 2006).

49. Franz Faul et al., "G*Power 3: A Flexible Statistical Power Analysis Program for the Social, Behavioral, and Biomedical Sciences," *Behavior Research Methods* 39(2007).

50. Jacob Cohen, *Statistical Power Analysis for the Behavioral Sciences*, 2nd ed. (Hillsdale, NJ: Erlbaum, 1988).

51. Shaughan A. Keaton and Christopher C. Gearhart, "Measuring Self-Reported Perceptions of Sport Fandom: Evaluating, Developing, and Refining Existing Scales in Sport Psychology" (paper presented at the Seventh Summit on Communication and Sport, Brooklyn, NY, 2014).

52. Keaton, "Sport Team Fandom, Arousal, and Communication: A Multimethod Comparison of Sport Team Identification with Psychological, Cognitive, Behavioral, Affective, and Physiological Measures"; Keaton and Gearhart, "Identity Formation, Identity Strength, and Self-Categorization as Predictors of Affective and Psychological Outcomes: A Model Reflecting Sport Team Fans' Responses to Highlights and Lowlights of a College Football Season"; "Measuring Self-Reported Perceptions of Sport Fandom: Evaluating, Developing, and Refining Existing Scales in Sport Psychology"; Shaughan A. Keaton, Christopher C. Gearhart, and James M. Honeycutt, "Fandom and Psychological Enhancement: Effects of Sport Team Identification and Imagined Interaction on Self-Esteem and Management of Social Behaviors," *Imagination, Cognition and Personality* 33 (2014); Shaughan. A. Keaton and James M. Honeycutt, "The Effects of Sport Fandom on Physiology, Communication, and Mental Health," in *The Influence of Communication in Physiology and Health*, ed. J. M. Honeycutt, C. Sawyer, and S. A. Keaton (New York: Peter Lang, 2014); Keaton, Watanabe, and Gearhart, "A Comparison of College Football and Nascar Consumer Profiles: Identity Formation and Spectatorship Motivation."

53. Keaton, "Sport Team Fandom, Arousal, and Communication: A Multimethod Comparison of Sport Team Identification with Psychological, Cognitive, Behavioral, Affective, and Physiological Measures"; Keaton and Gearhart, "Identity Formation, Identity Strength, and Self-Categorization as Predictors of Affective and Psychological Outcomes: A Model Reflecting Sport Team Fans' Responses to Highlights and Lowlights of a College Football Season"; Keaton and Honeycutt, "The Effects of Sport Fandom on Physiology, Communication, and Mental Health."

54. Keaton, "Sport Team Fandom, Arousal, and Communication: A Multimethod Comparison of Sport Team Identification with Psychological, Cognitive, Behavioral, Affective, and Physiological Measures"; Keaton and Gearhart, "Identity Formation, Identity Strength, and Self-Categorization as Predictors of Affective and Psychological Outcomes: A Model Reflecting Sport Team Fans' Responses to Highlights and Lowlights of a College Football Season"; "Measuring Self-Reported Perceptions of Sport Fandom: Evaluating, Developing, and Refining Existing Scales in Sport Psychology."

55. Ibid.

56. Ibid.

57. Morris Rosenberg, *Society and the Adolescent Self-Image*, Revised ed. (Collingdale, PA: Diane Publishing Company, 1999).

58. Virginia Braun and Victoria Clarke, "Using Thematic Analysis in Psychology," *Qualitative Research in Psychology* 3 (2006).

59. John W. Creswell, *Qualitative Inquiry & Research Design: Choosing among Five Approaches*, 2nd ed. (Thousand Oaks, CA: Sage, 2007).
60. Ibid, 239–40.
61. Anselm Strauss and Juliet M Corbin, *Basics of Qualitative Research: Grounded Theory Procedures and Techniques* (Thousand Oaks, CA: Sage, 1990); Creswell, *Qualitative Inquiry & Research Design: Choosing among Five Approaches*.
62. Robert B. Cialdini et al., "Basking in Reflected Glory: Three (Football) Field Studies," *Journal of Personality and Social Psychology* 34 (1976).

REFERENCES

Anderson, Chris. *The Long Tail: Why the Future of Business Is Selling Less of More*. Hachette Digital, Inc., 2006.

boyd, danah m., and Nicole B. Ellison. "Social Network Sites: Definition, History, and Scholarship." *Journal of Computer-Mediated Communication* 13, no. 1 (2007): 210–30.

Braun, Virginia, and Victoria Clarke. "Using Thematic Analysis in Psychology." *Qualitative Research in Psychology* 3 (2006): 77–101.

Bryant, Jennings. "Viewers' Enjoyment of Televised Sports Violence." In *Media, Sports, and Society*, eds. Lawrence A. Wenner, 270–89. Newbury Park, CA: Sage, 1989.

Bryant, Jennings, Paul Comisky, and Dolf Zillmann. "The Appeal of Rough-and-Tumble Play in Televised Professional Football." *Communication Quarterly* 29, no. 4 (1981): 256–62.

Byrne, Barbara M. *Structural Equation Modeling with Amos*. 2nd ed. Ottawa, Ontario, Canada: Routledge, 2010.

Cialdini, Robert B., Richard J. Borden, Avril Thorne, Marcus Randall Walker, Stephen Freeman, and Lloyd Reynolds Sloan. "Basking in Reflected Glory: Three (Football) Field Studies." *Journal of Personality and Social Psychology* 34 (1976): 366–75.

Clavio, Galen. "Social Media and the College Football Audience." *Journal of Issues in Intercollegiate Athletics* 4 (2011): 309–25.

Clavio, Galen, and Patrick Walsh. "Dimensions of Social Media Utilization among College Sport Fans." *Communication & Sport* (2013): 2167479513480355.

Cohen, Jacob. *Statistical Power Analysis for the Behavioral Sciences*. 2nd ed. Hillsdale, NJ: Erlbaum, 1988.

Creswell, John W. *Qualitative Inquiry & Research Design: Choosing among Five Approaches*. 2nd ed. Thousand Oaks, CA: Sage, 2007.

Ervin, Laurie H., and Sheldon Stryker. *Theorizing the Relationship between Self-Esteem and Identity. Extending Self-Esteem Theory and Research: Sociological and Psychological Currents*. New York: Cambridge University Press, 2001.

Faul, Franz, Edgar Erdfelder, Albert-Georg Lang, and Alex Buchner. "G*Power 3: A Flexible Statistical Power Analysis Program for the Social, Behavioral, and Biomedical Sciences." *Behavior Research Methods* 39 (2007): 175–91.

Funk, Daniel C., and Jeffrey D. James. "The Psychological Continuum Model: A Conceptual Framework for Understanding an Individual's Psychological Connection to Sport." *Sport Management Review* 4 (2001): 119–50.

Gantz, Walter, and Lawrence Wenner. "Fanship and the Television Sports Viewing Experience." *Sociology of Sport Journal* 12 (1995): 56–74.

Gearhart, Christopher C., and Shaughan A. Keaton. "The Influence of Motives for Selecting a Favorite Team on Sport Team Identification and Fan Behavior." Paper presented at the National Communication Association, New Orleans, LA, 2011.

Hambrick, Marion E., Jason M. Simmons, Greg P. Greenhalgh, and T. Christopher Greenwell. "Understanding Professional Athletes' Use of Twitter: A Content Analysis of Athlete Tweets." *International Journal of Sport Communication* 3, no. 4 (2010): 454–71.

James, Jeffrey D., and Stephen D. Ross. "Comparing Sport Consumer Motivations across Multiple Sports." *Sport Marketing Quarterly* 13 (2004): 17–25.

Jensen, Jonathan A., Shaina M. Ervin, and Stephen W. Dittmore. "Exploring the Factors Affecting Popularity in Social Media: A Case Study of Football Bowl Subdivision Head Coaches." *International Journal of Sport Communication* 7, no. 2 (2014).

Kasavana, Michael L., Khaldoon Nusair, and Katherine Teodosic. "Online Social Networking: Redefining the Human Web." *Journal of Hospitality and Tourism Technology* 1, no. 1 (2010): 68–82.

Keaton, Shaughan A. "Sport Team Fandom, Arousal, and Communication: A Multimethod Comparison of Sport Team Identification with Psychological, Cognitive, Behavioral, Affective, and Physiological Measures." dissertation, Louisiana State University, 2013.

Keaton, Shaughan A., and Christopher C. Gearhart. "Measuring Self-Reported Perceptions of Sport Fandom: Evaluating, Developing, and Refining Existing Scales in Sport Psychology." Paper presented at the Seventh Summit on Communication and Sport, Brooklyn, NY, 2014.

———. "Identity Formation, Identity Strength, and Self-Categorization as Predictors of Affective and Psychological Outcomes: A Model Reflecting Sport Team Fans' Responses to Highlights and Lowlights of a College Football Season." *Communication & Sport* (2013).

Keaton, Shaughan A., Christopher C. Gearhart, and James M. Honeycutt. "Fandom and Psychological Enhancement: Effects of Sport Team Identification and Imagined Interaction on Self-Esteem and Management of Social Behaviors." *Imagination, Cognition and Personality* 33 (2014): 251–69.

Keaton, Shaughan. A., and James M. Honeycutt. "The Effects of Sport Fandom on Physiology, Communication, and Mental Health." in *The Influence of Communication in Physiology and Health*, eds. J. M. Honeycutt, C. Sawyer and S. A. Keaton. New York: Peter Lang, 2014.

Keaton, Shaughan A., Nicholas M. Watanabe, and Christopher C. Gearhart. "A Comparison of College Football and Nascar Consumer Profiles: Identity Formation and Spectatorship Motivation." *Sport Marketing Quarterly* 24, no. 1 (2015).

Kline, Rex B. *Principles and Practice of Structural Equation Modeling.* 2nd ed. New York: Guilford, 2005.

Madrigal, Robert. "Cognitive and Affective Determinants of Fan Satisfaction with Sporting Event Attendance." *Journal of Leisure Research* 27 (1995): 205–27.

Mahony, Daniel F., Robert Madrigal, and Dennis R. Howard. "Using the Psychological Commitment to Team (Pct) Scale to Segment Sport Consumers Based on Loyalty." *Sport Marketing Quarterly* 9 (2000): 15–25.

Pegoraro, Ann. "Sport Fandom in the Digital World." *Routledge Handbook of Sport Communication* (2013): 248–58.

Raykov, Tenko, and George A. Marcoulides. *A First Course in Structural Equation Modeling.* 2nd ed. Mahwah, NJ: Erlbaum, 2006.

Rosenberg, Morris. *Society and the Adolescent Self-Image.* Revised ed. Collingdale, PA: Diane Publishing Company, 1999.

Rowe, David. "Following the Followers Sport Researchers' Labour Lost in the Twittersphere?" *Communication & Sport* 2, no. 2 (2014): 117–21.

———. *Global Media Sport: Flows, Forms and Futures.* New York: Bloomsbury Academic, 2011.

Sapolsky, Barry S. "The Effect of Spectator Disposition and Suspense on the Enjoyment of Sport Contests." *International Journal of Sport Psychology* (1980).

Shilbury, David, Shayne Quick, and Hans Westerbeek. *Strategic Sport Marketing.* Sydney, Australia: Allen and Unwin, 1998.

Sloan, Lloyd R. "The Motives of Sports Fans." In S*ports, Games, and Play: Social and Psychological Viewpoints*, ed. J. H. Goldstein, 175–240. Hillsdale, NJ: Erlbaum, 1989.

Smith, Garry J., Brent Patterson, Trevor Williams, and J. Hogg. "A Profile of the Deeply Committed Male Sports Fan." *ARENA Review* 5, no. 2 (1981): 26–44.

Stavros, Constantino, Matthew D. Meng, Kate Westberg, and Francis Farrelly. "Understanding Fan Motivation for Interacting on Social Media." *Sport Management Review* (2013).

Stewart, Robert K., Aaron C. T. Smith, and Matthew Nicholson. "Sport Consumer Typologies: A Critical Review." *Sport Marketing Quarterly* 12 (2003): 206–16.

Strauss, Anselm, and Juliet M Corbin. *Basics of Qualitative Research: Grounded Theory Procedures and Techniques.* Thousand Oaks, CA: Sage, 1990.

Sutton, William A., Mark A. McDonald, George R. Milne, and John Cimperman. "Creating and Fostering Fan Identification in Professional Sports." *Sport Marketing Quarterly* 6 (1997): 15–22.

Tabachnick, Barbara G., and Linda S. Fidell. *Using Multivariate Statistics.* 5th ed. Boston: Pearson Education, 2007.

Trail, Galen T., Dean F. Anderson, and Janet S. Fink. "Consumer Satisfaction and Identity Theory: A Model of Sport Spectator Conative Loyalty." *Sport Marketing Quarterly* 14, no. 2 (2005): 98–111.

Trail, Galen, T., Matthew J. Robinson, Ronald J. Dick, and Andrew J. Gillentine. "Motives and Points of Attachment: Fans Versus Spectators in Intercollegiate Athletics." *Sport Marketing Quarterly* 12 (2003): 217–27.

Wann, Daniel L. "The Causes and Consequences of Sport Team Identification." In *Handbook of Sports and Media,* eds. A. A. Raney and J. Bryant, 331–52. Mahway, NJ: Erlbaum, 2006.

———. "Preliminary Validation of the Sport Fan Motivation Scale." *Journal of Sports and Social Issues* 19 (1995): 377–96.

Wann, Daniel L., and Nyla R. Branscombe. "Sports Fans: Measuring Degree of Identification with Their Team." *International Journal of Sports Psychology* 24 (1993): 1–17.

———. "Die-Hard and Fair-Weather Fans: Effects of Identification on BIRGing and CORFing Tendencies." *Journal of Sports and Social Issues* 14 (1990): 103–17.

Wann, Daniel L., and Christi L. Ensor. "Family Motivation and a More Accurate Classification of Preferences for Aggressive Sports." *Perceptual and Motor Skills* 92, no. 2 (2001): 603–5.

Wann, Daniel L., Frederick G. Grieve, Ryan K. Zapalac, and Dale G. Pease. "Motivational Profiles of Sport Fans of Different Sports." *Sport Marketing Quarterly* 17 (2008): 6–19.

Wann, Daniel L., Kathleen B. Tucker, and Michael P. Schrader. "An Exploratory Examination of the Factors Influencing the Origination, Continuation, and Cessation of Identification with Sports Teams." *Perceptual and Motor Skills* 82 (1996): 995–1001.

Watanabe, Nicholas M., Shaughan A. Keaton, and Kozo Tomiyama. "A Cross-Cultural Examination of Sport Fandom: Japanese and Us Team Identity Formation, Spectatorship Motives, and Psychological Effects." Paper presented at the Seventh Summit on Communication and Sport, Brooklyn, NY, 2014.

Weiller, Karen H., and Catriona T. Higgs. "Fandom in the 40's: The Integrating Functions of All American Girls Professional Baseball League." *Journal of Sport Behavior* 20, no. 2 (1997): 211–31.

Williams, Jo, and Susan J. Chinn. "Meeting Relationship-Marketing Goals through Social Media: A Conceptual Model for Sport Marketers." *International Journal of Sport Communication* 3, no. 4 (2010): 422–37.

Chapter Six

The Online Community

Fan Response of Community's *Unlikely Fifth Season*

Matthew R. Collins and Danielle M. Stern

In May 2012 *Community* fans received the news they had been anxiously waiting for since the show's third season came to a close: confirmation that the show had been renewed and there would be a fourth season.[1] *Community* has achieved a cultish, fan- and critic-favorite status as a smart, cutting edge television show that does not adhere to the rules of a normal sitcom. When season four of community was announced, fans were overjoyed. Months of campaigning and raising awareness seemed to have saved their show, but there was a catch. The show's creator and executive producer, Dan Harmon, had been fired by NBC, leaving what fans believed would be a void in the heart of the show, and it appeared they were not wrong. The fourth season was ill received, with each individual episode of its shortened, thirteen-episode season receiving the lowest ratings the series had ever seen, with multiple episodes dropping below a seven out of ten rating based on user reviews from IMDB.com.[2] This study focuses on sectors of *Community* fandom online to gauge and analyze fan responses to two renewals of *Community* to answer the question: how do viewers use social online spaces to communicate their fandom following the saving of their show?

This study is an exercise in fandom studies, a branch of cultural and communication studies prominently brought into the mainstream by theorist Henry Jenkins.[3] Fans actively participate in the media they consume, from dressing up as their favorite characters, writing fan fiction, sharing scene remixes and reenactments online, or simply posting about the show online to connect to other fans. As audience studies have evolved over time, so has the media that audiences consume. With the advent of the Internet, television has changed drastically from its original form, with practices such as binge watching and delayed viewings becoming increasingly normal. This chapter establishes how fans use their agency to save a show, and that much of their agency is facilitated by the Internet. Similarly, the Internet becomes a hub of discussion for these fans, creating a vast wealth of information for scholars to tap into, as it is full of well-documented and candid responses to episode summaries and reviews.

This study was narrowed to specific avenues of online fan discussion. Two Web sites were selected for their vast popularity and a passionate user base. Combined, the A.V. Club and the *Community* subreddit /r/community contain thousands upon thousands of comments and users, offering a variety of different viewpoints and opinions of fans to analyze. The subreddits for the season 4 finale and the season 5 premiere of *Community* were chosen because the season 4 finale subreddit has comments pertaining to the reactions of the entire fourth season, as well as that specific episode. The season 5 finale subreddit includes discussion of that specific episode, as well as expectations for the rest of the fifth season. Reading and analyzing these two rich hubs of online fan discussion revealed how online fans react after a successful "save our show" campaign.

The analysis revealed three trends, which will be discussed in more depth later in the chapter. The first trend came as little surprise. Many of the comments were almost unanimously supportive of the fifth season and the return of Dan Harmon, exclaiming that their show was finally back to its normal self. The opposite of this praise existed when examining the fourth season's finale, which was universally panned by fans and was incredibly ill received. However, with the return of the series for a fifth season, the fans began to express changes in their views toward the fourth season, discussing it in a more positive, almost nostalgic light; an attitude which only appeared once the fifth season was released. The last trend involved attacks on other shows (notably CBS's *The Big Bang Theory*) as inferior to *Community*. The analysis reveals the enhanced impact television audiences have over production in a social media era that encourages direct response from viewers to content producers. The agency of the socially wealthy audience, connected online in myriad spaces, must not only be heard, but also responded to by the financial executives whose influence continues to disperse.

ACTIVE AUDIENCES IN THE SOCIAL MEDIA ERA

In recent decades, theories have shifted from passive audience perspectives and one-to-many approaches to gradually increasing attention to the agency audiences operate over the media they consume. Fan studies emerged in the era of participatory culture, presented by Henry Jenkins, who built upon the works of Michel de Certeau. According to de Certeau, "Everyday life invents itself by poaching in countless ways on the property of others."[4] De Certeau and Jenkins provide a framework for understanding how fans create meaning from negotiating and selecting media elements and practices they enjoy.

However, one important critique of the genre of fan studies arose with the publication of *Textual Poachers*. The problem was that Jenkins identified himself as a fan, and did so intentionally to separate himself from other scholars of fans who had looked at the genre in too "sensationalistic" ways, which "foster misunderstandings about this subculture."[5] This presented a problem, as Jenkins was called under question for being too celebratory of fans. Jenkins argued that the rise of the "aca-fan" has allowed for a happy marriage of the two identities of the academic and the fan.[6] Hills argued that the aca-fan must create a balancing act to appease both of these worlds, saying the aca-fan must be "careful not to present too much of their enthusiasm while tailoring their accounts of fan interest and investment to the norms of 'confessional' (but not too confessional) academic writing."[7]

But the debate about the aca-fan's need to balance these two realms does not end with Hills and Jenkins. More recently, Paul Booth acknowledged that the "aca-fan" does not do enough to involve fans in the process of research. He claimed that although fans are very aware of their identity as fans, they know little to nothing about the academic work that goes on by the aca-fans who study them. Booth continues to suggest that fan scholars should "be more engaged with fan communities," allowing fans to "enter the academic discourse on fandom more openly."[8] Doing so would create a richer experience for fans to identify with the research being conducted about them, and allow academics to feel more connected to the fan groups they belong to. In order to accomplish this, Booth suggested academics become more involved with the social media sites the fans use, sharing their articles and essays with the fans on a one-on-one level.[9] Certainly, understanding one's role as an aca-fan must be taken into consideration when conducting fan studies.

Like most aspects of media studies, the social landscape has impacted how researchers conduct their research. For instance, a study from George Mason University about prevailing trends in online video watching versus television watching discovered that out of a sample group of 1,500 U.S. adults, "the proportion of people who use both television and the Internet to watch video content (55.4 percent) is larger than the proportion of people who solely rely on television (42.3 percent)."[10] Although this data is not indicative of audiences as a whole, it still speaks to the nature of the new media landscape, particularly, that the larger group of audiences studied are using the Internet, compared to the smaller group, which does not. Sonia Livingston explained that researchers cannot approach studying audiences in the same way as attempted decades ago. Livingston explained that ethnographic research is losing its place in audience studies, as the audience begins to increasingly turn to the Internet to showcase their reactions and feelings about a television program over a "real world" reaction, creating, for an ethnographic researcher, an audience that was "inaccessible."[11] However, it is hard to ignore that the Internet fosters opportunities for audiences to connect in important, rich ways.

Jenkins dubbed the evolving media landscape one of media convergence: "Consumers are learning how to use these different media technologies to bring the flow of media more fully under their control and to interact with other users. . . . They are fighting for their right to participate more fully in their culture."[12] Jenkins furthered his research on convergence culture by reminding scholars that just because we are being thrown into a convergence media landscape, it does not mean that old media are now obsolete, rather that they are being repurposed and reappropriated, or converged with new media.[13] This argument is supported by Cha Jiyoung's findings that while a majority of adults use the Internet and television, only 1.5 percent claimed they only used the Internet to watch video content, television shows or otherwise.[14]

The past decade has seen a natural rise in fandom studies focused on the Internet and social media. Anne Marie Todd wrote about how fans create meaning and identity within themselves based on the characters on the show *Friends*, but focused on how this occurred within online message boards during the final episode of *Friends*. Todd discusses that the finale of *Friends* took place on television, but much of the discussion and reactions were found online, further supporting the idea of convergence as not displacing old media, but converging within it.[15] Although Todd's main argument was not intrinsically about online fan communities, her study would have been impossible without considering the role the Internet had in *Friends* fandom. Rebecca Williams analyzed fans' self-identification with *The West Wing*, making it clear that her study was focused only on "visible groups of Internet fans," because the "isolated nature" of fans not on the Internet was problematic.[16] Williams's work is just one example of the emerging importance of the role of the Internet and social media in fan studies.[17]

In this evolving academic landscape, the stories with which fans engage have also changed. Although not explicitly tied to the Internet, Jenkins defined transmedia storytelling as "a new aesthetic that has emerged as a response to media convergence," which "depends on the active participation of knowledge communities."[18] Transmedia storytelling identifies a story from any medium that is supplemented by a continuation of the story from another medium. The novelization of a film or a movie based on a television show would be considered a transmedia story. Married to the idea of transmedia storytelling is the concept of participatory culture, defined by Jenkins, Purushotma, Weigal, Clinton, and Robison as "a culture with relatively low barriers to artistic expression and civic engagement," and a "strong support for creating and sharing creations."[19] The idea is that fans gain more enjoyment from participating with the content they enjoy, and transmedia storytelling offers an outlet to do so. Neil Perryman examined *Dr. Who* as one of the most successful examples of modern transmedia storytelling. He explained how the BBC successfully created a transmedia storytelling experience with *Dr. Who* through primarily Internet-based events such as mini-episodes and cryptic Web sites where the audience could engage and participate with more content in the *Dr. Who* universe outside of watching the show.[20]

However, transmedia storytelling is a symbiotic, not unidirectional relationship between content producers and fans. The fans take original content of producers, and the producers sometimes borrow ideas from fans. Scholars have begun looking at how fans participate in transmedia storytelling without the official sanction of the content producers. Jason Mittell analyzed a fan site focused on the television show *Lost* called "Lostpedia," a wiki site with over 25,000 registered users and 150 million page views. Mittell examined the site to see "how the wiki platform enables fan engagement, structures participation, and distinguishes between various forms of content, including canon, fanon, and parody."[21] Mittell's examination of Lostpedia demonstrates that fans are creating, on their own accord, a transmedia event over which the content producers have no control. This enables many of the fans of *Lost* to accept what other fans have written about the universe of *Lost* as primary, even though the content producers have never addressed these scenarios.

In his study of *Lost* fans' participation with the text, Carlos Scolari argued that the role of user generated video (such as YouTube) has transformed the fan experience. He called for "a more flexible definition of transmedia storytelling" to allow these types of videos to be considered part of the transmedia storytelling experience.[22] One expansion of this participation is the distinction between a fan and cult fan, which Hills defined as an "intensely felt fan experience" that is often wrapped up with identity.[23] Sometimes this intensity contributes to fan narcissism, which Cornel Sandvoss described as fans watching television shows to see themselves in the shows due to a need to "build an intense identification with their object of fandom," which is why fans of a certain sports team refer to themselves as "we."[24]

It is important to note how cult fans see themselves not only as part of the group of fans they belong to, but how they view non-fans outside of their group. According to Jenkins, fans were "nervous what would happen if their underground culture was exposed to public scrutiny."[25] Although this has much to do with debate over how to handle being an aca-fan, it draws interesting ideas about how cult fans see outsiders. For instance, Costello and Moore interviewed television fans who were actively separating themselves from other fans who might consume television in "an unstructured and habitual fashion," quoting one person they researched who spoke about a certain show they enjoyed as saying, "It's the only network show which doesn't insult our intelligence. It is a thinking person's TV."[26]

Certainly, there is a level of elitism within some cult television fandom, as well as a withdrawn nature and negative response to outsiders as Jenkins identified. But even more compelling is the learning curve that might be associated with becoming interested in a new television show. For instance, Booth wrote about *Inspector Spacetime*, a fictional show within a show on *Community*, which parodies *Dr. Who*. Booth explained that with only a few seconds of screen time on *Community*, *Inspector Spacetime* has garnered an explosive community of fans who created an entire universe for this show within a show that exists entirely online.[27] In order to fully understand *Community*, a fan must know the origins and story surrounding *Inspector Spacetime*. As such, to be a fan of *Community* means committing to understanding paratexts[28] and engaging more than the average television viewer, producing cult fandom. In turn, this chapter investigates the presence of cult fandom in the social media experience of *Community* fans. Intentional or not, these examples show that cult fans may appear weird, elitist, or overly confusing, creating barriers against other potential fans, possibly driving them away from enjoying a show. However, for the fans of the show that participate in the loosely defined transmedia storytelling, Internet forums and fan sites become a way for any fan to be able to catch up, or fill in the gaps of any aspect of an episode they might have missed. Clearly, cult fandom in the intersection of social media landscape and transmedia storytelling is a complex process that deserves further study.

EXAMINING FANS QUALITATIVELY

This project analyzed online reactions to the latest season of the NBC sitcom *Community* (season 5) in order to gauge fan response to the return of show runner Dan Harmon, who was not involved with season 4, but after much protest was asked back to the show by NBC. The point of this study was to discover trends in fan reactions once they receive what they want from a series, such as the return of the original show runner. The same concept could be applied to the return of a canceled show, a cast member returning who might have left, or any other variety of fan-demanded change. Similar studies that analyze sites of online fandom and user comments have been conducted, such as Miriam Greenfield's study of online *Lord of the Rings* fandom, which focused on analyzing forum posts of online *Lord of the Rings* fansites and how users gossip about the stars of the movie series.[29] In the current study, the authors followed in Greenfield's footsteps by reading through and analyzing user comments contributed to social media sites of fandom.

To find these reactions, the authors looked at online discussions from two highly trafficked sites focused on *Community*. The first was the A.V. Club, which garnered more than four thousand comments on the two *Community* episode reviews included in the analysis.[30] The second is the *Community* subreddit of reddit.com, /r/community, which has more than 130,000 unique subscribers and thousands of comments on the discussion threads for the two episodes in the analysis.[31] These two sites provided a wealth of fan commentary on the reactions surrounding the premiere of season 5, as well as fan's comparison of the episodes to previous seasons. This study looked at two episodes of the show in particular, the first episode of the current fifth season titled "Re-Pilot" and the last episode of the fourth season, titled "Advanced Introduction to Finality." These episodes were chosen because they provided a good basis for when fans considered the show at its worst (the finale of the fourth season without Dan Harmon) and what fans campaigned for with the "repiloting" of the series' fifth season and Dan Harmon's return. "Re-Pilot" gave insight into fan's reactions on the return of Dan Harmon, and was directly compared to the fourth-season episodes. These sites and the discussions contained within them offered a wealth of information based on these two episodes.

In order to manage the massive amounts of comments from both sites, the analysis of A.V. Club was limited to comments that generated response threads, that is, comments about the comments, as this provided the best discourse between fans and allowed the more popular opinions to be more noticeable. On the *Community* subreddit, the authors read all the comments, but once again, focused the majority of the analysis on comments that generated comment threads, and the comments within those threads.

Constructionist grounded theory as explained by Kathy Charmaz is central to the research.[32] First, after analysis, the reactions and data were separated into three different themes (two major and one minor): those who enjoyed Harmon's return, hailing the show with Harmon as a return to form; those that not only enjoyed Harmon's return, but found a new appreciation for the fourth season; and those who gossiped about other shows. Although there was no direct discussion with the participants while gathering the research, the authors interacted with fans based on Charmaz's encouragement for the academic to share writings with the participants to further "co-construct" the analysis.[33] Similarly, the study followed Paul Booth's suggestion that the "aca-fan" allow the fans they are studying to be more involved with the research process by posting their writing on the online areas the fans are talking about the show and the research is being conducted.[34] To accommodate both of these ideas, drafts of this work were submitted to the *Community* subreddit to see if what was gathered accurately reflected how the fans felt.

What makes this method of online research valuable is supported by Mark Andrejevic's research, which shows that fans take pleasure in discussing the shows they watch in online, and how show producers can use these online texts to shape the show's direction.[35] Online discussions have become an instant and a more thorough review process of fan reactions than any rating system or ethnographic research can really be, so much so that producers actively check them to gauge fan response. Adding to this discussion is Sonia Livingston's idea that as consumers are shifting toward online discussion and consumption of television, the researcher must follow them in order to avoid studying an audience that is "inaccessible."[36] This method provided a rich amount of data from a variety of sources. This study cultivated a complex data set that the authors hope furthers the importance of academic studies of online fandom in the social media era.

Analysis

The news of the return of Dan Harmon as show runner of *Community* for season 5, after being fired following season 4, proved to be as shocking as it was exciting for fans of the show. The fans of *Community* demonstrated, similar to fans in past "save our show" campaigns, that television audiences can significantly impact production choices, especially in the social media environment. After the social media campaigns, emails or letters to network executives, and enormous amounts of online discussions successfully served their purpose, what replaces this unified front of fans?

After reading thousands of reactions and comments, three prevailing themes emerged. The first trend was overwhelmingly positive praise for season 5 and an overall consensus of quality increase from season 4. The second trend identified was those that had sympathy for season 4, which included fans either expressing gratitude that the season existed or simply stating that it did not deserve the criticism it receives, while also praising the season as a means to keep the show alive while it went through difficult times (cancellation rumors and the firing of Dan Harmon). The third and smallest trend centered on the fans gossiping and insulting the show *The Big Bang Theory* instead of commenting directly on *Community*.

A Return to Form

The first and most popular trend that was apparent and held by the majority of fans were intensely positive reactions to the return of Dan Harmon to *Community* with an overwhelming amount of comments expressing joyous praise of how much better the season 5 premiere was from season 4. The types of responses in this trend varied. The first was amateur critical praise of the show. For instance, user That TV Nerd commented on the A.V. Club review, "'Repiloting' was a glorious exercise in self restraint. It stripped away all of *Community*'s excesses—both lame and hilarious—and dug right down to the emotional roots of each character. Precisely what the doctor ordered after season 4."[37] The other types of comments which gave praise to season 5 were more simplistically enthusiastic—quotes from the episode that the commenters enjoyed the most, followed by a thread of more quotes that other fans enjoyed.

Both of these comments as well as their similar counterparts on the page created a stark contrast between the reactions on the A.V. Club for the season 4 finale, of which there were very few responses offering praise. Some comments on season 4 involved direct criticisms of the writer of the episode, Megan Ganz. As user Nathan Ford's Evil Twin said, evoking a *Star Wars* quote, "We trusted you Megan Ganz! You were the chosen one! It was said you would destroy the pandering writers not join them! Bring balance back to the show not leave it in darkness!"[38] Many more comments on the season 4 finale followed a similar suit, with nearly all criticizing the lack of quality and expressing utter dissatisfaction with the show. Similarly, as season 5 comments contained quotes that portrayed a celebratory attitude in order to share laughs with fellow commenters, the quotes found in the comment section of season 4 were stated in a sarcastic manner that highlighted the intense dissatisfaction fans had with the episode. For example, "But they finally found a way to make paintball cool again,"[39] followed by a comment thread with others that stated, "They did not. And that line was icky,"[40] "*That* was when the pandering got to me,"[41] and "If I was Danny Pudi, I'd feel violated only because I had to say that."[42]

Fans of season 4 felt pandered to, insulted, and even violated by the way the show was treating them without the direction of Harmon. Fans blamed show producers such as Megan Ganz for the low the quality of the programming, while naming Harmon as the key to the original great the quality of the series. Fans also used quotes from the episode to demonstrate how disappointing the show was, as opposed to using quotes to highlight how enjoyable the show was. These fans' reactions to a certain line from season 4 are relatively personal, such as attacks made on Ganz, but also having sympathetic feelings for the actor who had to say the line. This sort of identification with the show and the actors within them showcase the personal connection many fans have with the show, which is similar to Todd's study of *Friends* fandom online when the show went off the air. Todd described the fans as blurring the distinctions between the actors who played characters on the show and the characters on the show they identify with.[43] In the case of the line spoken by Danny Pudi, fans were insulted that they were taken out of the scene and felt embarrassed for not only the character they know, but the actor who plays him.

Turning to an analysis of /r/community's positive reactions toward the show, similar trends emerged. For instance, /r/community's reaction was overwhelmingly positive praise for Harmon's return. A large portion of the comments were direct quotes from the show, showcasing an excitement for the new season and Harmon's return, but many made direct compliments toward Harmon. For example, "Take a bow, Harmon,"[44] "God Bless you, Dan Harmon,"[45] and, "That was perfect. . . . I love you Dan Harmon."[46] When commenters were not directly praising, or quoting, they were professing their excitement for how great the new season began, with comments like, "That was the best hour of my life,"[47] and "One thing is for sure. This really is a reboot of the series."[48]

Different from The A.V. Club, however, there were few overwhelmingly negative comments directed toward the season 4 finale and especially few directed toward the showrunners during season 4. In fact, most were incredibly positive, with a few dissenters commenting on the paintball quotes, saying "What's wrong with rehashing tired old plotlines in substantially worse ways,"[49] and "This whole season just felt like fan fiction to me."[50] However, comments such as these were few and far between. One possible explanation for the general sense of enthusiasm toward the season 4 finale generated on /r/community rests in the type of discussion encouraged compared to A.V. Club. The /r/community exists as a simple discussion of the episode, whereas the A.V. Club review comments section shares a review and a grade (which in the case of the season 4 finale, was a D), which might influence the course the discussion takes in the comments section.

Gratitude for a Bridge Season

One of the most prevailing and unexpected trends in the comments of online fans was the overwhelming support of season 4 showcased with the premiere of season 5. The shocking aspect of this support comes from the fact that even though the majority of reactions to the season 4 finale were negative, no fans firmly showed appreciation that the show was allowed a fourth season or for the people who kept the show afloat against the negative criticism. Certainly, there was a somber nostalgic vibe since nobody knew or expected for the show to return for a fifth season or for Harmon's return to run it. However, the commenters offered nothing in the way of appreciating the fourth season so much as they were appreciating the show as a whole. This mood changed with the premiere of season 5, with comments that ranged from appreciative to apologetic.

For instance, user TheASDF commented on A.V. Club about the season 5 premiere, which spurred a comment thread of similar reactions: "I feel bad for season 4. People shit on it like it was the worst season of a thing to happen. It *wasn't* that bad, you guys."[51] Within that comment thread, others agreed with this outlook, stating that the season needed to happen without Harmon in order to get the show back on track because it was faltering regardless. Comments that expressed a need for Harmon to be removed from the show, in order to improve it, included, "I think it gave Harmon a chance to recharge his batteries,"[52] and "Harmon seemed to have a wiser, more mature perspective of his show and his characters, and I think the firing and the time off are likely responsible for that."[53] These comments suggest that some fans felt the show needed to reach its nadir in order for it to reach the level of its former quality. One commenter went as far back as the Harmon-led season 3 to discuss why his absence from season 4 was a good thing: "I honestly didn't really think a Harmon-led season 4 would be much better than the season 4 we got. I thought the latter half of season 3 spun out of control . . . so even if season 4 was bad, I think a Harmon-led season 4 would be a different kind of bad."[54] It is important to remember that while these comments do exist, they were the minority on A.V. Club, as opposed to negative comments about season 4 or praise for season 5.

The situation was similar on /r/community, which was unexpected due to the general support (or at least lack of negative reactions) to the season 4 finale. On /r/community, a comment thread was started in the discussion of the season 5 premiere when one user shared, "Let's take a minute to thank the season 4 showrunners, no matter your opinion of them, for keeping the show going enough to get us a fifth season."[55] This comment was followed with a few replies in the same vein such as, "I'll come out with you as a person who kinda liked season 4."[56] However, other than a few comments in support of season 4, many fans reacted harshly toward season 4 and its showrunners. For example, "I don't think I'm going to thank someone for doing the job they got paid to do, especially when I think they did a bad job of it."[57]

Attacks on Other Shows

The last emergent theme was the *Community* subreddit's animosity toward another series, *The Big Bang Theory*. Occasionally fans of *Community* have become embittered when their show is in danger of cancellation, while a show such as *The Big Bang Theory* continues to be one of the most popular on network television.[58] This relates to Costello and Moore's study on elitism within online fan communities, who quoted one participant as claiming their show was "thinking person's TV."[59] The situation on the *Community* subreddit was no different, as there were more than a few comments throughout the episode discussion thread exemplifying this aspect of elitism. One commenter noted that a family member walked in on a joke that aired during the season 5 premiere and left the room dismissively. Annoyed and bothered with their family member, the user stated, "They're all BBT fans whose basis for comedy is the context-free joke, I have to remind myself."[60] The comment that followed agreed with the user's plight, "Some people can't be bothered to have to put more than minimal brain power into TV watching."[61]

This animosity toward other programs is somewhat surprising since these comments were shared on the season 5 premiere, a thread that contained almost unanimously positive comments. These negative commenters still seemed to have an air of hostility about them, even though their show had been renewed, regardless of *The Big Bang Theory*'s status.

These findings coincide with Sandvoss's comments about narcissistic fans, watching a show to see themselves within it. These fans feel that *Community* is a smart show, and that they are smart people, and they have to tolerate or put up with anyone else who does not enjoy the show. It also exemplifies Sandvoss's idea of fans viewing themselves as part of a team, having to fight against fans of other shows.[62] This shared identity pairs them up against other sitcoms, which they feel they need to attack and diminish so their show can win.

Although many of the comments analyzed were amateur reviews, a vast majority of the comments were essentially online gossip. As Greenfield wrote in her study of *Lord of the Rings* fandom, online fan communities have a tendency to shift toward gossip as when they originate. Gossiping can be used as form of "casual leisure." However, over time, online fan gossip can shift and take a more prominent role, even becoming "a major focus that uses a variety of gossip techniques to survive."[63] This can most certainly be applied to the fan revelations about *Community*, as many of the comments analyzed and mentioned within this study related to gossip in some capacity. Whether it was directed toward actors, characters, show producers, network executives or other shows, *Community* fans made large work of gossip as a focal point to connect with each other via social media spaces of A.V. Club and reddit.

CONCLUSION

This study looked at reactions of fans to the renewal of *Community* as well as the renewal of Dan Harmon as showrunner for a few reasons. Chiefly, it is important to note the reactions and opinions of fans using social media spaces such as blogs and reddit, as this is where most fan discussion exists. Fans may still congregate in a friend's living room or meet up at a convention to discuss and participate with their media, but these are few and far between, not to mention are often entirely inaccessible or limiting to the academic who might be interested in studying the fans and their culture. That is why online fan studies are so important to the fields of communication and media studies. They offer an avenue not only for fans to congregate and participate, but for academics to become fully immersed in the world of fans and better be able to study fan practices. Since media consumption in the digital age is so focused around the Internet and social media, it only makes sense that these spaces are the destination to study media consumption.

This chapter highlights that even when audiences got what they wanted, they were not always happy about it. As seen in the analysis, fans reacted poorly to the fourth season of *Community*. They felt that the return of the show without Dan Harmon was a cheap trick, almost a bait and switch. To get the show back was not enough, they needed the show creator back as well or it would not be worth the time and effort they had put in. As expected, these fans discussed this online, generating thousands of comments spanning multiple discussion threads. However, not all was lost for *Community* fans, with the renewal of *Community* for a fifth season, as well as the return of Dan Harmon, everything seemed right in the community, and the online discussion reflected that notion. Fans were thrilled to have Harmon's *Community* back, which is not surprising. However, what was unexpected was the way certain fans' perceptions of the fourth season changed. With the return of the fifth season, many people began discussing the fourth season again, but with a different attitude. Instead of bashing the show, they reflected upon it relatively positively. These comments appear to operate to make the fourth-season "save this show" campaign worth the effort.

While working on this study, limitations emerged within the process. First, as fans of *Community*, the authors did not like the fourth season. As such, a limitation could be the initial process of the authors gravitating toward comments that echoed that sentiment. In the end, these were the majority, and if fans had said they had liked the fourth season in substantial ways, these expressions would have been included in the analysis. However, the authors did feel compelled to look over comments that did not reflect a preconceived vision of the series. This is part of the problem Booth identifies with being an aca-fan and also the reason the authors chose to post early versions of the analysis on the *Community* subreddit. The goal was grounded analysis that was not diffused solely through researcher bias. Unfortunately, researcher posts were not as successful as intended since they did not garner any engaging or helpful responses. For this reason, a different approach to accessing and sharing work with fans might be more beneficial rather than simply posting the work online and asking for responses and reflections.

The potential for research in the reactions of fans after a successful save our show campaign can be taken in a variety of directions. First, while engaging with this material, the authors noticed certain attitudes both of the sites exhibited, which were touched on earlier in this chapter. For example, saying that since the A.V. Club was a review site that gave the season 4 finale a D–, it possibly framed the attitudes the commenters on that site had toward the episode, which turned out to be more hostile than in reddit, where comments were a bit more amicable, while still remaining negative. Further research could address these competing attitudes of online fandom, or find other areas of fandom in social online spaces to discover how different Web sites react to the same events surrounding media texts.

Similar to investigating different Web sites, it might be important to investigate other shows that have successfully been saved. For instance, *Arrested Development* was renewed by Netflix in 2013 for a fourth season (seven years after being canceled by Fox). *Arrested Development*'s production deal secured with an instant streaming Internet company broke new ground for television and fandom. In fact, as of this chapter writing, Yahoo announced it would begin production on a sixth season of *Community*. This creates an interesting research topic as the Web allows for a different presentation of television content, which can impact how fans react. Finally, it would be interesting to investigate the return of the canceled CW series *Veronica Mars* via a Kickstarter campaign initiated by creator Rob Thomas.

Audience studies have come a long way since their inception. Gradually, audiences have been understood to enact more agency with television texts and within scholarly debate. The current study contributes to this growing body of research. *Community* fans were able to successfully save their show and their showrunner, showcasing their large amount of agency over the production. Moreover, the fans actively discussed, critiqued, and praised the show online with each other upon *Community*'s return, continuing their agency in social media spaces.

NOTES

1. A year later, fans were faced with the same situation, wondering if the show would be picked up for a fifth season. Fans once again entered the same anxious stage, when the fifth season came to a close with no confirmation of a sixth season by NBC. A few months after NBC finally declared *Community*'s cancelation, an unexpected announcement by Yahoo to take over production of the program saved it once again in summer 2014.

2. Community episode list. http://www.imdb.com/title/tt1439629/episodes?season=4&ref_ =tt_eps_sn_4, accessed July 2, 2014.

3. Henry Jenkins first looked at the subculture that was created around fans of *Star Trek* and how they poached aspects of the show, remixing and reimaging them for their own purposes. Henry Jenkins, *Textual Poachers: Television Fandom and Participatory Culture* (New York: Routledge, 1992).

4. Michel de Certeau, "The Practice of Everyday Life," in *Critical Cultural Theory* (4th ed.), ed. J Storey. (Essex, England: Pearson, 1984), 545.

5. Jenkins, *Textual Poachers*, 8.

6. Jenkins, *Textual Poachers*, 12.

7. Matt Hills, *Fan Cultures* (New York: Routledge, 2002), 1–2.

8. Paul Booth, "Augmenting Fan/Academic Dialogue: New Directions in Fan Research," *Journal of Fandom Studies*, 1 (2013): 121, doi: 10.1386/jfs.1.2.119_1.

9. Booth, "Fan Research," 133.

10. Cha Jiyoung, "Do Online Video Platforms Cannibalize Television?: How Viewers Are Moving from Old Screens to New Ones," *Journal Of Advertising Research*, 53 (2013): 8. doi:10.2501/JAR-53-1-071-082.

11. Sonia Livingston, "The Challenge of Changing Audiences: Or, What Is the Researcher to Do in the Age of the Internet?" *European Journal of Communication,* 19 (2004): 3, 10, accessed July 2, 2014, http://dx.doi.org/10.1177/0267323104040695.

12. Henry Jenkins, "The Cultural Logic of Media Convergence," *International Journal Of Cultural Studies,* 7 (2004): 37, doi:10.1177/1367877904040603.

13. Henry Jenkins, *Convergence Culture: Where Old and New Media Collide* (New York: New York University Press, 2008), 15–16."

14. Jiyoung, "Old Screens to New Ones."

15. Ann Marie Todd, "Saying Goodbye to Friends: Fan Culture as Lived Experience," *The Journal of Popular Culture,* 44 (2011): 856, doi: 10.1111/j.1540-5931.2011.00866.x.

16. Rebecca Williams, "'This Is the Night TV Died': Television Post-Object Fandom and the Demise of *The West Wing,*" *Popular Communication,* 9 (2011): 268, doi:10.1080/15405702.2011.605311.

17. Whiteman compared the identities of online communities for the show *City of Angels* and the video game *Silent Hill Heaven,* or Soukup's research on how fan Web sites about celebrities allow online communities to further identify with the celebrity by "controlling the representation" of that celebrity. Many more examples exist of fan studies not explicitly about Internet fandom, (these studies could exist without the Internet, but would be much more difficult or near impossible) but about other aspects of fandom that use the Internet as a supplement to the study. However this is not the only avenue of research convergence media has made available for fan studies, as there is an entire sect of fan research in which the Internet is the fundamental aspect. Natasha Whiteman, "The De/Stabilization of Identity in Online Fan Communities," *Convergence: The Journal of Research into New Media Technologies,* 15 (2009): 391–410, doi:10.1177/1354856509342341; Charles Soukup, "Hitching a Ride on a Star: Celebrity, Fandom, and Identification on the World Wide Web," *Southern Communication Journal,* 71 (2006): 321, doi:10.1080/10417940601000410.

18. Henry Jenkins, *Fans, Bloggers, and Gamers* (New York: New York University Press, 2006), 20.

19. Henry Jenkins et al., *Confronting the Challenges of Participatory Culture: Media Education for the 21st Century* (Cambridge, MA: The MI Pre.), xi.

20. Neil Perryman, "*Doctor Who* and the Converge. e of Media," *Convergence: The Journal of Research into New Media Technologies,* 14 (2008): 21–39, doi:10.1177/1354856507084417.

21. Jason Mittell, "Sites of Participation: Wiki Fandom and the Case of Lostpedia," *Transformative Works and Culture,* 3 (2009): (section 1.3, section 0.1), accessed July 2, 2014, doi:10.3983/twc.2009.0118. Fanon, for reference, is the writing of fan fiction, which has become widely regarded by the fan community as canon, while canon is what is considered true in a fictional universe as the content producers have created and intended.

22. Carlos A. Scolari, "Lostology: Transmedia Storytelling and Expansion/Compression Strategies," *Semiotica,* 195 (2013): 48, doi:10.1515/sem-2013-0038.

23. Hills, *Fan Cultures,* 2.

24. Cornel Sandvoss, *Fans: The Mirror of Consumption* (Cambridge: Polity Press, 2005), 101.

25. Jenkins, *Fans, Bloggers,* 1.

26. Victor Costello and Barbara Moore, "Cultural Outlaws: An Examination of Audience Activity and Online Television Fandom," *Television New Media,* 8 (2007): 130, doi: 10.1177/1527476406299112.

27. Paul Booth, "Reifying the Fan: *Inspector Spacetime* as Fan Practice," *Popular Communication,* 11 (2013): 148–49.

28. Paratexts are the extras that surround existing media texts, such as promotional materials, trailers, and podcasts.

29. Miriam Greenfeld, "Serious Play: How Gossiping Is Accomplished in an Online Fandom Forum," (paper presented at the annual meeting for the International Communication Association, New York, 2005).

30. Todd VanDerWerff, "Community: 'Advanced Introduction to Finality,'" A.V. Club. Available online at http://www.avclub.com/tvclub/community-advanced-introduction-to-

finality-97134; Todd VanDerWerff, "Community: 'Repilot'/'Introduction to Teaching.'" A.V. Club. Available online at http://www.avclub.com/tvclub/repilotintroduction-to-teaching-106548.

31. "Season 4 Episode 13 (Finale) Discussion Thread," accessed July 2, 2014, http://www.reddit.com/r/community/comments/1e1azm/season_4_episode_13_finale_discussion_thread/; "Discussion Thread for Community S05E01 – 'Repilot,'" accessed July 2, 2014, http://www.reddit.com/r/community/comments/1u9pg0/discussion_thread_for_community_s05e01_repilot/.

32. Kathy Charmaz, *Constructing Grounded Theory: A Practical Guide Through Qualitative Analysis* (London: Sage, 2006).

33. Charmaz, *Constructing Grounded Theory*, 366.

34. Booth, "Fan Research," 133.

35. Mark Andrejevic, "Watching Television Without Pity: The Productivity of Online Fans," *Television & New Media*, 9 (2008): 1.

36. Livingston, "Age of the Internet," 10.

37. Poor Sheltered Homeschooler, January 2, 2014 (09:46 AM), commenter on Todd VanDerWerff, "Community: 'Repilot'/'Introduction to Teaching,'" A.V. Club (media review Web site), April 2014, http://www.avclub.com/tvclub/repilotintroduction-to-teaching-106548.

38. Nathan Ford's Evil Twin, May 9, 2013 (10:10 PM), commenter on Todd VanDerWerff, "Community: 'Advanced Introduction to Finality,'" A.V. Club (media review Web site), April 2014, http://www.avclub.com/tvclub/community-advanced-introduction-to-finality-97134.

39. *Lt. Broccoli*, May 9, 2013 (10:11 PM), commenter on Todd VanDerWerff, "Community: 'Advanced Introduction to Finality,'" A.V. Club (media review Web site), April 2014, http://www.avclub.com/tvclub/community-advanced-introduction-to-finality-97134. *Community* is well known for its paintball concept episodes, in which the entire campus becomes engulfed in a massive paintball match.

40. Scrawler, May 9, 2013 (10:33 PM), commenter on Todd VanDerWerff, "Community: 'Advanced Introduction to Finality,'" A.V. Club (media review Web site), April 2014, http://www.avclub.com/tvclub/community-advanced-introduction-to-finality-97134.

41. Nathan Ford's Evil Twin, May 9, 2013 (10:33 PM), commenter on Todd VanDerWerff, "Community: 'Advanced Introduction to Finality,'" A.V. Club (media review Web site), April 2014, http://www.avclub.com/tvclub/community-advanced-introduction-to-finality-97134.

42. Kumagoro, May 10, 2013 (12:39 PM), commenter on Todd VanDerWerff, "Community: 'Advanced Introduction to Finality,'" A.V. Club (media review Web site), April 2014, http://www.avclub.com/tvclub/community-advanced-introduction-to-finality-97134.

43. Todd, "Lived Experience," 863.

44. no1partyanthem, January 3, 2014 (01:22 UTC), commenter on roger_ "Discussion thread for Community S05E01 – 'Repilot,'" /r/community (discussion forum), April 2014, http://www.reddit.com/r/community/comments/1u9pg0/discussion_thread_for_community_s05e01_repilot/.

45. fluxuation, January 3, 2014 (01:28 UTC), commenter on roger_ "Discussion thread for Community S05E01 – 'Repilot,'" /r/community (discussion forum), April 2014, http://www.reddit.com/r/community/comments/1u9pg0/discussion_thread_for_community_s05e01_repilot/.

46. Stabbedinthefoot, January 2, 2014 (08:27 PM), commenter on roger_ "Discussion thread for Community S05E01 – 'Repilot,'" /r/community (discussion forum), April 2014, http://www.reddit.com/r/community/comments/1u9pg0/discussion_thread_for_community_s05e01_repilot/.

47. Classic_Wingers, January 2, 2014 (08:57 PM), commenter on roger_ "Discussion thread for Community S05E01 – 'Repilot,'" /r/community (discussion forum), April 2014, http://www.reddit.com/r/community/comments/1u9pg0/discussion_thread_for_community_s05e01_repilot/.

48. m0rris0n_hotel, January 2, 2014 (08:43 PM), commenter on roger_ "Discussion thread for Community S05E01 – 'Repilot,'" /r/community (discussion forum), April 2014, http://www.reddit.com/r/community/comments/1u9pg0/discussion_thread_for_community_s05e01_repilot/.

49. LookingForAlaska, May 9, 2013 (09:29 PM), commenter on seytonmanning "Season 4 Episode 13 (Finale) Discussion Thread," /r/community (discussion forum), April 2014, http://www.reddit.com/r/community/comments/1e1azm/season_4_episode_13_finale_discussion_thread/?limit=500.

50. Serllyn, May 9, 2013 (10:32 PM), commenter on seytonmanning "Season 4 Episode 13 (Finale) Discussion Thread," /r/community (discussion forum), April 2014, http://www.reddit.com/r/community/comments/1e1azm/season_4_episode_13_finale_discussion_thread/?limit=500.

51. TheASDF, January 2, 2014 (9:27 PM), commenter on Todd VanDerWerff, "Community: 'Repilot'/'Introduction to Teaching,'" A.V. Club (media review Web site), April 2014, http://www.avclub.com/tvclub/repilotintroduction-to-teaching-106548.

52. Jerk-AssHomer, January 3, 2014 (04:43 AM), commenter on Todd VanDerWerff, "Community: 'Repilot'/'Introduction to Teaching,'" A.V. Club (media review Web site), April 2014, http://www.avclub.com/tvclub/repilotintroduction-to-teaching-106548.

53. Wad VanDerTurf, January 3, 2014 (10:49 AM), commenter on Todd VanDerWerff, "Community: 'Repilot'/'Introduction to Teaching,'" A.V. Club (media review Web site), April 2014, http://www.avclub.com/tvclub/repilotintroduction-to-teaching-106548.

54. Feed The Collapse, January 3, 2014 (08:57 AM), commenter on Todd VanDerWerff, "Community: 'Repilot'/'Introduction to Teaching,'" A.V. Club (media review Web site), April 2014, http://www.avclub.com/tvclub/repilotintroduction-to-teaching-106548.

55. DaLateDentArthurDent, January 2, 2014 (07:16 PM), commenter on roger_ "Discussion thread for Community S05E01 – 'Repilot,'" /r/community (discussion forum), April 2014, http://www.reddit.com/r/community/comments/1u9pg0/discussion_thread_for_community_s05e01_repilot/.

56. latam9891, January 2, 2014 (08:27 PM), commenter on roger_ "Discussion thread for Community S05E01 – 'Repilot,'" /r/community (discussion forum), April 2014, http://www.reddit.com/r/community/comments/1u9pg0/discussion_thread_for_community_s05e01_repilot/.

57. Imnotgoodwithnames, January 3, 2014 (12:47 AM), commenter on roger_ "Discussion thread for Community S05E01 – 'Repilot,'" /r/community (discussion forum), April 2014, http://www.reddit.com/r/community/comments/1u9pg0/discussion_thread_for_community_s05e01_repilot/.

58. Author Matthew Collins is drawing here from his experience as a fan and active user on the *Community* subreddit.

59. Costello and Moore, "Online Television Fandom," 130.

60. 12ichmond, January 2, 2014 (07:48 PM), commenter on roger_ "Discussion thread for Community S05E01 – 'Repilot,'" /r/community (discussion forum), April 2014, http://www.reddit.com/r/community/comments/1u9pg0/discussion_thread_for_community_s05e01_repilot/.

61. Rydog814, January 2, 2014 (07:49 PM), commenter on roger_ "Discussion thread for Community S05E01 – 'Repilot,'" /r/community (discussion forum), April 2014, http://www.reddit.com/r/community/comments/1u9pg0/discussion_thread_for_community_s05e01_repilot/.

62. Sandvoss, *Fans*, 96.

63. Greenfield, "Serious Play."

REFERENCES

Andrejevic, Mark. "Watching Television Without Pity: The Productivity of Online Fans," *Television & New Media* 9 (2008): pp. 24–46.

Booth, Paul. "Augmenting Fan/Academic Dialogue: New Directions in Fan Research," *Journal of Fandom Studies* 1, no. 2 (2013): 119–37.

———. "Reifying the Fan: Inspector Spacetime as Fan Practice," *Popular Communication* 11, no. 2 (2013): 146–59.

Charmaz, Kathy. *Constructing Grounded Theory: A Practical Guide Through Qualitative Analysis.* London: Sage, 2006.

Costello, Victor, and Moore, Barbara. "Cultural Outlaws: An Examination of Audience Activity and Online Television Fandom," *Television New Media* 8, no. 2 (2007): 124–43.

de Certeau, Michel. "The Practice of Everyday Life," In *Critical Cultural Theory* (4th ed.), ed. John Storey, 545–55. Essex, England: Pearson, 1984.

Greenfeld, Miriam. (2005). "Serious Play: How Gossiping Is Accomplished in an Online Fandom Forum." Paper presented at the annual meeting for the International Communication Association, New York, May, 2005.

Hills, Matt. *Fan Cultures.* New York: Routledge, 2002.

Jenkins, Henry. *Convergence Culture: Where Old and New Media Collide.* New York: New York University Press, 2008.

———. *Fans, Bloggers, and Gamers.* New York: New York University Press, 2006.

———. "The Cultural Logic of Media Convergence." *International Journal Of Cultural Studies* 7, no. 1 (2004): 33–43.

———. *Textual Poachers: Television Fandom and Participatory Culture.* New York: Routledge, 1992.

Jenkins, Henry, Purushotma, Ravi, Weigal, Margaret, Clinton, Katie, and Robison, Alice J. *Confronting the Challenges of Participatory Culture: Media Education for the 21st Century.* Cambridge, MA: MIT Press, 2009.

Jiyoung, Cha. "Do Online Video Platforms Cannibalize Television?: How Viewers Are Moving from Old Screens to New Ones," *Journal of Advertising Research* 53, no. 1 (2013): 71–82.

Livingston, Sonia. "The Challenge of Changing Audiences: Or, What Is the Researcher to Do in the Age of the Internet?" *European Journal of Communication* 19, no. 1 (2004): 75–86.

Mittell, Jason. "Sites of Participation: Wiki Fandom and the Case of Lostpedia," *Transformative Works and Culture*, no. 3 (2009): http://dx.doi.org/10.3983/twc.2009.0118.

Perryman, Neil. "*Doctor Who* and the Convergence of Media," *Convergence: The Journal of Research into New Media Technologies* 14, no. 1 (2008): 21–39.

Reddit. "Discussion Thread for Community S05E01 – 'Repilot.'" Available online at http://www.reddit.com/r/community/comments/1u9pg0/discussion_thread_for_community_s05e01_repilot/.

———. "Season 4 Episode 13 (Finale) Discussion Thread." Available online at http://www.reddit.com/r/community/comments/1e1azm/season_4_episode_13_finale_discussion_thread/.

Sandvoss, Cornel. *Fans: The Mirror of Consumption.* Cambridge: Polity Press, 2005.

Scolari, Carlos A. "Lostology: Transmedia Storytelling and Expansion/Compression Strategies," *Semiotica* 195 (2013): 45–68.

Soukup, Charles. "Hitching a Ride on a Star: Celebrity, Fandom, and Identification on the World Wide Web," *Southern Communication Journal* 71, no. 4 (2006): 319–37.

Todd, Ann Marie. "Saying Goodbye to Friends: Fan Culture as Lived Experience," *Journal of Popular Culture* 44, no. 4 (2011): 854–71.

VanDerWerff, Todd. "Community: 'Advanced Introduction to Finality.'" A.V. Club. Available online at http://www.avclub.com/tvclub/community-advanced-introduction-to-finality-97134.

———. "Community: 'Repilot'/'Introduction to Teaching.'" A.V. Club. Available online at http://www.avclub.com/tvclub/repilotintroduction-to-teaching-106548.

Whiteman, Natasha. "The De/Stabilization of Identity in Online Fan Communities," *Convergence: The Journal of Research into New Media Technologies* 15, no. 4 (2009): 391–410.

Williams, Rebecca. "'This Is the Night TV Died': Television Post-Object Fandom and the Demise of The West Wing," *Popular Communication* 9, no. 4 (2011): 266–79.

Chapter Seven

Game(s) of Fandom

The Hyperlink Labyrinths That Paratextualize Game of Thrones *Fandom*

Garret Castleberry

Premiering in March 2010, HBO's *Game of Thrones* (or *GoT*) fashioned a reputation for layered narrative hybridity that challenges critics and fans alike. Audiences were mandated to follow not only multiple characters expanding across medieval fantasy realms but also to successfully navigate a nontraditional storytelling style that destabilizes central protagonists while repositioning antiheroes and villains as sympathetic signifiers for audience identification. Based upon fiction writer George R. R. Martin's (GRRM) now synonymous *A Song of Ice and Fire* book series (*ASoIaF*),[1] *GoT* grows in viewership at a time where HBO works to expand its brand legitimacy to foreign and thus global markets.[2] As familiarity with the cultural construct gains popularity beyond niche readers and initial inquisitors to the HBO series, networks of fan sites stimulate an aggressively expansive form of cultural capital. At the same time, the girth of online TV critics (often a hybrid between pensive paid-bloggers and reformed academics) and criticism explores the televisual implications that *GoT* depicts. These paratexual critical forums both legitimate and interrogate the cultural capital now required in order for popular entertainments to expand and retain audience attention. Such forums reinforce what some call "the new intertextual commodity."[3] Also in play is the staying power necessary to establish if not exploit franchises and fandom toward exhausting the symbiotic units' extended narrative-consumer potential.

On one hand, paratextual online discourse legitimates a certain formerly highbrow association with what constitutes not only must-see TV but also must-review, discuss, judge, weigh in, binge, and so on. Ongoing discussions constitute a cornerstone process of what critics identify as the prestige cable drama or appointment television for those incapable of determining such exclusivity. Yet beneath the corporate team-cheers and jeers of *Salon*, *Slate*, or *AV Club* Web sites, the raw materials gathered by "purist" Internet enthusiasts come together at *WinterIsComing.net* (WIC), a subsite created in 2008 that functions as a kind of catch-all for *GoT*-related news, rumors, reactions, and speculations. *WIC* is not the only *GoT*-focused Web site—indeed other sites like *Westeros.org* and *WatchersOnTheWall* each function like meta-kingdoms warring for online fan legitimacy and *GoT* cultural currency—but *WIC* does offer a tragic case study, mimetic of *GoT*, wherein too great of legitimacy can ultimately yield conformity. As cultural clout for *GoT* grows dramatically each season, even the independently created cannot stave the off the attentive hegemony of media conglomeration.[4]

This chapter focuses on examining contrasting fandom ontologies[5] between these two concurrent fandom online sourcing styles (TV recappers versus *WIC*'s catch-and-release info-scroller) in relation to GRRM and Benioff/Weiss's HBO adaptation. First regarding TV critics that specialize in recapping, this chapter looks into how they weigh critiques across a spectrum of reactions that inevitably feeds the show's already existing reputation for layering and depth. Recapping emerges as an institutional method for fan-critics, fan-tagonists and industry-sponsored promoters to extend paratextual discourses and discursive battles over context and content.[6]

Second, examining the monolithic hyperlink gatherer *WinterIsComing.net* as another unique hybrid demonstrates how fans can legitimate appreciation for a given brand or property via organizational accumulation and cooperative online community.[7] Whereas the former attempts objectivism through criticism, the latter adopts objectivity through surveillance-accumulation or the constant gathering of *GoT*-related data. Together these two modes of Internet fandom mirror the franchise property they reflect through the three concepts of world-building, nonlinear storytelling, and strategic ambiguity. These conclusions aid scholars and audiences in understanding how televisual properties navigate both inside and outside the control of their producers while they collect and establish both legitimate and imitative archives throughout developmental, broadcast, and reception phases of cultural production.

GOT AS A METAPHORICAL EXPLANATION TO DE-CONTEXTUALIZE LABYRINTHINE FAN FACTION DISCOURSES

The use of metaphor posits one of the most successful ways in which storytellers distill the complex values and emotions of the human experience into consumable narratives. Lay audiences might call this transformational process "magical" while some scholars may identify it as "spectacle" just as industry producers maintain a host of veritable terms not the least of which would be "marketable." Contemporary post-industrial/post-global capitalism places both a premium and stringent demands on the value of metaphorical language as a commodity exchange rich not only in cultural value among audience-consumers but also economic value in an increasing transmedia global market. This chapter embraces creative metaphor in ways that synthesize interwoven relationships between texts and paratexts, the narrative world building of GRRM and now Benioff/Weiss with the dialogic worldbuilding among niche fan populations that comprise professional TV recappers. Thus identification and theorization of liminal fandom behaviors are made possible by drawing upon recognizable examples for those familiar with the *GoT* or *ASoIaF* series.

I use the HBO dramas as a primary reference for two reasons. First, the HBO television series captures global fascination arguably more so than GRRM's literary texts. Second, because larger audiences and wider demographics sample the show over the in-production book series, it is logical to draw upon the cultural impact that the TV drama holds over fans as well as paratextual discourse generated through online news venues. This chapter argues that structuring a metaphorical context demonstrates the text's metaphorical potency, as well as the text's reflective value, which thus communicates the rhetorical power and transmedia cultural resonance generated from GRRM's creative vision.

This chapter identifies and investigates the hyperlink labyrinths that contextualize GoT fandom. The aforementioned recapper culture and *WIC* database comprise two forms of oppositional engines for *GoT* paratextual discourse that compliment one another in their distinct designs. The phrase oppositional engine proposes a term signifying the dialectic distinction between these two industry brands of fandom. The professional recapping culture in mainstream news sites both reacts and responds to texts like *GoT*. Recappers acknowledge competing and agreeing voices that promote the similar reactionary experiences. In contrast, the *WIC* fan site functions as a "catch-all" data mine or perhaps communal hub for any/all *GoT* information. Both exude similar hyperlinking technologies that posit them as engines for information. As blog-like, both function politically to produce powerful discourses that encourage fandom and extend brand awareness.[8] Yet often these engines run in oppositional directions.

The following sections unpack the first of these two oppositional engines previewed as "recapping culture." Recap culture assumes massive popularity online in an era where industries require "clicks" to justify "costs," hence the advent of *clickbait* as a term to describe how Web sites now lure readers.[9] The rhythmic nature of recapping functions like a labyrinthine enabler that clutters together combinations of description and reaction but varies in depth when it comes to interpretation and evaluation.[10] Following the cultural resonance of recappers and reapping, this essay then scales the metaphorical walls of *WIC* and examines the communal and industrial cache that catchall engines offer fan communities. Along these oppositional engine routes, balance between genre theory oppositions of imitation and innovation ground and extend this chapter's contribution to fan studies. Theoretical balance includes incorporating and introducing several old and new terms fan studies scholars will recognize. Implicating the theoretical language of fan studies speaks to the discipline's short but growing history as well as interdisciplinary dexterity and utility. Ideally, this chapter will introduce terms fan studies and media scholars may find useful for future works. Ultimately, this chapter seeks to extend past-present-future conversations through a combination of critical-cultural and original insights.

GOT AS LABYRINTH TEXT—GOT FANDOM AS LABYRINTHINE PARATEXT

The world-building mechanics that construct and communicate GRRM's fantasy universe of *ASoFaI* and *GoT* evoke one of the most aggressive creative accomplishments that rival if not exceed similar creations like J. R. R. Tolkein's *Lord of the Rings/The Hobbit* book series/film franchise and George Lucas's expansive and lucrative *Star Wars* universe. Unlike other globally popular (and massively populated) fiction worlds such as the comic book universes of DC and Marvel Comics, respectively, Martin's world constitutes original content drawn together by a single mind,[11] unlike the hundreds of production staff that scatter a dozen countries across several continents to create HBO's prestige fantasy epic.[12] The scale of GRRM's vision and Benioff/Weiss' HBO adaptation boasts the largest cast in the history of television.[13] Each season before *GoT* returns from hiatus, fan sites flutter with charts detailing volumes of characters, motivations, alliances, and predicaments so that audiences can reacquaint themselves to the show's narrative convolution.[14] The books feature increasingly complex maps that detail the fictional geographies of "Westeros" and "Essos." For the TV series, the HBOGO app features digitally interactive maps that expand territories, kingdoms, and such, as the show grows in intricacy each year. Aside from GRMM's ever-expanding characters and locations, Benioff/Weiss continue to innovate new production approaches and storytelling stylistics that suggests diverse genres beyond epic fantasy.[15]

Themes range the gauntlet of televisual-cinematic emotional expression and visual storytelling, from patriarchal issues to gender troubles, from sidekick comedies to ancient horrors, from war allegories to courtroom trials, from romance novels to torture porn. *GoT* revels in the kinds of narrative density that transcends conventional characters, locations, genres, or expectations. That said transcending (and transgressing) audience and fan expectations comes at great cost. But while labyrinthine production design stylistics[16] engage viewers in ways few TV series have, R-rated creative freedom and HBO-encouraged exploitative adultness posits *GoT* in the crossfire of fans and anti-fans alike.[17] Thus emerges a paratextual labyrinth that exceeds *GoT* in size, scope, space, and temporal elasticity.

Jonathan Gray draws upon the work of Gerard Genette for his definition of paratexts. Gray classifies paratexts as "those semi-textual fragments that surround and position the work."[18] Paratexts traditionally encircle a master text without necessarily penetrating its diegetic space. For contemporary TV texts, a series' popularity may dictate heavy or light volumes of paratextual discourse. In the case of *GoT*, paratextual fan factions compete and unite in ways that herald and challenge the course of the series. Discourse may include issues of translation from literary to TV text, the role violence or sexual violence plays as contextual versus gratuitous, and the taboo ritual of discussing spoilers in advance of *GoT* airings. Fan fervor amplifies through the progression of digitized devices couples with the onset of multimedia conglomerations, convergence cultures, and transmedia storytelling. Fan-scholars recognize the intertwined challenges that accompany media texts, fan factions, and paratextual discourses. Gray names negative fan factions "anti-fans" and speaks to their distinction:

> Although the fan is positively charged, what of those who are negatively charged? What of anti-fans? This is the realm not necessarily of those who are against fandom per se, but of those who strongly dislike a given text or genre, considering it inane, stupid, morally bankrupt and/or aesthetic drivel.[19]

In some ways, Gray highlights the utility of viewing fandom as a democratic process wherein disagreeing bodies offer variant perspectives.

Derek Johnson offers a theoretical variation with his advent of the term *fantagonism*. Johnson theorizes how fantagonism denotes the "ongoing, competitive struggles between both internal factions and external institutions to discursively codify the fan-text-producer relationship according to their respective interests."[20] For *GoT*, the situation becomes thrice layered between original book author GRRM, showrunners Benioff/Weiss, and each episode's director. While the series receives praise for narrative consistency and artful televisual translation,[21] certain interpretive ambiguities demonstrate how diverting visions complicate already complex production processes and narrative densities. One example of diverting vision can be observed through the aging of the Stark family/children on *GoT*. The book series runs much closer together temporally, whereas because of the behemoth production schedule,[22] casting demands, and narrative complexity, the child actors originally cast experience varying growth spurts that must alter the timeline of the series or risk diegetic disconnection from viewers.[23]

A second hypothetical issue that inverts the child actor-age problem is the sexualization of child actors, or rather, shifting character ages on the show depicted in sexual situations *as* children in the books.[24] This kind of sociocultural navigation thrust enormous moral-ethical responsibility upon the various collaborators as well as audiences. Thus, for texts like *GoT* and their respective fan factions, Johnson's fan-text-producer relationship could be extended to reflect the author-adaptor-interpreter motif in place of "producer." Furthermore, the literary fan factions posit an additional layer wherein two simultaneous textualities coexist in a temporal race toward the texts' conclusions. Following this extension, perhaps an indefatigable equation that mitigates dueling textualities may adjust the fan-text-producer equation for the appropriate layers: (Literary fans +/– TV fans +/– Literary/TV fans) → (Literary Text/TV Text) → (Original Author +/– TV Showrunners +/– Director Vision). This almost comical extension of Johnson's original equation helps identify at least one area of contemporary complexity that labyrinthine texts like *GoT* face. But what of labyrinthine fandom?

Gray, Sandvoss, and Harrington agree that "the changing cultural status of fans is probably best illustrated by the efforts of those in the public gaze, such as celebrities and politicians seeking to connect with consumers and voters by publicly emphasizing their fan credentials."[25] The role of celebrity fans (e.g., celebrities *as* fans) and even celebrity meta-fans (e.g., celebrities famous for a text that are also fans of the text) stresses the significant role that social media plays in the proliferation of transmedia texts. President Barack Obama made notorious headlines in his 2012 *People* magazine interview—in which he boasts about Saturday afternoon TV binges of Showtime's *Homeland* and HBO's *Boardwalk Empire*[26] —and with the president's 2014 request to HBO Studio Chief Richard Plepler for advanced copies of *Game of Thrones* and *True Detective* for personal use during an extended holiday weekend.[27] In recent years, the TV serial drama steadily rose in prominence over film and comic book industries at San Diego's Comic Con media-cultural convention.

In 2014, numerous publications pronounced *GoT* the convention's defacto "winner" in terms of fan buzz and producer-actor cooperation.[28] In some ways, the *GoT* fan wagon circles itself as the show's social-media savvy self-aware actors and actresses evoke and perform what we might call meta-fandom moments of intertextual reflexivity. For example, in 2014 *GoT* actress Lena Headey defiantly posted cryptic spoilers through playfully concocted photos on her Instagram page. Reactions were decidedly mixed among fans and critics but her social media use-as-breaking the fourth wall points to a new hybrid between brand extension and celebrity-fans.[29] This is what Jenkins means when he assesses that, "media companies act differently today because they have been shaped by the increased visibility of participatory culture: they are generating new kinds of content and forming new kinds of relationships with their consumers."[30] In other words, HBO doesn't want to risk sounding like a cultural grandfather when, for instance, *GoT* gains cultural clout as the number one pirated television show four consecutive years on air.[31] Instead, HBO's personnel frequently plays it cool (like the Fonz or perhaps "the Dude") and allows such piracy to *increase* their prestige status and global currency as a soft-after brand, especially in the wake of Netflix's rival success.[32]

Meanwhile the ontology of recap culture provides unique spaces to retort and reform production and institutional issues. Recappers legitimize discursive sights/sites that debate ethical and sociocultural questions. Insolently these discursive entanglements identify topics that underlie HBO's self-aware gratuitous branding.[33] On the other hand, textual shortcomings like instances of *sexposition*[34] throughout the series and season 4's "rape versus consent" debate[35] fuel actual discourse that increases the paratextual validity fandom fuels and then demonstrates a recourse impact on perceptions of the text, reflections of how society should engage taboo subject matter, and in effect elevate legitimacy for fan studies. This issue engages what Pam Wilson describes as "narrative activism,"[36] where active fans campaign to eliminate conflicting agents—for example, between producers and writers or book authors and TV adaptors—from causing further harm to the diegetic text or the shared interpretive vision beheld by fans.[37]

RECAP CULTURE AS DISCURSIVE PARATEXTUAL WORLD-BUILDING AND REAL WORLD DEBATING

In *Textual Poachers*, Jenkins theorizes how viewing strategies "made possible by the technology's potentials, extend the fans' mastery over the narrative and accommodate the community's production of new texts from the series materials."[38] In its context, Jenkins's commentary on viewing strategies emphasizes the technological innovation that "VCRs, Reruns, and Rereading" have on the historical evolution of fan participation and meaning making. While Jenkins original intent works within its historically dated cultural context, his commentary toward the fandom nature of viewing strategies extends contemporary discussion. Certainly VCRs enable the kinds of rereading that affords mass audiences larger accessibility to not only text(s) but also the kinds of functions that temporally displace texts so that fans can absorb, slow down, fast-forward, pause, and generate close readings if not entirely new interpretations and uses of/for content. The onset of Internet proliferation, DVD/Blu-ray and recording apparatuses like TiVo/DVR/etc., as well as more recent transmedia technologies and culturally popular practices like binge watching and live-tweeting provide a plurality of viewing processes and a multiplicity of interpretive-consumptive avenues.

One avenue of particular interest is the emergent proliferation of *recapping*, or the typed summarization of newly aired TV content, primarily dramas and sitcoms, with an infusion of personal reaction and cultural commentary. Recapping reflects the diverse nature of TV criticism in that it varies in degrees of interpretation, reader accessibility, informed or educated insight, and thus the rhetorical power to persuade readers. Anyone can recap a program, but a unique craft or stylized rhetorical recap dominates the arts and entertainment sections of some of the most prominent online American news sources. Among the "elite" staffs of TV bloggers, I argue their respective discourses flavor distinct cultural sway among the interpretational fan communities across televisual genres. In turn a recapper's qualitative and quantitative fervor or dismissal acts to enable increased *textual legitimacy* for purported traditional readers, while enabling fan legitimacy through mainstream identification with certain preferred readings and texts. On one hand, introducing the word "elite" conjures notions of cultural privilege. On the other hand, professional bloggers wield potent sociopolitical power regarding degrees of transmedia trafficking and thus exist rhetorically as persuasive pendulums of critical retort and liminal discourse. Recappers straddle lines between traditional lay TV criticism and pseudo-academic mini-lectures, yet such temporal constraint relieves some criticisms of their intellectual heft. [39]

Ultimately, this chapter highlights how recap culture functions as a new kind of fandom, an informed fandom drawing upon Internet resources and technologies in ways previous generations and iterations of fan communities have not. Indeed, scholars of audience and fan studies mine the roles Internet technology plays and the ways in which fans play with these technologies. [40] This chapter argues the distinct cultural flavor mitigated by professional recapping—and perhaps more importantly the vast replication of secondary freelance or hobbyist recappers that saturate mainstream and independent blogospheres—function to reflect through diegetic mimesis the world-building process that GRRM and Benioff/Weiss extend but struggle to maintain. In effect, such cross-comparison demonstrates how this distinct "brand" of fan culture vested in *GoT/ASoIaF* posits as difficult a labyrinthine gauntlet as those depicted within GRRM's fantasy fiction world Westeros.

MAGIC FINGERS, RECAP SCREENERS, AND PARATEXTUAL ILLUSIONS

Johnson theorizes fan factions exercise powerful political persuasion in the realm of paratextual discourse. Johnson insists audiences "can also challenge corporate producers by constructing interpretive consensuses that delegitimize institutional authority over the hyperdiegetic text."[41] TV recappers, as active audience and active fan, shift the discursive authority of the text into their digital dialogic arenas once a show concludes its initial "live" airing. From this point, industry-mining (and mindful) recappers take to the Web to craft their insta-feed responses to media. A certain recapper segment holds privileged position as these recappers preload their responses due to advanced access to content. "Screeners" posit an industry term comprising (formerly) tapes, DVDs, or digital codes that allow credentialed critics media materials ahead of time. Regarding *temporal privilege*, this industry technique ensures a relative saturation quota for textual discourse circulates immediate if not prior impact to an episodic airing. Advanced access thus shrinks the elastic distance between a show airing and post-episode recap publication. In effect, recappers then publish so quickly that bloggers appear magical if not super-human in their typing-processing-publishing capabilities. This technique pads recapper efficiency and thus boosts persuasive appeal. Through this institutional method the screener process arguably adds a layer of mysticism for online readers, which increases the illusionary spectacle that strengthens paratextual discourse.

On the other hand, magic finger recapping inoculates certain kinds of privileged readings that gain online access and thus discursive legitimacy faster than the typical lay fans and audiences. Furthermore, professional recappers catering to TV fans channel messages through *Vulture* or *AV Club* or even slightly more mainstream news outlets like *Slate* and *Salon*. These recappers embed rich discourses and textual meanings into commentaries but their corporate-professional privilege arguably paints recap close readings with shades of industry influence.[42] For example, consider writer Scott Erik Kaufman, a former academic at the University of California, Irvine who offered courses on popular entertainments like *Game of Thrones*. Kaufman often produces unfiltered criticisms with salty R-rated analyses for the political-cultural blog *Lawyers, Guns, & Money*. While Kaufman still contributes to *LG&M*, he now works as associate editor for *The Raw Story* and less occasionally for the *Onion*'s sister Web site the *AV Club*. Yet when comparing the wily outlaw approaches that innovate *LG&M* analyses[43] to corporate pressures adhering softer communicative and critical approaches,[44] the results while subtle should not be understated. This effect might be identified as a *corporate softening* (to shield myself from harsher criticism) inoculated by industry influence.

One positive aspect of industrial influence comes through the use of professional recaps as a method for generating cultural interest toward a bevy of enriching texts overlooked by nontraditional/post-millennial media consumers. That said one notable mode of give/take negotiation—brought on by advanced screeners and early industry-sponsored recap releases—occurs through delimitation. Johnson explains how, "corporate producers' creative choices often delimit the range of interpretation possible within fan meta-texts, authorizing some but denying others."[45] For example, Matt Weiner frequently clarifies moments of artistic ambiguity on critical fave *Mad Men*[46] and Vince Gilligan also debunked Walter White's "inferred interiority"[47] for the series finale "Felina" during a live Q&A on AMC's torturously forced *Talking Bad* post-mortem show. In many ways, the advent and unappealing onset of numerous post-mortem talk shows diffuse the richness and cultural agency that is both alluring and charming about fan communities. In effect, when post-mortems like *Talking Dead* or *Talking Bad* or *Sons of Anarchy: Post-Mortem* or *Bates Motel: Post-Mortem* create an immediate live feed-back loop with series showrunners, episode writer-producers and actors, such programs relieve the text's polysemic and polyvalent potency, the creative currency that fuels fandom.

HBO is slightly guilty but takes a much preferred *less is more* stance with their OnDemand services. For audiences that access programs like *GoT*, *Boardwalk Empire*, or *True Blood* on HBO OnDemand, the conclusion of a given episode's credits typically follows with a pre-produced 2–3 minute "Inside the Episode" short. These shorts cut between select footage of the just-ended episode, accompanied with key insights into the narrative motiva-tions at stake from a writer or producer's perspective. I argue that these media packages hold greater rhetorical value through their process of limited use. Notably, the "Inside the Episode" shorts do not air after live or subse-quent cable channel airings nor do they run on HBO's online HBOGO ser-vice. The shorts do appear on DVD/Blu-ray box set releases along with a host of alternative materials that attract fan attention. Arguably the brevity of HBO's "Inside the Episode" helps protect the program's textual mystique. The textual mystique affectively describes a cable serial's *je ne sais quoi* or intangible qualities. Often the textual mystique forms through a combination of creative and culturally significant inputs, a televisual alchemy of sorts.

Thus, recappers and recapping function as potent "clickables" or Internet eye-catching materials that translate into cultural currency via "likes" and "shares" and eventually transform into advertiser dollars through corporate data-mining processes that translate views into statistics and statistics into demographic ratios and demographic ratios into concentrated marketing niches. If such clickable data is resold, say to HBO's subsidiary merchandize companies (HBO is owned by TimeWarner no less), then those secondary merchandizing companies can go on to buy the advertizing spaces strategi-cally placed above, below, and to the sides of a given Twitter/Facebook/Instagram news feed. Thus, the fans who already click and read *GoT* recaps looking for new perspectives on episodes find doubly rewarding adverts listing *GoT* T-shirts, beer tumblers, lapel pens, and so on. Of course this process identifies contemporary marketing practices among numerous shows and creative properties, particularly those with strong corporate backing. The branding process links product and audience via commoditization and consumption. Tracing this marketing strategy does not denote cynical conspiracy against hidden recapper agendas but instead acknowledges the industrial role(s) performed in the manufacturing of paratextual content. Words like "corporate softening" or "industrial influence" may generate mixed responses in how fans feel about the texts they covet and the paratexts they participate in. Jenkins refers to this phenomenon as "colonial cringe" or shared negative reaction to a cultural-industrial output.[48] Following this logic of corporate sponsorship, the recent history of *WinterIsComing.net* posits a rich contrast to the spread out discourses constitute recap culture.

WINTERISCOMING AS A "WATCHER ON THE [PARATEXTUAL] WALL"

If recap culture constitutes one end of the oppositional engine spectrum among *GoT* fandom's labyrinthine paratext, then *WIC* might best be understood as one of several possible labyrinth subcenters powered by panoptic vantage points. In *GoT*, "The Wall" separates Westeros from harsh weather and harsher dangers that lie to the North. The Wall posits a strategic vantage point in *GoT* that provides dualistic panoptic surveillance between those protected and those feared. Notably, the Watchers who occupy the Wall reuse the House Stark credo "Winter is Coming," which doubles as an ominous philosophy for grim preparedness. Underneath this one-thousand-foot structure, those of "the Night's Watch" gain access through strategic corridors that provide privileged access. While the fictitious Wall purports thousands of years in age, according to *WIC*'s "about" page, the Web site on the other hand was founded in 2008 around the time HBO first commissioned *GoT*'s pilot.[49] This section overviews key features showcasing *WIC*'s versatility as an "ideal" fan site and describes how media conglomeration processes dilute and homogenize the individuality that makes sites like *WIC* function as desirable panoptic engines among fans.

In essence, *WIC* functions as a nexus for *GoT* web content. *WIC*'s Web design shows no prejudice in the abundance and directions of both textual and paratextual online content relating strictly to *GoT* and *ASoIaF*. One might question whether a Web site devoted to a TV series can generate enough content when the TV drama only runs for ten weeks out of the year. Yet *WIC* practices a no-discrimination policy and finds ways to gain and spread access throughout *GoT*'s pre- and post-production phases. In comparison, consumers purchasing HBO Blu-ray collector's editions unpack caches of unaired production content designed to expand upon economic limitations the TV series faces. These features range from digitized storyboards to character/world histories voiced by series actors to select episode commentaries. Yet the DVD/Blu-ray content exhibits a narrow-focused offering. Meanwhile *WIC* encompasses transmedia conglomeration innovated by the Internet, which provides an economically flexible tool temporally elastic in scope and accessibility.

One week *WIC* may embed insider interviews with the design firm behind the popular and immersive *GoT* opening credits sequence. The next day or week *WIC* offers copious reviews, interviews with various personnel, updates on GRRM's personal blog, word of mouth from the production locations, and all the post-millennial buzzworthy pop ideology that proliferates alongside billion-dollar franchises and Hollywood synergy. It is significant to note *WIC*'s initial grassroots-style design and focus. *WIC* was presented without advertising interruptions, noisy margins, or unnecessary spam windows begging users to sign in through social media account. Fans could simply visit at will and gain access to random content that ranges from screen caps of how to make an actress appear nine-months pregnant while unclothed to recurring segments like "Curtain Calls" (whenever a character dies on-screen) or "Dame of Thrones" (highlighting female character intricacies) or "The North Remembers," a genteel breakdown of each month's highlights and hyperlinks.

In so many ways, *WIC* planners extend the site beyond what fans gain from DVD/Blu-ray extras or the *L.A. Times*. For fans, Web surfing or data mining for *GoT* answers, conversations, and infotainment functions as personal labor (e.g., *fan-chores*) that may narrow interpretation but also engage individual fulfillment via metaphorical journeys toward fictional ⇔ real "truth." *WIC* features an active, passionate fan community that uses comments sections as soundboards. But just as the site's ominous meta-mantra forecasts, change is inevitable. Announced in January 2014, the once-intimate fan site reached an industrial epoch that captures a double bind between what gives paratextual spaces a magical *je ne sais quoi* and how social media conglomeration compounded by consumer demand for creative industries come to homogenize indigenous online culture.

According to the January 2014 announcement, *WIC* was purchased and currently exists wholly redesigned by the un-ironically titled conglomerate network, *Fansided*. *Fansided* started as a small-market sports-based Internet fan site that has gained traction and burgeoned into a booming entertainment engine that corrals over three hundred diverse fan-related sites under its corporate umbrella.[50] The justifiable logic behind such strategies is merging multiple sites with separate but highly devoted fan-users creates dynamic potential for crossover appeal. In other words, fan-users become fan-used. Formerly *WIC* communicated quiet constraint and close adherence to *GoT*-related color palettes in its previous web design. Now the homogenized "sports blue" of *Fansided*'s homepage template saturates *WIC* borders, head-er and footer space. Instead of the woodsy colors and moody grays that evoke fan fervor and epic fantasy, the long generic scrolling wall could easily be mistaken for someone's [mom's] Facebook feed. While much remains, including the clever titles that encourage sporty role-play, corporate softening and industry influence dominate *WIC*'s unspoken qualities, its subtext, its paratextual soul. Like the undead White Walkers that roam North of the Wall, *WIC* functions in a way that's culturally lifeless and drolly homogenous. A paratextual labyrinth built into other paratextual labyrinths built into other paratextual labyrinths, *WIC* captures content and now constitutes captured content. The site—while still very useful—is hardly soulful.

CONCLUDING THOUGHTS ON THE OPPOSITIONAL ENGINES OF INDUSTRIAL FANDOM

The previous sections presented two oppositional engines of online paratextual fandom designed to survey the cultural resonance circling GRRM's *ASoIaF* series now translated into *GoT* by HBO's Benioff and Weiss. Engaging fan studies theories and theorists ideally continues their rich merit while extending new possibilities. This evaluation highlights ways in which professional recappers and *WIC* capture and recast texts, create paratextual discourse, and undermine originality in favor of media conglomeration and convergence culture. Ultimately, this chapter seeks to provide critical insight into the motivations that drive these economic and sociocultural oppositional engines. In the case of recappers and recapping, these industrial-professional fans produce and circulate important critical dialogues and legitimate social issues. In the case of *WIC*, its originators created a savvy product that resonates culturally to such a high degree that it becomes an economically viable investment, arguably fulfilling a key tenant of the American Dream and postglobal Western capitalism at large. These two paratexts mimic their *GoT* source material in ways that challenge audiences-readers-fans. First, the paratexts perpetuate labyrinthine discourses that spread throughout hyperlinked corridors, winding down ontological hallways and into alternating corners of knowledge and perspective. These practices digitize debate and thus capitalize on potential for broader social reflexivity among other critics, bloggers, and fans. Second, the *WIC* fan site highlights values of creativity, ingenuity (or is it *engine*-uity?), and community but also functions as cautionary allegory for extremes relating to convergence culture, corporate homogenization, and even cultural hegemony. Future fan studies would benefit from long-term exploration into these respective oppositional engines, particularly as HBO's *GoT* continues to grow in popularity with each successive year.

NOTES

1. The frequent use of several long names and titles requires heavy use of shorthand abbreviations, which is customary for televisual media texts and appropriate considering the social media nature of the fan faction paratexts I explore.

2. Amol Sharma (2014), Troy Dreier (2013), and Lisa Richwine (2012) collectively summarize HBO's contemporary foreign market expansion strategies. In effect, contemporary convergence markets offset costs where HBO arguably loses U.S. momentum while it sustains newfound profitability oversees. Retrieved from http://online.wsj.com/articles/hbo-weighs-more-web-tv-services-overseas-1407106582 and http://www.streamingmedia.com/Articles/Editorial/Featured-Articles/How-HBO-Go-Expanded-South-with-a-Latin-America-Rollout-88584.aspx.

3. Henry Jenkins, (2004), "The Cultural Logic of Convergence Culture," *International Journal of Cultural Studies* 7(1), 39, and "The new intertextual commodity," (P. David Marshall, 2002) in *The New Media Book*, Dan Harries (ed.). London: British Film Institute.

4. I examine the *WIC* media buyout in greater detail during later analysis.

5. Tom Gruber's shortened definition of *ontology* as a "specification of a conceptualization" suffices nicely for my brief use of the term (Tom Gruber, 1992). "What Is an Ontology?" Retrieved online from Stanford's listserve, http://www-ksl.stanford.edu/kst/what-is-an-ontology.html).

6. Derek Johnson, (2007) "Fan-tagonism: Factions, Institutions, and Constituitive Hegemonies of Fandom," In *Fandom: Identities and Communities in a Mediated World*, 285–300, eds. Jonathan Gray, Cornel Sandvoss, and C. Lee Harrington. New York: New York University Press, 287. I address these interrelated terms in greater detail in subsequent sections.

7. Sara Gwenllian Jones notes "online fan cultures are more symbiotic than they are antagonistic" in "Web Wars: Resistance, Online Fandom and Studio Censorship" (London: British Film Institute, 2003), 171. Jenkins sees "new kinds of cultural power emerging as fans bond together within larger knowledge communities, pool their information, shape each other's opinions, and develop a greater self-consciousness about their shared agendas and common interests" (New York: New York University Press, 2007), 362–63.

8. Ibid. See the political power that blogging offers convergence culture (Jenkins, 2004), 36–37.

9. Lexi Hansen, *The Clickbait Phenomenon* (June 26, 2014).

10. So many recappers fail to uphold minimalist academic standards of the "DIET" model (*describe, interpret, evaluate, theorize*) that sorting through recaps becomes an exercise in grading but without the fear of teacher evaluations recourse. That said, prescribing academic grading qualities in recapping provides another quick and easy method for identifying which recapping/recapper paratexts offer the strongest research potential in their ability.

11. Contrasting ideas of *originality*, many genre theorists and critical scholars prescribe that no wholly original concept or idea exists but instead generates through a combination of imitation and innovation. For expansions on these epistemologies, consult Jean Baudrillard, *Simulacra and Simulation* (Ann Arbor: University of Michigan); Guy Debord, *Society of the Spectacle* (Detroit, MI: B&R, 1983); Rick Altman's *Film/Genre* (London: BFI, 1999); and Cawelti's *Adventure, Mystery, and Romance* (Chicago: University of Chicago, 1976).

12. Kate Byron, "These three countries are winning the 'Game of Thrones,'" *CNBC* (5 April, 2013).

13. Arlene Paredes, "Game of Thrones Spoilers: 'We have the largest cast on television right now.'—David Benioff," *International Business Times*, May 30, 2012. http://au.ibtimes.com/articles/346654/20120530/game-thrones-spoiler-video-valar-morghulis.htm#.U9wye1ZrrRo.

14. Jessica Toomer (*Huffington Post*, 2013); Jerry Mosemack (*USAToday*, 2014); Jace Lacob (*Daily Beast*, 2011); Stuart Jeffries (*Guardian*, 2013) offer a minor sampling of perhaps a hundred or more textual fluffers that function as intertextual eye candy while attempting to simplify *GoT* complexity for new audiences.

15. The show falls in line with Lemke's theorizations of multimodal genres in transmedia storytelling (Jay Lemke, 2009). "Multimodal genres and transmedia traversals: Social semiot-

ics and the political economy of the sign," *Semiotica*, 1(4), 283–97. Castleberry samples *GoT*'s genre elasticity, which helps proliferate cross-cultural identification with the text. "Creating Game of Thrones Cross-Demographic Appeal through Genre-Mixing Iconicity," *In Media Res* (online) (Castleberry, 2014).

16. Kristy Barkan, "Visual Effects: The True Magic of *Game of Thrones*," *ACMSIGGRAPH* (2014, online); Bob Bricken, "*Game of Thrones*' season 4 sets are so detailed they'll blow your mind," *io9* (2014, online).

17. For inflammatory accusations against HBO executives, see Kyle Buchannon "This is why *Game of Thrones* has so much nudity," (2012, Web site); and Ali Plumb's *Empire* magazine podcast (2012, Web site).

18. In Jonathan Gray (2003). "New audiences, new textualities: anti-fans and non-fans," *International Journal of Cultural Studies*, 6(1), 72.

19. Ibid., 70.

20. Johnson (2007), "Fan-tagonism: Factions, institutions, and constituitive hegemonies of fandom," In *Fandom*, 287.

21. Mark Harris discusses the broad shift in TV landscaping with special kudos for *GoT* complexity (*New York Magazine* online, 2014), while Gavin Polone boldly claims "All TV should be more like *Game of Thrones*" (*Vulture* online, 2012) and RogerEbert.com editor-in-chief Matthew Zoller Seitz notes even the most gratuitous production choices in "Season Two Is Artful and Adult" (*Vulture* online, 2012).

22. Twelve-month production cycles for ten-episode seasons based on interviews from Maureen Ryan, "Game of Thrones' Third Season: How Many Episodes Will There Be?" (*Huffpost TV*, 2012).

23. Margaret Lyons surfaces this considerable TV problem in "*Game of Thrones*' Kid-Actor Problems" (*Vulture* online, 2012) and Jennifer Vineyard performs a follow-up report two years later, "*Game of Thrones*' Showrunner D. B. Weiss on How the Show Will Handle Its Aging Child Stars" (*Vulture* online, 2014).

24. Would-be Westeros princess Daenerys Targaryen is bartered by her sexually abusive brother Visyris to become the sex-slave "Queen" to barbarian Dothraki King Khal Drogo. In Martin's book Dany is in her middle teens while sexually active and abused but the show vaguely updates her to appear somewhat of-age. On the TV show, the Stark sons encounter consensual sexual relationships at ages increased from the books, while the Stark daughters face habitual threats of rape while underage in both the books and TV series.

25. Gray, Jonathan, Cornel Sandvoss, and C. Lee Harrington, (eds.), "Introduction: Why Study Fans?" 1–18, in *Fandom: Identities and Communities in a Mediated World* (New York: New York University Press, 2007), 5.

26. Sandra Sobieraj Westfall, "President Obama Talks Facebook and TV Habits," *People*, December 11, 2012. http://www.people.com/people/article/0,,20553487,00.html.

27. David Carr and Ravi Somaiya, "Punching above Its Weight, Upstart Netflix Pokes at HBO," *New York Times*, February 16, 2014. http://www.nytimes.com/2014/02/17/business/media/punching-above-its-weight-upstart-netflix-pokes-at-hbo.html?_r=2.

28. David Bloom, "Power Lifters: Five Brands That Got a Big Social Media Lift from Comic Con." *Deadline*, July 30, 2014. http://www.deadline.com/2014/07/comic-con-social-media-brands-game-of-thrones-the-walking-dead-mad-max-fury-road-wwe/; "The Women of 'Game of Thrones' Dominated Comic Con," *StyleList*, July 28, 2014. http://www.stylelist.com/view/the-women-of-game-of-thrones-dominated-comic-con/.

29. "A *Game of Thrones* Actress May Be Giving Away Huge Spoilers on Instagram" (Frank Pallotta, *Business Insider* [Web site], 2014); "Game of Thrones Season Four: Lena Headey Sparks Calls of 'Spoiler!' with Instagram Post" (Jess Denham, *The Independent UK* [Web site], 2014); "*Game of Thrones* actress Lena Headey Cheekily Refuses to Apologize for Instagram 'Spoilers'" (Joanna Robinson, *Vanity Fair* [Web site], 2014).

30. In Jenkins, *Fandom*, 2007, 362.

31. Mandi Bierly, "'Game of Thrones' Retains Crown as Most Pirated TV show," *EW*, December 27, 2013. http://www.cnn.com/2013/12/27/showbiz/tv/game-of-thrones-most-pirated-show-ew/; Jay McGregor, "Game of Thrones Season Finale Becomes Most Pirated

Show in History," *Forbes.com*, June 17, 2014. http://www.forbes.com/sites/jaymcgregor/2014/
06/17/game-of-thrones-season-finale-becomes-most-pirated-show-in-history/.

32. AAP, "'Game of Thrones' Director Says 'Cultural Buzz' More Important Than Ratings
for Survival," *Sidney Morning Herald*, February 26, 2013. http://www.smh.com.au/
entertainment/tv-and-radio/downloads-dont-matter-20130226-2f36r.html#ixzz2LywE7AZ2.

33. Ibid.; Buchannon, 2012; and Plumb, 2012.

34. The term "sexposition" arose from hybrid TV critic/academic Myles McNutt on his TV
criticism blog *Cultural Learnings* (McNutt, 2011). Device-naming practices thus initiate hyperlinked streams of online consciousness, in this case relating to *GoT* discourse. In "HBO, You're
Busted," Mary McNamara notoriously criticized HBO and *GoT*'s incessant penchant for gratu-
ity (*L.A. Times*, 2011), which led to Matthew Zoller Seitz's cautious defense rebuttle "In
defense of HBO's 'unnecessary' nudity" (*Salon*, 2011), and then spilled into the blogosphere
with contributions like "Game of Tits" from Lady T. of *funnyfeminist.com* (online, 2011). As a
potent idiom, sexposition now inhabits the Internet lexicon with mainstream regularity among
TV critics and fans alike—a testimony to the political virility blogs still possess—and even
boasts its own detailed Wikipedia history to boot.

35. "A Frank Discussion of *Game of Thrones*' Rape Scene and Its Epidemic of Sexual
Violence" (Jill Pantozzi, Rebecca Pahle, and Victoria McNally, *The Mary Sue* [Web site],
2014); "George R. R. Martin Defends Sexual Violence in 'Game of Thrones' (Daisy Wyatt,
The Independent [Web site], 2014); "For 'Game of Thrones,' Rising Unease Over Rape's
Recurring Role" (Davd Itzkoff, *New York Times* [Web site], 2014).

36. See Pam Wilson, *Reality TV: Remaking Television Culture* (New York: New York
University Press, 2008), 337.

37. Ibid., Johnson (2007) richly synthesizes Wilson's use of *narrative activism* in a case
study of fan reaction against incumbent showrunner Marti Noxon during *Buffy the Vampire
Slayer*'s sixth season.

38. From Henry Jenkins, *Textual Poachers: Television Fans and Participatory Culture*
(New York: Routledge, 1992), 73.

39. There is much to say on the notion of *liminal discourse* and the role of professional
recappers later given their *betwixt and between* status, as Victor Turner would say.

40. Theorists and industry insiders alike tackle the potencies that convergence technologies
enable. For an introductory sampling, see Berger (2012) *Media and Society: A Critical Perspective* (New York: Rowman & Littlefield); Jenkins (2006) *Convergence Culture* (New York:
New York University Press) and Jenkins, Ford, and Green's (2013) *Spreadable Media: Creating Value and Meaning in a Networked Culture* (New York: New York University Press.), as
well as industry perspectives like Frank Rose's (2012) *The Art of Immersion* (New York: W.
W. Norton & Co.) and legal ramifications as detailed in Lawrence Lessig's (2004) *Free Culture: The Nature and Future of Creativity* (New York: Penguin Books.).

41. Johnson, "Fan-tagonism," in *Fandom*, 291.

42. Zukin and Maguire (2004) on the significant role that the institutional field plays in
creating and maintaining consumer culture, where "Consumer culture is produced, as well, by
agents who work directly in the corporate economy as managers, marketers, and advertising
'creatives'; by independent 'brokers' who analyze and criticize consumer products; and by
dissidents who initiate alternative responses to the mass consumption system. This broad
framework allows us to consider consumption as *an institutional field*, that is, a set of intercon-
nected economic and cultural institutions centered on the production of commodities for individual demand" (Sharon Zukin and Jennifer Maguire, 2004). "Consumers and consumption,"
Annual Review of Sociology, 30, 175.

43. See Kaufman's [SEK] habitual favor for drawing "laser eyes" on screen caps to analyze
the eye lines in *Game of Thrones* production style: (Scott Erik Kaufman, "I see that you've seen
that I saw you: miscommunication in 'Second Sons' (*Game of Thrones*)," *Lawyers, Guns &
Money*, May 25, 2013. http://www.lawyersgunsmoneyblog.com/2013/05/i-see-that-youve-
seen-that-i-saw-you-miscommunication-in-second-sons-game-of-thrones.

44. Compare the former example to the straightforwardness of *The Raw Story* recap from
season four (Scott Kaufman, "Recap: *Game of Thrones*: Season 4, Episode 7, 'Mockingbird,'

The Raw Story, May 19, 2014. http://www.rawstory.com/rs/2014/05/19/recap-game-of-thrones-season-four-episode-seven-mockingbird/.
    45. Johnson, "Fan-tagonism" In *Fandom,* 2007, 291.
    46. "Matthew Weiner Talks About Mad Men's Mid-Season Finale, *2001,* and Why Joan Is So Mad at Don." (Denise Martin, *Vulture* [Web site], 2014).
    47. Jason Mittell, "Serial Characterization and Inferred Interiority," *In Media Res,* December 14, 2011. http://mediacommons.futureofthebook.org/imr/2011/12/14/serial-characterization-and-inferred-interiority.
    48. Henry Jenkins, "Afterword: The Future of Fandom." In *Fandom,* 2007, 363.
    49. "Winter Is Coming: About Us," *Winteriscoming.net* [Web site] Retrieved from http://winteriscoming.net/about/.
    50. "Fansided: About Us," *Fansided.com* [Web site]. Retrieved from http://fansided.com/about/.

REFERENCES

Barkan, Kristy. (2014). "Visual effects: The true magic of *Game of Thrones.*" *ACMSIGGRAPH* [Web site]. Retrieved from http://www.siggraph.org/discover/news/visual-effects-true-magic-game-thrones.
Benioff, D., and Weiss, D. B. (Producers). (2011–present). *Game of Thrones* [TV series]. Los Angeles: Home Box Office.
Berger, Arthur Asa. (2012). *Media and Society: A Critical Perspective,* 3rd ed. New York: Rowman & Littlefield.
Bricken, Bob. (2014). "*Game of Thrones*' season 4 sets are so detailed they'll blow your mind," *io9* [Web site]. Retreived from http://io9.com/game-of-thrones-season-4-sets-are-so-detailed-theyll-1516881897.
Buchannon, Kyle. (2012). "This is why *Game of Thrones* has so much nudity," *Vulture* [Web site]. Retrieved from http://www.vulture.com/2012/06/game-of-thrones-nudity-nude-scenes.html.
Byron, Kate. (2013). "These three countries are winning the *Game of Thrones,*" *CNBC* [Web site]. Retrieved from http://www.cnbc.com/id/100619559#.
Castleberry, Garret. (2014). "Creating *Game of Thrones* Cross-Demographic Appeal through Genre-Mixing Iconicity," *In Media Res* (online). Retrieved from http://mediacommons.futureofthebook.org/imr/2014/09/23/creating-game-thrones-cross-demographic-appeal-through-genre-mixing-iconicity.
Cawelti, John G. (1976). *Adventure, Mystery, and Romance.* Chicago: The University of Chicago.
Danesi, Marcel. (2012). *Popular Culture: Introductory Perspectives,* 2nd ed. New York: Rowman & Littlefield.
Denham (2014). "Game of Thrones Season Four: Lena Headey Sparks Calls of 'Spoiler!' with Instagram Post," *The Independent UK* [Web site]. Retrieved from http://www.independent.co.uk/arts-entertainment/tv/news/game-of-thrones-season-four-lena-headey-sparks-calls-of-spoiler-with-instagram-post-9401582.html.
Dreier, Troy. (2013). "How HBO Go expanded South with a Latin America rollout." Retrieved from http://www.streamingmedia.com/Articles/Editorial/Featured-Articles/How-HBO-Go-Expanded-South-with-a-Latin-America-Rollout-88584.aspx.
Fansided. [Web site]. http://fansided.com.
Gray, Jonathan. (2003). "New Audiences, New Textualities: Anti-Fans and Non-Fans." *International Journal of Cultural Studies,* 6(1), 64–81. DOI: 10.1177/1367877903006001004.
Gray, Jonathan, Cornel Sandvoss, and C. Lee Harrington. (2007). "Introduction: Why study fans?" 1–18. In *Fandom: Identities and Communities in a Mediated World,* eds. Jonathon Gray, Cornel Sandvoss, and C. Lee Harrington. New York: New York University Press.
Hansen, Lexy. (2014). The clickbait phenomenon. *iTracking Research Inc.,* Retrieved from http://itrackingresearch.com/clickbait-phenomenon/.

Harris, Mark. (2012). "TV is not TV anymore: A revolution in how we watch was just a start. Now the good stuff." *New York Magazine* [Web site]. Retrieved from http://nymag.com/arts/tv/upfronts/2012/mark-harris-tv-2012-5/.

HBOGO. [Web site]. http://www.hbogo.com/#whatis/.

Itzkoff, Dave. (2014). "For *Game of Thrones*, Rising unease over rape's recurring role." *New York Times* [Web site]. Retrieved from http://www.nytimes.com/2014/05/03/arts/television/for-game-of-thrones-rising-unease-over-rapes-recurring-role.html.

Jeffries, Stuart. (2013). "*Game of Thrones*—A beginner's guide." *The Guardian* [Web site] Retrieved from http://www.theguardian.com/tv-and-radio/2013/mar/25/game-of-thrones-beginners-guide.

Jenkins, Henry. (2007). "Afterword: The Future of Fandom." In *Fandom: Identities and Communities in a Mediated World*, 357–64. Eds. Jonathan Gray, Cornel Sandvoss, and C. Lee Harrington. New York: New York University Press.

———. (2004). "The Cultural Logic of Media Convergence." *International Journal of Cultural Studies, 7*(1). DOI: 10.1177/1367877904040603.

Jenkins, Henry, Sam Ford, and Joshua Green. (2013). *Spreadable Media: Creating Value and Meaning in a Networked Culture*. New York: New York University.

Johnson, Derek. (2013). *Media franchising: Creative license and collaboration in the culture industries*. New York: New York University.

———. (2007). "Fan-tagonism: Factions, Institutions, and Constituitive Hegemonies of Fandom." In *Fandom: Identities and Communities in a Mediated World*, 285–300. Eds. Jonathan Gray, Cornel Sandvoss, and C. Lee Harrington. New York: New York University Press.

Jones, Sara Gwenllian. (2003) "Web Wars: Resistance, Online Fandom and Studio Censorship." In *Quality Popular Television*. Eds. Mark Jancovich and James Lyons. London: British Film Institute.

Lacob, Jace. (2011) "*Game of Thrones* for Dummies." *Daily Beast* [Web site] Retrieved from http://www.thedailybeast.com/articles/2011/04/13/game-of-thrones-for-dummies.html.

Lemke, Jay. (2009). "Multimodal Genres and Transmedia Traversals: Social Semiotics and the Political Economy of the Sign," *Semiotica*, 1(4), 283–97. DOI 10.1515/SEMI.2009.012.

Lyons, Margaret. (2012). "*Game of Thrones*' kid-actor problem." *Vulture* [Web site]. Retrieved from http://www.vulture.com/2012/04/game-of-thrones-kid-actor-problem.html.

Marshall, P. David. (2002). "The New Intertextual Commodity." In *The New Media Book*, Ed. Dan Harries. London: British Film Institute.

Martin, Denise. (2014). Matthew Weiner talks about *Mad Men*'s mid-season finale, *2001*, and why Joan is so mad at Don." *Vulture* [Web site]. Retrieved from http://www.vulture.com/2014/05/matthew-weiner-interview-mad-men-mid-season-7-finale-joan-hates-don-bert-cooper-dance.html?mid=facebook_vulture.

Martin, George R. R. (1996). *A Game of Thrones*. New York: Bantam.

McNamara, Mary. (2011). "HBO, You're Busted." *Los Angeles Times* [Web site]. Retrieved from http://articles.latimes.com/2011/jul/03/entertainment/la-ca-hbo-breasts-20110703.

McNutt, Myles. (2011). "*Game of Thrones*—'You Win or You Die!'" *Cultural Learnings* Online blog. Retrieved from http://cultural-learnings.com/2011/05/29/game-of-thrones-you-win-or-you-die/.

Mosemack, Jerry. (2014). "Get Familiar with *Game of Thrones*' Family Trees." *USAToday* [Web site]. Retrieved from http://www.usatoday.com/story/news/nation-now/2014/04/13/game-of-thrones-family-trees/7399299/.

Murray, Simone. (2004). "'Celebrating the Story the Way It Is': Cultural Studies, Corporate Media, and the Contested Utility of Fandom," *Continuum: Journal of Media & Cultural Studies*, 18(1), 7–25. DOI: 10.1080/1030431032000180978.

Pallotta, Frank. (2014). "A '*Game of Thrones*' actress may be giving away huge spoilers on her Instagram." *Business Insider* [Web site]. Retrieved from http://www.businessinsider.com/game-of-thrones-lena-headey-instagram-spoilers-2014-6.

Pantozzi, Jill, Rebecca Pahle, and Victoria McNally. (2014). "A frank discussion of *Game of Thrones*' rape scene and its epidemic of sexual violence. *TheMarySue.com* [Web site]. Retreived from http://www.themarysue.com/game-of-thrones-rape-discussion/.

Phillips, Kendall R. (2005). *Projected Fears: Horror Films and American Culture*. Westport, CT: Praeger.

Plumb, Ali. (2012). "Neil Marshal *Game of Thrones* podcast." *Empire* [Web site]. Retrieved from http://www.empireonline.com/news/story.asp?NID=34164.

Polone, Gavin. (2012). "Polone: All TV should be more like *Game of Thrones*." *Vulture* [Web site]. Retrieved from http://www.vulture.com/2012/04/game-of-thrones-best-show-on-tv-polone.html.

Richwine, Lisa. (2012). Netflix, HBO to battle for Nordic viewers. *Reuters*. Retrieved from http://www.reuters.com/article/2012/08/15/us-netflixinc-scandinavia-idUSBRE87E0JL20120815.

Robinson, Joanna. (2014). "*Game of Thrones* actress Lena Headey cheekily refuses to apologize for Instagram 'spoilers.'" *Vanity Fair* [Web site]. Retrieved from http://www.vanityfair.com/vf-hollywood/game-of-thrones-lena-headey-wont-apologize-for-spoilers.

Rose, Frank. (2012). *The Art of Immersion: How the Digital Generation is Remaking Hollywood, Madison Avenue, and the Way We Tell Stories*. New York: Norton, 2012.

Ryan, Maureen. (2012). "*Game of Thrones* third season: How many episodes will there be?" *Huffington Post—TV* [Web site]. Retrieved from http://www.huffingtonpost.com/2012/04/11/game-of-thrones-third-season_n_1416386.html.

Seitz, Matthew Zoller. (2012). "TV review: *Game of Thrones* season two is artful and adult." *Vulture* [Web site]. Retrieved from http://www.vulture.com/2012/03/tv-review-game-of-thrones-season-two-review.html.

———. (2011). "In defense of HBO's 'unnecessary' nudity." *Salon* [Web site]. Retrieved from http://www.salon.com/2011/07/06/game_of_thrones_sex_and_nudity/.

Sharma, Amol. (2014). "HBO Weighs More Web-TV Services Overseas: The Time-Warner Unit is Exploring Countries to Expand Its Web-TV Model." *Wall Street Journal*. Retrieved from http://online.wsj.com/articles/hbo-weighs-more-web-tv-services-overseas-1407106582.

Toomer, Jessica. (2013). "This *Game of Thrones* chart is all you need to get caught up before the season finale." *Huffpost TV* [Web site]. Retrieved from http://www.huffingtonpost.com/2014/06/10/game-of-thrones-chart-_n_5480101.html.

Vineyard, Jennifer. (2014). "*Game of Thrones*' showrunner D. B. Weiss on how the show will handle its aging child stars." *Vulture* [Web site]. Retrieved from http://www.vulture.com/2014/03/db-weiss-on-game-of-thrones-kids-aging.html.

WinterIsComing.net. [Web site]. http://winteriscoming.net/about/.

Wyatt, Daisy. (2014). "George R. R. Martin defends sexual violence in Game of Thrones." *Independent* [Web site]. Retrieved from http://www.independent.co.uk/arts-entertainment/tv/news/george-rr-martin-defends-sexual-violence-in-game-of-thrones-9342569.html.

Zukin, Sharon, and Jennifer Smith Maguire. (2004). "Consumers and Consumption," *Annual Review of Sociology*, 30, 173–97. DOI: 10.1146/annurev.soc.30.012703.110553.

Chapter Eight

Be Original

Examining Fan Comments on A&E's Duck Dynasty Facebook Page After the Robertson Suspension[1]

Michel M. Haigh

In April 2013, *Duck Dynasty* became *A&E*'s most successful program. The fourth season premiere was the "most watched nonfiction telecast to ever air on cable television."[2] However, by June 2014, *Duck Dynasty*'s sixth season premiered to its lowest ratings. It drew 4.6 million viewers, which was down 46 percent from the season 5 premiere.[3] Kissell states "docudramas in general tend to have a relatively short shelf life. The series seemed to lose steam in the back half of its fourth season, and its ratings decline accelerated post-controversy."[4]

The controversy Kissell was referring to occurred in December 2013 when Phil Robertson, the family's patriarch, was featured in a *GQ* profile. In the 4,000 word-article written by Drew Magary, Robertson states "We're Bible-thumpers who just happened to end up on television. . . . You put in your article that the Robertson family really believes strong that if the human race loved each other and they loved God, we would just be better off."[5] However, when asked, "What, in your mind, is sinful?" Robertson responded, "Start with homosexual behavior and just morph out from there."[6] This statement and additional comments Robertson made about African Americans led to Robertson being suspended by *A&E* on December 18, 2013.

More than 250,000 fans signed a petition to express dissent against *A&E* and Phil's suspension. Nine days after suspending Robertson, on December 27, *A&E* announced he would return to taping. Nine of the ten episodes of season 5 had already been filmed, and *A&E* was not going to edit Robertson out of the episodes.[7]

The current study coded fan comments posted to *A&E*'s *Duck Dynasty* Facebook page in December of 2013. It examined the comments made when Robertson was suspended as well as the comments made after he was reinstated by *A&E*. What happens when a network acts contrary to their slogan "Be Original?" Cavalcante points out fan posts had an "us" versus "them" divide when examining the *Here Comes Honey Boo Boo* Facebook page. The "us" fans often posted comments that were positive and celebrated the family, and the "them" (or the haters) posted negative and derogatory comments. The "us" fans would then comment on the authenticity of the family to counter the "haters."[8] The "us" versus "them" scenario could play out on the *Duck Dynasty* Facebook page. The "us" could be fans of the show that post messages of support for Phil. The "them" could be dissidents motivated to express their unhappiness with the show, those expressing dissent with *A&E* for suspending Phil, or those expressing dissent with *A&E* after Phil was reinstated. The following study examined Facebook posts for an "us" versus "them" theme to see if fans felt a connection with the Robertson family.

LITERATURE REVIEW

Disposition Theory

Duck Dynasty blurs the lines between scripted and reality television. *Duck Dynasty* episodes are formatted much like a sitcom because each follows two story lines. *A&E*'s executive vice president for programming said the show quickly found its "natural voice."[9] "What you see on television is what you get in real life," said *New York Times* reporter Neil Genzlinger who spent time with the family.[10] Viewers of reality TV say they enjoy the genre because it provides authenticity, intimacy, and focuses on "real" people.[11] Barton found one of the main reasons people watched reality television was for personal utility. Personal utility includes viewers watching reality television because the personalities in the show make the viewers less lonely. The viewers relax and forget about their own problems when watching the show, and viewers watch because the show is different than other shows on television.[12] Another reason viewers watch reality television is for enjoyment. Enjoyment "the pleasure derived from consuming media entertainment," is an emotional response to the content, as well as a judgment of "interaction of user, content, and environmental variables."[13] Enjoyment increases when characters the viewers like experience positive outcomes in the show, or when characters the viewers dislike experience negative outcomes.[14]

Disposition Theory describes the process of forming attachments with characters. "Audiences develop emotional alliances with characters they encounter in fictional media . . . audiences make moral judgments about the characters in movies, television shows, and the like."[15] The moral judgments formed are the basis of viewers having an emotional reaction to a character. Granted, the viewer will only show positive affect toward characters behaving in harmony with the viewer's moral code.[16] Raney stated disposition-based theories are concerned with viewers' emotional response to media content. Viewers' emotional response will depend on their level of empathy and their feelings toward the characters.[17] Raney suggests viewers form relationships with characters on an affective continuum. The continuum extends from extremely positive to indifference.[18]

Weber, Tamborini, Lee, and Stipp state individuals' dispositions toward characters are based on the characters' actions. Viewers favor characters that act in an appropriate manner and dislike characters that act inappropriately. Viewers enjoy narratives where good characters are rewarded and bad characters are punished. These two variables (character morality and behavioral outcomes) shape viewers' enjoyment levels.[19] Raney suggests viewers make judgments about characters without scrutinizing their behavior by relying on the schema, or mental networks of information they have formed to story lines they have been exposed to in the past.[20] For example, viewers have schema for the characters in a romantic comedy. There will be the heroine, "Mr. Wrong," and "Mr. Right." Viewers have expectations for the characters' actions based on watching romantic comedies in the past.

The emotional attachments viewers are forming with characters in scripted television can also be formed with the TV personalities presented in reality television. Another way to look at this emotional attachment is by examining it as a relationship the viewer has with a character or a parasocial relationship.

Parasocial Relationships

Horton and Wohl argued there is an emotional impact media personas have on viewers' lives.[21] "People come to feel intimately related to performers whom they have never met, but with whom they feel a special and close relationship."[22] These relationships can be similar to social relationships. Research found viewers were attracted to characters because the traits they exhibited were similar to traits that attracted people in social relationships.[23] The intensity of the parasocial relationship was related to adult attachment patterns, and similar to the intensity of romantic and close social relationships.[24]

Perse and Rubin found as parasocial relationships develop, viewers feel more familiar with the performers and their uncertainty about the characters' is reduced.[25] A parasocial relationship occurs over time through repeated interactions. "Performers create a sense of history to relationships and open up the possibility for "inside jokes" that make repeated viewers/listeners feel that they have unique knowledge about the performer and a shared history."[26]

Cohen differentiates between fandom and parasocial relationships. He states "fandom is based on keeping one's idols at a psychological distance that allows being awed by them, and parasocial relationships are predicated on developing more intimate feelings toward a performer; identification involves an intense, if temporary, merging of the self with a media character."[27] However, in the case of reality television, one could argue the line between fans and parasocial relationships is blurred. Reality television allows the audience to form parasocial relationships and identify with the personalities. If audience members can have empathy toward fictional characters than they should have empathy toward reality television personalities.

Research shows parasocial relationships occur between reality television viewers and the personalities on the show. The comments on social media sometimes mimic comments friends would make with other friends. For example, when examining the Facebook page for *Here Comes Honey Boo Boo*, fans often posted about the family's realness and authenticity. However, Cavalcante points out fan posts had an "us" versus "them" divide. The "us" fans often posted comments that were positive and celebrated the family, and the "them" (or the haters) posted negative and derogatory comments. The "us" fans would then comment on the authenticity of the family to counter the "haters."[28]

In a study of the official message board affiliated with *The Real Housewives* series, Cox also found an "us" versus "them" divide."[29] However, the distinction was "between 'us' as working-class viewers and 'them' as upper-class housewives, such distinctions are not always proffered in a productive way."[30]

Duck Dynasty is another show that can provide glimpses into an "us" versus "them" divide. Examining the types of posts on the Facebook page during the suspension will provide some insight into the fans' perspectives. The "them" in this case could be *A&E* or the "them" could be supporters of *A&E*'s decision to suspend Phil. The "us" would be the fan supporters of the Robertson family and Phil.

For example, the following post was made on the *Ducky Dynasty* Facebook page in December of 2013: "If it involves money, a person can get away with anything in America. *Duck Dynasty* is *A&E*'s golden goose . . . *A&E* cancel *DD*!" Another "them" post stated, "I will not watch anything on this network until the bearded redneck is permanently suspended." These are clearly "them" posts meaning the fans posting sided with *A&E*.

Posts that indicated "us" showed support for Phil. Examples of these posts included: "About time y'all put Phil back on. He should have never been suspended. I guess my boycott of *A&E* is over now. . . . Go Phil!"

Another fan pointed out the "us" versus "them" comments appearing on the *Duck Dynasty* page, "What I don't get is why [are] ALL the *Duck Dynasty* haters wasting their time & energy on a *Duck Dynasty* fan page?"

Duck Dynasty also provides a unique opportunity to examine another type of conflict Hernandez identified.[31] Hernandez discusses the Southern masculinity portrayed in *Duck Dynasty*. She states "representations of masculinity in *Duck Dynasty* have larger implications for how the American psyche understands Southern femininities, masculinities, and gender roles and relations."[32]

Hernandez identified different forms of Southern masculinity that can sometimes be at odds with each other. For example one type is the Southern Christian gentleman, the masculine marital ideal, the hunter, or the white-trash redneck.[33] The current study examines the Southern Christian gentleman. Craig Thompson Friend stated the Christian gentleman and the masculine marital ideal are both based on the values of honor, mastery, and dedication to family.[34] However, the Southern Christian gentleman "performed his masculinity via religious and civic dedication."[35] Some examples can be found in the fan comments on the *Duck Dynasty* Facebook page. For example, one supporter commented, "One of the reasons that I watch this show is for the family values and the family beliefs in the Lord our God. This family is the best and the show is the best. I don't miss an episode but I will not watch any of the *A&E* programs if Phil or *Duck Dynasty* is taken off the air." Another supporter commented, "So proud of you all for sticking together like a real and true family, way to go Phil, God bless." Another post said, "Their family is faith based and funny as heck . . . thank you *A&E*." However, a dissenter posted, "Woohoo! The multi-millionaires pretending to be camo-wearing, hillbillies for their fake reality show will be back. It's almost like the suspension was a fake stunt to stir up their redneck fans."

The following research questions were posed to understand how fans responded when *A&E* suspended Robertson from filming. The study wanted to examine the "us" versus "them" theme apparent in the posts, as well as identify the different ways the viewers related to the Southern masculinity depicted in the show.

RQ1: What was the most common way fans showed support for Robertson?

RQ2: What was the most common way Southern Christian masculinity was discussed via fans' posts?

RQ3: How often did fans post about getting back at "them" (*A&E*) for placing Robertson on suspension?

RQ4: Did the topic of the posts change before and after Robertson was reinstated?

Method

Facebook posts from the *Duck Dynasty* page were collected for the first twenty-four hours after the suspension was announced (December 18, 2013, n = 3,365). Additional data was collected from the page when Robertson was reinstated (December 27, 2013, n = 422). *A&E* maintains the *Duck Dynasty* page. The unit of analysis was each Facebook post made to the *A&E Duck Dynasty* Facebook page (N = 3,787). The unit of analysis was the original post, not the comments made to the post. *A&E* did not respond to the fan posts on the page.

Demographics

The number of "likes" (M = 1.78, SD = 1.84), shares (M = 0.03, SD = 0.12), and comments (M = 0.80, SD = 2.26) varied for each post. Very few posts included an external link (3.4 percent), or a photo of Robertson (7.7 percent). Usually the photo of Robertson was a "meme" (7.0 percent). Males (49.8 percent) tended to post more frequently than females (42.9 percent).

Coder Training

Two undergraduate students enrolled at a Mid-Atlantic university conducted the content analysis. A written coding instrument was developed to code the sample. Coding norms were established during a supervised training session. Seven percent (7 percent) of the sample was coded during the training phase. Coders established a high degree of standardization during the training phase, resulting in effective inter-coder reliabilities between 0.99 and 1.00 when employing Cohen's Kappa.

Categories Coded

The posts were either marked "present" or "absent" if the post included information about the topic. See Table 8.1 for a complete list of categories and the frequency each occurred. Posts could fit more than one category based on each thought unit.

Support for Phil

Posts discussing freedom of speech were coded using the following categories: First Amendment rights, political correctness, support for Phil, and comments to "bring back Phil." Some example posts from the *Duck Dynasty* Facebook page that fit this category included: "Boo to *A&E*. Everyone is entitled to their opinions and beliefs. I love *Duck Dynasty* and love living in America, where we have freedom of speech." Another fan posted, "Freedom of speech and freedom to worship is what America was founded on. If you don't agree with these standards, then don't watch the program." And finally, another fan stated, "Welcome to the land of political correctness, where the freedom of speech is reserved for only liberals and persecution is reserved for Christians and conservatives!"

Southern Christian Values

Posts discussing topics related to Southern family values were coded using the following categories: "Praying for you/your family," "Bible verses," referencing "Lord/God/Jesus/Holy Spirit, etc.," "Christians," "sin," "forgiveness," "Robertson family members are role models," and GLBT (Gay, Lesbian, Bisexual, Transgender) statements. For example, one fan posted "I would like to give the *Duck Dynasty* my support and prayers. Phil was just standing up for his Christian beliefs. I also believe homosexuality is a sin. . . . I will keep the Robertson family in my prayers. This will pass." Another fan posted, "The Robertson family is on point with morals and family values. I love to watch their show and it lightens my heart and soul." Another example stated, "I am a Christian and I love the show *Duck Dynasty*. I support Phil Robertson and everything that he said to the *GQ*. Phil we are praying for you and your family."

A&E

Posts focused on the following topics: "Boycotting *A&E*," and *A&E*'s motto "Be Original." Additional comments in this category included: "unliking" the *A&E Duck Dynasty* Facebook page, *A&E* needing to apologize to Phil, and "shame on *A&E*." One fan posted, "*A&E*'s motto 'Be Original' apparently doesn't apply to all." Another stated, "Love the cover photo!!! Should come with asterisk. . . . Cause *A&E* wants you to be original *but just don't let anyone know about it. Especially *GQ*. Boo *A&E*. Boooo."

RESULTS

Descriptive statistics and Chi-Square analyses were used to answer the research questions. Frequencies were used to answer Research Questions 1–3. Research Question 1 asked how fans showed support for Phil. Categories coded to examine this included categories discussing the First Amendment rights (26.3 percent), political correctness (12.4 percent), support for Phil (14.1 percent), and comments stating "bring back Phil" (16.6 percent).

Research Question 2 asked if fans would post about Southern Christian masculinity. Posts discussing topics related to Southern family values were coded using the following categories: "Praying for you/your family" (11.6 percent), "Bible verses" (12.0 percent), referencing "Lord/God/Jesus/Holy

Table 8.1.

Coding Categories, Cohen's Kappa, Percent Present	Cohen's Kappa	Percent Present
Boycotting A&E	0.99	32.2%
"I support Phil"	1.0	14.1%
First Amendment rights	0.99	26.3%
"Praying for you/your family"	0.99	11.6%
Reference specific Bible verses	1.0	12.0%
Reference "Lord/God/Jesus/Holy Spirit/etc."	0.99	16.9%
Reference Christians	1.00	17.9%
Reference "sin"	1.00	12.7%
Forgiveness	0.99	11.2%
Political correctness	1.00	12.4%
Robertson family members are role models	0.99	11.7%
A&E's motto: "Be Original"	0.99	11.6%
Quote or reference the GQ article	1.00	11.9%
GLBT statements	0.99	16.7%
A&E needs to apologize to the Robertson family	0.99	13.0%
Unlike the Duck Dynasty Facebook page	0.99	15.2%
"Shame on A&E"	1.00	16.6%
"Bring back Phil"	1.00	16.6%

Spirit, etc." (16.9 percent), "Christians" (17.9 percent), "sin" (12.7 percent), "forgiveness" (11.2 percent), "Robertson family members are role models" (11.6 percent), and GLBT statements (16.7 percent).

Research Question 3 asked how often fans posted about getting back at "them" (*A&E*). Posts discussing the channel discussed the following topics: "Boycotting *A&E*" (32.2 percent), *A&E*'s motto "Be Original" (11.6 percent), *A&E* needing to apologize to the Robertson family (13.0 percent), fans "unliking" the *Duck Dynasty* page because it was run by *A&E* (15.2 percent), as well as fans stating "shame on *A&E*" (16.6 percent).

Research Question 4 asked if the topic of the post would change before and after Robertson's reinstatement. Chi-Square analyses found significant differences for all the coded categories including: First Amendment rights χ^2 (1, N = 3,787) = 87.84, $p < 0.001$; political correctness χ^2 (1, N = 3,787) = 62.41, $p < .001$; support for Phil χ^2 (1, N = 3,787) = 45.44, $p < 0.001$; "Praying for you/your family" χ^2 (1, N = 3,787) = 46.03, $p < 0.001$; Bible verses χ^2 (1, N = 3,787) = 45.85, $p < 0.001$; referencing "Lord/God/Jesus/

Holy Spirit" χ^2 (1, N = 3,787) = 19.95, p < 0.001; "Christians" χ^2 (1, N = 3,787) = 53.68, p < 0.001; "sin" χ^2 (1, N = 3,787) = 63.88, p < 0.001; "forgiveness" χ^2 (1, N = 3,787) = 57.52, p < 0.001; "Robertson family members are role models" χ^2 (1, N = 3,787) = 50.66, p < 0.001; GLBT statements" χ^2 (1, N = 3,787) = 29.94, p < 0.001; "Boycotting *A&E*" χ^2 (1, N = 3,787) = 167.53 p < 0.001; and *A&E*'s motto 'Be Original'" χ^2 (1, *N = 3,787*) = 59.91, p < 0.001.

Discussion

This study examined the fans' reaction after *A&E* announced Robertson was going to be suspended for comments made in a *GQ* article. It wanted to examine if fans had parasocial relationships with Robertson, how they showed support, and if an "us" versus "them" theme was present on the fan Facebook page.

Research Question 1 examined the most common way fans showed support on the *Duck Dynasty* Facebook page. The most common way fans supported Robertson was by discussing his First Amendment rights, followed by statements discussing "I support Phil," or statements about Robertson not adhering to political correctness and being honest in the *GQ* article.

Research Question 2 examined the most common way Southern Christian masculinity was discussed via fans' posts. The most common way fans discussed Southern Christian masculinity was by discussing Robertson being Christian. The next most common topic in this type of post was "Lord/God/Jesus/Holy Spirit," followed by posts discussing GLBT issues, "sin," "Bible verses," "Robertson family members are role models," "Praying for you/your family," and mentions of "forgiveness." These types of posts indicate the intensity of the parasocial relationship was related to adult attachment patterns similar to those in social relationships.[36] Fans were able to identify with Robertson and showed empathy toward him.[37] The different forms of Southern masculinity identified in the posts support Craig Thompson Friend's concept of the masculine marital ideal for honor, mastery, and dedication to family.[38] And, the Southern Christian gentleman "performed his masculinity via religious and civic dedication."[39]

The pattern of posts supports the pattern found when examining the *Here Comes Honey Boo Boo* Facebook page. On both pages, fans often posted about the family's realness and authenticity.

And there was an "us" versus "them" divide. The "us" fans often posted comments that were positive and celebrated the family, and the "them" (or the haters) posted negative and derogatory comments. The "us" fans would then comment on the authenticity of the family to counter the "haters" and in this case, the "them" could also be *A&E*.[40]

Research Question 3 examined how often fans posted about getting back at "them" (*A&E*). Actually, the most common post fans made was boycotting *A&E* until Robertson was reinstated. Fans also discussed *A&E* being hypocrites by acting contrary to the channel's slogan "Be Original." Fans thought *A&E* should apologize to Robertson for suspending him, and posted comments shaming *A&E*. The most common ways fans were going to get back at *A&E* and show support for Robertson was to stop watching the channel and "unfriend"/"unlike" the *Duck Dynasty* page. Posts about *A&E* acting contrary to its slogan were almost as common as the posts about praying for the family, forgiveness, and the family being role models.

Research Question 4 asked if the topic of the post would change before and after Robertson was reinstated. The topics did change before and after Robertson was reinstated. Additionally, more people posted when *A&E* first announced the suspension than after Robertson was reinstated.

Limitations and Future Directions

All studies have limitations. One of the biggest limitations for this study was the timing of the *A&E* reinstatement of Robertson. December 27 coincides with the winter holidays and family time, so the number of people posting in support of *A&E*'s decision to reinstate Robertson may have been impacted. Additional data could have been collected at a later period of time to see if more fans posted after the winter holidays.

Several things became apparent while the research was being conducted. The organization was not likely to respond to or delete the negative fan posts. As Haigh and Wigley point out (in this volume), when Facebook updates how the page looks, none of the posts from the fans appear on the main page. The only way an observer would know the organization was going through a crisis is by clicking on the "posts by others" section of the page.

The current findings support the idea that when a celebrity or persona has a crisis, particularly one that threatens an established parasocial relationship, those invested in the relationship will use social media to voice their support and take action against the "haters" or parties that threaten the relationship. The results also support the idea that an audience identifies with reality TV personas just as they do fictional characters.[41] This was evident in the number of fan posts that included sentiments of empathy.[42]

Future research should continue to examine how fans use social media to further their parasocial relationships with reality TV personas. There is still more to learn about the "us" versus "them" interactions on these types of pages, what motivates fans to take the time to post comments on the fan pages, as well as what motivates them to show support for celebrities in times of crisis. There is still limited research applying Disposition Theory to reality TV such as *Duck Dynasty*. Perhaps some reality TV fans don't form attachments with reality stars; it is also worth figuring out what type of personality one needs to succeed in reality TV.

It is unclear how the fans' support during times of crisis impacts the decisions made by those in power. For example, did the petition started by the fans of *Duck Dynasty* and the posts on Facebook influence *A&E* to reinstate Phil? In the case of Paula Deen, fans were willing to boycott products that had dropped Deen as a spokeswoman; the same pattern was starting to occur during the Robertson suspension. Many fans posted to keep a lookout for products manufactured by the Robertson family and not those produced in connection with *A&E*. How much does this impact sponsors' return on investments? The Deen and Robertson scandals are similar in some ways; therefore, it would be interesting to compare the two cases. Both involved First Amendment issues and both included accusations of prejudice and racism. One caveat is that Deen's use of the N-word happened more than thirty years earlier, while Robertson's discussion of homosexuality occurred in real time.

In conclusion, it appears fans tend to rally around reality TV celebrities in crisis. They are motivated to post support on social media as well as start petitions. They are willing to boycott products and channels in support of the reality TV star in crisis.

NOTES

1. Thanks to: Haley Burnside and Yixuan Li, undergraduates in the college, for their help in coding the Facebook posts.
2. P. Nash Jenkins. "How Low Will Duck Dynasty's Ratings Go?" *Time*, June 13, 2014. *Time*, accessed September 9, 2014, http://time.com/2868620/duck-dynasty-ratings-season-6-robertson/.
3. Rick Kissell. "A&E's 'Duck Dynasty' Opens Season 6 with More Declines." *Variety*, June 12, 2014, accessed September 9, 2014, http://variety.com/2014/tv/ratings/aes-duck-dynasty-opens-season-6-with-more-declines-1201219469/.
4. Kissell, "A&E's 'Duck Dynasty' Opens Season 6 with More Declines."
5. Drew Magary. "What the Duck," *GQ*, January 2014, accessed Sept. 9, 2014, http://www.gq.com/entertainment/television/201401/duck-dynasty-phil-robertson.
6. Magary, "What the Duck."
7. Tim Kenneally. "What Does Phil Robertson's 'Duck Dynasty' Suspension Really Mean for Season 5? Not Much." December 20, 2013, accessed September 9, 2014, http://www.thewrap.com/duck-dynasty-phil-robertson-suspension-ae/.
8. Andre Cavalcante, "You Better 'Reneckognize'!: Deploying the Discourses of Realness, Social Defiance, and Happiness to Defend *Here Comes Honey Boo Boo* on Facebook," in *Reality Television: Oddities of Culture*, ed. Alison F. Slade et al. (Lanham, MD: Lexington Books, 2014), 39–58.
9. Neil Genglinger. "Lured in by a Family Just Being Itself on TV." *New York Times*, October 7, 2012, accessed September 9, 2014, http://www.nytimes.com/2012/10/08/arts/television/duck-dynasty-lures-a-growing-audience-on-ae.html?pagewanted=all&_r=0.
10. Genglinger, "Lured in by a Family Just Being Itself on TV."
11. Cavalcante. "You Better 'Reneckognize'!: Deploying the Discourses of Realness, Social Defiance, and Happiness to Defend *Here Comes Honey Boo Boo* on Facebook."

12. Kristin Barton. "Reality Television Programming and Diverging Gratifications: The Influence of Content on Gratifications Obtained," *Journal of Broadcasting & Electronic Media* 53, no. 3 (2009): 460–76.

13. Arthur Raney. "The Effects of Viewing Televised Sports," in *The Sage Handbook of Media Processes and Effects*, ed. Robin L. Nabi and Mary Beth Oliver (Thousand Oaks, CA: Sage, 2009), 441.

14. Art Raney. "The Psychology of Disposition-Based Theories of Media Enjoyment," in *Psychology of Entertainment*, ed. Jennings Bryant and Peter Vorderer (Mahwah, NJ: Lawrence Erlbaum Associates, 2006), 137–50.

15. Nancy Rhodes and James C. Hamilton. "Attribution and Entertainment: It's Not Who Dunnit, It's Why," in *Psychology of Entertainment*, ed. Jennings Bryant and Peter Vorderer (Mahwah, NJ: Lawrence Erlbaum Associates, 2006), 125.

16. Rhodes and Hamilton, "Attribution and Entertainment: It's Not Who Dunnit, It's Why."

17. Raney, "The Psychology of Disposition-Based Theories of Media Enjoyment."

18. Art Raney, "Expanding Disposition Theory: Reconsidering Character Liking, Moral Evaluations, and Enjoyment." *Communication Theory* 14, no. 4 (2004): 348–69.

19. Rene Weber, Ron Tamborini, Hye Eun Lee, and Horst Stipp, "Soap Opera Exposure and Enjoyment: A Longitudinal Test of Disposition Theory." *Media Psychology* 11 (2008): 462–87.

20. Raney, "Expanding Disposition Theory: Reconsidering Character Liking, Moral Evaluations, and Enjoyment."

21. Donald Horton, R. Richard Wohl. "Mass Communication and Parasocial Interaction." *Psychiatry* 19, no. 3 (1956): 215–29.

22. Jonathan Cohen. "Parasocial Interaction and Identification" in *The Sage Handbook of Media Processes and Effects*, ed. Robin L. Nabi and Mary Beth Oliver (Thousand Oaks, CA: Sage, 2009), 225.

23. Cynthia Hoffner. "Children's Wishful Identification and Parasocial Interactions with Favorite Television Characters." *Journal of Broadcasting & Electronic Media* 40, no. 3 (1996): 389–402. Cynthia Hoffner and Joanne Cantor, "Perceiving and Responding to Mass Media Characters," in *Responding to the Screen: Reception and Reaction Processes*, ed. Jennings Bryant and Dolf Zillmann (Hillsdale, NJ: Erlbaum, 1991): 63–103. John R. Turner, "Interpersonal and Psychological Predictors of Parasocial Interaction with Different Television Performers." *Communication Quarterly* 41, no. 4 (1993): 443–53.

24. Jonathan Cohen. "Parasocial Relations and Romantic Attraction: Gender and Dating Status Differences." *Journal of Broadcasting & Electronic Media* 41, no. 4 (1997): 516–29. Trim Cole and Laura Leets. "Attachment Styles and Intimate Television Viewing: Insecurely Forming Relationships in a Parasocial Way," *Journal of Social and Personal Relationships* 16, no. 4 (1999): 495–511.

25. Elizabeth M. Perse and Rebecca Rubin. "Attribution in Social and Parasocial Relationships." *Communication Research* 16, no. 1 (1989): 59–77.

26. Cohen, "Parasocial Interaction and Identification," 227.

27. Cohen, "Parasocial Interaction and Identification," 229.

28. Cavalcante. "You Better 'Reneckognize'!: Deploying the Discourses of Realness, Social Defiance, and Happiness to Defend *Here Comes Honey Boo Boo* on Facebook."

29. Nicole B. Cox. "Bravo's '*The Real Houswives*': Living the (Capitalist) American Dream?" in *Reality Television: Oddities of Culture*, ed. Alison F. Slade, et al. (Lanham, MD: Lexington Books, 2014), 77–99.

30. Cox, "Bravo's '*The Real Houswives*': Living the (Capitalist) American Dream?" 87.

31. Leandra H. Hernandez, "'I Was Born This Way': The Performance and Production of Southern Masculinity in A&E's *Duck Dynasty*," in *Reality Television: Oddities of Culture*, ed. Alison F. Slade et al. (Lanham, MD: Lexington Books, 2014): 21–37.

32. Hernandez, "'I Was Born This Way': The Performance and Production of Southern Masculinity in A&E's *Duck Dynasty*," 29.

33. Hernandez, "'I Was Born This Way': The Performance and Production of Southern Masculinity in A&E's *Duck Dynasty*."

34. Craig Thompson Friend, "From Southern Manhood to Southern Masculinities: An Introduction," in *Southern Masculinity: Perspectives on Manhood in the South Since Reconstruction,* ed. Craig Thompson Friend (Athens: The University of Georgia Press, 2009), vii–xxvi.

35. Hernandez, "'I Was Born This Way': The Performance and Production of Southern Masculinity in A&E's *Duck Dynasty*," 24.

36. Cohen, "Parasocial Relations and Romantic Attraction: Gender and Dating Status Differences." Cole and Leets, "Attachment Styles and Intimate Television Viewing: Insecurely Forming Relationships in a Parasocial Way."

37. Cohen, "Parasocial Interaction and Identification," 230.

38. Craig Thompson Friend, "From Southern Manhood to Southern Masculinities: An Introduction," in *Southern Masculinity: Perspectives on Manhood in the South Since Reconstruction,* ed. Craig Thompson Friend (Athens: The University of Georgia Press, 2009): vii–xxvi.

39. Hernandez, "'I Was Born This Way': The Performance and Production of Southern Masculinity in A&E's *Duck Dynasty*," 24.

40. Cavalcante, "You Better 'Reneckognize'!: Deploying the Discourses of Realness, Social Defiance, and Happiness to Defend *Here Comes Honey Boo Boo* on Facebook."

41. Cohen, "Audience Identification with Media Characters."

42. Cohen, "Parasocial Interaction and Identification."

REFERENCES

Barton, Kristin. "Reality Television Programming and Diverging Gratifications: The Influence of Content on Gratifications Obtained." *Journal of Broadcasting & Electronic Media* 53, no. 3(2009): 460–76.

Cavalcante, Andrew. "You Better 'Reneckognize'!: Deploying the Discourses of Realness, Social Defiance, and Happiness to Defend Here Comes Honey Boo Boo on Facebook." In *Reality Television: Oddities of Culture,* edited by Alison F. Slade, Amber J. Narro, and Burton P. Buchanan, 39–58. Lanham, MD: Lexington Books, 2014.

Cohen, Jonathan. "Parasocial Interaction and Identification." In *The Sage Handbook of Media Processes and Effects,* edited by Robin L. Nabi and Mary Beth Oliver, 223–36. Thousand Oaks, CA: Sage, 2009.

———. "Parasocial Relations and Romantic Attraction: Gender and Dating Status Differences." *Journal of Broadcasting & Electronic Media* 41, no. 4 (1997): 516–29.

Cole, Trim and Laura Leets. "Attachment Styles and Intimate Television Viewing: Insecurely Forming Relationships in a Parasocial Way." *Journal of Social and Personal Relationships* 16, no. 4 (1999). 495–511.

Cox, Nicole B. "Bravo's 'The Real Housewives': Living the (Capitalist) American Dream." In *Reality Television: Oddities of Culture,* edited by Alison F. Slade, Amber J. Narro, and Burton P. Buchanan, 77–99. Lanham, MD: Lexington Books, 2014.

Eyal, Keren, and Jonathon Cohen. "When Good Friends Say Goodbye: A Parasocial Breakup Study." *Journal of Broadcasting & Electronic Media* 50, no. 3 (2006): 502–23.

Genglinger, Neil. "Lured in by a Family Just Being Itself on TV." *New York Times,* October 7, 2012, accessed Sept. 9, 2014, http://www.nytimes.com/2012/10/08/arts/television/duck-dynasty-lures-a-growing-audience-on-ae.html?pagewanted=all&_r=0.

Hernandez, Leandra, H. "'I Was Born This Way': The Performance and Production of Southern Masculinity in A&E's Duck Dynasty." In *Reality Television: Oddities of Culture,* edited by Alison F. Slade, Amber J. Narro, and Burton P. Buchanan, 21–37. Lanham, MD: Lexington Books, 2014.

Hoffner, Cynthia. "Children's Wishful Identification and Parasocial Interactions with Favorite Television Characters." *Journal of Broadcasting & Electronic Media* 40, no. 3 (1996): 389–402.

Hoffner, Cynthia and Joanne Cantor. "Perceiving and Responding to Mass Media Characters." In *Responding to the Screen: Reception and Reaction Processes,* edited by Jennings Bryant and Dolf Zillmann, 63–103. Hillsdale, NJ: Erlbaum, 1991.

Horton, Donald and R. Richard Wohl. "Mass Communication and Parasocial Interaction." *Psychiatry* 19, no. 3 (1956): 215–29.

Jenkins, P. Nash. (2014). How Low Will Duck Dynasty's Ratings Go?" *Time*, June 13, 2014. *Time*, accessed September 9, 2014, http://time.com/2868620/duck-dynasty-ratings-season-6-robertson/.

Kenneally, Tim. "What Does Phil Robertson's 'Duck Dynasty" Suspension Really Mean for Season 5? Not Much." December 20, 2013, accessed September 9, 2014, http://www.thewrap.com/duck-dynasty-phil-robertson-suspension-ae/.

Kissell, Rick. "A&E's 'Duck Dynasty' Opens Season 6 with More Declines." *Variety*, June 12, 2014, accessed September 9, 2014, http://variety.com/2014/tv/ratings/aes-duck-dynasty-opens-season-6-with-more-declines-1201219469/.

Magary, Drew. "What the Duck," *GQ*, January 2014, accessed September 9, 2014, http://www.gq.com/entertainment/television/201401/duck-dynasty-phil-robertson.

Perse, Elizabeth M. and Rebecca Rubin. "Attribution in Social and Parasocial Relationships." *Communication Research* 16, no. 1 (1989): 59–77.

Raney, Arthur. "The Effects of Viewing Televised Sports." In *The Sage Handbook of Media Processes and Effects*, edited by Robin L. Nabi, and Mary Beth Oliver, 439–53. Thousand Oaks, CA: Sage, 2009.

———. "The Psychology of Disposition-Based Theories of Media Enjoyment." In *Psychology of Entertainment*, edited by Jennings Bryant and Peter Vorderer, 137–50. Mahwah, NJ: Lawrence Erlbaum Associates, 2006.

———. "Expanding Disposition Theory: Reconsidering Character Liking, Moral Evaluations, and Enjoyment." *Communication Theory* 14, no. 4(2004): 348–69.

Rhodes, Nancy, and James C. Hamilton. "Attribution and Entertainment: It's Not Who Dunnit, It's Why." In *Psychology of Entertainment*, edited by Jennings Bryant and Peter Vorderer, 119–36. Mahwah, NJ: Lawrence Erlbaum Associates, 2006.

Thompson Friend, Craig. "From Southern Manhood to Southern Masculinities: An Introduction." In *Southern Masculinity: Perspectives on Manhood in the South Since Reconstruction*, edited by Craig Thompson Friend, vii–xxvi. Athens: University of Georgia Press, 2009.

Turner, John R. "Interpersonal and Psychological Predictors of Parasocial Interaction with Different Television Performers." *Communication Quarterly* 41, no. 4 (1993): 443–53.

Weber, Rene, Ron Tamborini, Hye Eun Lee, and Horst Stipp, "Soap Opera Exposure and Enjoyment: A Longitudinal Test of Disposition Theory." *Media Psychology* 11 (2008): 462–87.

Chapter Nine

"The Parents Have the Dream, but the Children Are in the Nightmare"

Digital Interactivity, Toddlers & Tiaras *Viewers, and Social Networking Sites*

Leandra H. Hernandez

Images of sexualized girls pervade today's media landscape, particularly in television shows. With the advent of reality television, television series such as *Dance Moms*, *Toddlers & Tiaras*, and *Little Miss Perfect* are showcased and exemplify the objectification and sexualization of young girls via hypersexualized pageantry; regimented, artificial femininity; and dance contests with gyrating, adult-like choreography. The series *Toddlers & Tiaras*, for example, chronicles the lives and pageantry experiences of young girls across the United States and showcases the maintenance and consumerism that go hand-in-hand with pageantry, including the thousands of dollars associated with tanning, fake teeth, nails, hair extensions, pageant coaches, and other components of the pageantry world.

Although research has explored the ways in which femininity is performed in *Toddlers & Tiaras*,[1] the relationship between pedophilia and *Toddlers & Tiaras*,[2] and arguments in favor of reforming the protections afforded to child participants in *Toddlers & Tiaras*,[3] what is missing is an exploration of the ways in which fans interact with *Toddlers & Tiaras* via social media pages and the ways in which the *Toddlers & Tiaras* show interacts with its fans. *Toddlers & Tiaras* is one of the most popular reality television shows in recent history because of its conflicting and problematic representations of young girls and pageantry,[4] with recent controversies including a mother dressing up her daughter as the Julia Roberts prostitute character from *Pretty Woman*, a mother forcing her daughter to "smoke" a fake cigarette during a costume routine, and another mother dressing her daughter as Dolly Parton with an inflatable bust and rear end pads.[5] Given the controversial nature of this television series and the fact that it has consistently been rated one of the top twenty primetime television series over the last few years,[6] research on this topic can speak to the *other* side of the television-viewer relationship—the way fans interact with the show and the conversations that viewers and nonviewers alike have online about the television series. As media scholar Mark Andrejevic has noted, this "mediated interactivity" relationship between producer, online medium, and fans allows fans to act as instant focus groups and promoters of television shows while simultaneously giving television producers a space to imbue their marketing strategies onto viewers.[7] Thus, this chapter seeks to critically explore what sorts of discourses are occurring about *Toddlers & Tiaras* on Tumblr, Facebook, and Twitter, the ways in which pageant participants that family members are discursively constructed in online conversations, and whether or not online discourse topics vary according to Web site. By analyzing *Toddlers & Tiaras* web pages and sites on Tumblr, Facebook, and Twitter, this chapter argues that fans utilize these online platforms to express their dedication to watching the series and, more importantly, debate the moral and ethical issues associated with "forcing" young girls into pageantry.

LITERATURE REVIEW: REALITY TV, AUDIENCES, AND SOCIAL NETWORKS

As many scholars have noted, reality television is an elusive genre that is often difficult to define.[8] Despite this difficulty, scholars have agreed on certain characteristics shared by reality television programming: it provides relatively inexpensive entertainment, it has a certain level of scripting, it is cheaper to produce, and it features "real" people instead of celebrities.[9] These characteristics have led to the genre's popularity, yet it has also led to scholars criticizing both the reality television genre *and* its viewers. Reality television has been critiqued for its seemingly low-quality television and, as media scholars Zizi Papacharissi and Andrew L. Mendelson have described, "reality TV poses a new low denominator for television content, promotes models of questionable social validity, and proliferates a culture of exhibitionism and voyeurism."[10] Moreover, Carolyn Michelle and Craig Hight have noted that critics of the genre and its viewers characterize reality television as tasteless television with its passive, uncritical, audience of dupes to follow.[11] Viewers of reality television have been described as "lower in intelligence, lacking in judgment and or/taste, unthinking voyeurs, unwitting dupes of commercialist broadcasters, [and] in danger of mistaking reality-TV programmes 'for reality.'"[12] Thus, early approaches to studying reality television programs and media audiences' interactions with these programs characterized audiences as passive dupes who took an uncritical eye to their favorite reality television series.

However, more recent approaches that explore the relationship between media audiences and reality television programs suggest that media audiences are more active and interactive than previously expected. Not only are media audiences cognizant of the reasons for which they watch certain reality television shows and of the values that draw them to certain television shows,[13] but the advent of Web 2.0 has created a new avenue where media audiences can vote for reality television shows, debate their favorite and least favorite aspects of certain character personalities and plots, and interact with series producers to have some sort of effect on plot twists and future season development.[14]

The first part of the relationship between media audiences and reality television programs focuses on media audiences' reasoning behind watching reality television shows. Research suggests that reality television audiences actively weigh their reasons for watching reality television programming and have a variety of personality characteristics and motivational reasons that contribute to why they watch reality television. For example, Steven Reiss and James Wiltz explored why adults enjoy watching reality television shows and found that reality television viewers primarily watch reality television to feel more important about themselves.[15] In other words, the more status-oriented a viewer is, the more likely he or she will report enjoyment and pleasure after watching reality television. Steven Reiss and James Wiltz argue that this self-fulfillment and self-gratification is caused by reality television in two ways:

> One possibility is that viewers feel they are more important (have higher status) than the ordinary people portrayed on reality television shows. The idea that these are "real" people gives psychological significance to the viewers' perceptions of superiority—it may not matter much if the storyline is realistic, so long as the characters are ordinary people. Further, the message of reality television—that millions of people are interested in watching real life experiences of ordinary people—implies that ordinary people are important.[16]

Thus, according to this study, reality television viewers are motivated to watch reality television because it makes them feel more superior about their lives and because they fantasize that they could potentially gain celebrity status someday. Motivated by vengeance and competition, certain reality television viewers thrive off of the interpersonal conflict and competition that characterize many reality television programs.

Another reason viewers watch reality television programming is because they enjoy watching what is perceived to be unique, unscripted interpersonal interactions (not necessarily riddled with conflict, per se) and because, at a base level, they find reality television appealing. For example, Robin L. Nabi and colleagues conducted two separate studies to assess on reality television viewers' motives for watching reality television and the role of voyeurism as an important or unimportant motivator of reality television consumption. [17] They found that, for the most part, viewers enjoy reality television because they are curious about other people's lives and because reality television is "perceived as relatively unique" due to its unscripted nature and the fact that viewers are watching "real people" like them on television. Similarly, Zizi Papacharissi and Andrew L. Mendelson explored reality television viewers' motives for watching reality television programming and found that viewers primarily watched reality television to pass time and because they were drawn to the novelty and entertainment aspect of the programming. [18] Thus, this research supports assertions made by Carolyn Michelle and other scholars, which argue that reality television viewers are indeed active, creative, and at some points critical. [19]

Reality television viewers are also active online on social networking sites, television blogs, and other Web sites dedicated to their favorite reality television shows. As Michael Stefanone and colleagues have noted, the advent of Web 2.0 radically changed the ways in which reality television audiences perceive their role and activity in the media environment. [20] They stated, "Rather than simply being the target of mediated messages, [reality television viewers] can see themselves as protagonists of mediated narratives and can integrate themselves into a complex media ecosystem." [21] Building upon Will Brooker's work, Su Holmes argues that the advent of Web 2.0 proliferates reality TV's overflow on to the Internet and "construct[s] a more fluid, flexible participatory interaction between text(s) and viewer." [22] This new ability to interact online and communicate via social networking sites about reality television programming and other media has been termed "digital interactivity."

Many scholars have noted that there is no simple way to define "interactivity." [23] Within a reality television context, it could be used to discuss viewers utilizing different types of technology to vote for reality television characters and participants. [24] It can also be used to describe reality television viewers acting as media producers of sorts. [25] Building upon Axel Bruns' work, S. Elizabeth Bird introduced the notion of digital interactivity, which focuses on audience members acting as "produsers" and participating in the collaborative viewer-show environment. [26] Here, audience members and fans can communicate online with each other about a particular television series, participate in the creation of digital content, and use certain technologies for voter interaction. [27]

Although the notion of "the creative, active online audience" has recently both inspired and dominated media audience research, there are few studies that explore how reality television audiences participate on social networking sites and discuss certain reality television programs. Deploying the term "mediated interactivity," Mark Andrejevic analyzed how online fan sites provide feedback to television writers, scriptwriters, and producers. He found that by participating in this digital interactivity process, fans were able to contribute their feedback to writers, thus affecting the future of certain reality television programming.[28] Fans also served as a form of unpaid labor by contributing time, energy, and feedback about their favorite reality television shows. Similarly, Victor Costello and Barbara Moore analyzed how online fans use the Internet to communicate about their favorite television programming and found that online fans fall on an activity continuum with "lurkers" on one end and thriving, active, communicative fans on the other end.[29] Contrary to studies that considered reality television fans to be "dumbed down" and "unsophisticated," Victor Costello and Barbara Moore found that fans were quite sophisticated, developed complex debates about television programming, and inspired meaningful exchanges on the Web site.[30]

Thus, considering previous research on reality television fans, motives for watching reality television, and how fans interact on social networking sites, it can be concluded that audience interaction is a variable concept, a spectrum upon which fans fall and interact online with various capacities. Audiences watch reality television for a variety of reasons, more commonly because they enjoy feeling more superior about their lives and watching what they perceive to be unique, real characters. Moreover, reality television audiences take an active role online as they communicate on official series Web sites and other social networking sites. Fans debate certain characters, vote to change series features, and even engage in meaningful, violent, or angry debates about various aspects of the series. However, one limitation of this research is that it does not segment certain types of reality television programming. Zizi Papacharissi and Andrew L. Mendelson's study analyzed viewers motives for watching reality television defined generally, as did Victor Costello's and Barbara Moore's study.[31] Costello and Moore noted that they specifically "attempted to cast a wider net" because they wanted to explore viewers' reactions to a wide range of programming.[32] However, doing so limits the ability to explore why and how reality television fans respond to certain programming online. Thus, this study seeks to explore *Toddlers & Tiaras* fans' digital interactivity on social networking sites such as Tumblr, Twitter, and Facebook. Given the series' peculiar controversial nature and the fact that it has consistently high viewer ratings, an analysis of fans' digital interactivity can shed light upon the various conversational topics dominating these social networking sites and perhaps viewers' justification for watching the series. Past research has analyzed the ways in which femininity is artificially constructed in the series, as well as the power struggles that occur between pageant parents and their children.[33] Does the controversial relationship between the "abusive parent" and the "oppressed pageant child" surface in online discourses about the show and if so, how? With this in mind, this study seeks to explore the following research questions:

1. What topics are *Toddlers & Tiaras* fans discussing online?
2. How are pageant participants and their family members discursively constructed on social networking sites?
3. Do online discourse topics vary according to Web site?
4. How do these discussions vary according to the type of site being used as the platform for discussion? In other words, what are the dispositions of the groups of fans according to the type of forum used?

METHOD

To answer these research questions, this essay employed a qualitative content analysis[34] of Tumblr Web sites dedicated to *Toddlers & Tiaras*, as well as the official *Toddlers & Tiaras* Facebook Web page and Twitter posts with the hashtag "#toddlersandtiaras." I searched for the official TLC *Toddlers & Tiaras* Web pages on Tumblr, Facebook, and Twitter, yet found that only Facebook featured an official TLC page dedicated to *Toddlers & Tiaras*. Thus, I chose to analyze the five active Tumblr accounts dedicated to *Toddlers & Tiaras* and the hashtag discussions associated with *Toddlers & Tiaras* on Tumblr and Twitter. Twitter did not have any user profiles dedicated solely to *Toddlers & Tiaras*. The Tumblr Web sites utilized for this analysis are below.

"Toddlers & Tiaras"	http://toddlersntiaras.tumblr.com/
"Toddlers & Tiaras"	http://toddlers-a-n-d-tiaras.tumblr.com/
"Toddlers & Tiaras"	http://toddlersandtiarasgirl.tumblr.com/
"Toddlers & Tiaras Confessions"	http://toddlersandtiarasconfessions.tumblr.com/
"Sparkle Babies"	http://sparklebabies.tumblr.com/

I chose to conduct a qualitative content analysis of the conversations occurring on the Web sites so that I could explore significant patterns that emerged, specifically pertaining to discourses about the role of pageant mothers in the series, potential motives for viewing the series, and any other major topics that might arise. Although this method has certain limitations pertaining to analyzing online texts, including potential inaccurate data interpretation and lack of context surrounding online users' posts, it is useful in this context because it allows the researcher to look at the "rough edges, special cases, and subtle peculiarities" of the online communicators' experiences.[35]

After the Tumblr Web sites, the official Facebook *Toddlers & Tiaras* Web page, and the hashtag #toddlersandtiaras conversations on Tumblr and Twitter were reviewed, a directed analysis was conducted where the statements were coded into initial codes and categories.[36] As Rebecca Curnalia notes, though directed analysis is "more pre-defined than typical grounded or thematic qualitative analysis, [it] allows the researcher to collapse and create categories while coding, and also includes identifying themes and trends in latent content after the initial coding."[37] Although the coding scheme initially utilized included categories pertaining to viewers' motives for watching the series and perceptions of the parent-pageant child relationship, it became clear that fans' viewing motives were far less pronounced in the data. Thus, the themes that emerged were viewers' feelings toward watching the series, viewers' identification with pageant participants, viewers' adoration for pageant participants, space for opposition/defense of the series and participants, Facebook users and the "Mom-ster" debate, and Twitter users and the "Abusive Parent" debate. All quotations included in the "Results" section were copied verbatim to preserve the meaning and individualities associated with the bloggers' and fans' posts.

RESULTS

Feelings about Watching the Series

The first theme that emerged during data analysis focuses on viewers' overall feelings about the series. While some bloggers and Facebook users expressed their joy about new series premieres, other bloggers and Facebook users discussed their guilt and shame associated with how much they like watching the series. The first Tumblr Web site, which is called "Toddlers & Tiaras,"[38] is run by a self-proclaimed Makenzie fan. On the site, the blog author posts multiple photos of the toddler beauty queens during preparation for the pageants and during the pageants, in addition to GIFs of various pageant participants performing and saying humorous statements during their on-screen interviews. The first theme exemplified in the Tumblr pages is the authors' commentary that expresses their love of the show. First, entries were posted that expressed the author's joy that the next season was about to air. For example, one image was posted that stated "Keep calm because T&T is back tomorrow" with the caption, "Toddlers & Tiaras is coming back to TLC!"[39] The next post states, "Toddlers and tiaras premiere comes on tomorrow! ☺ Are you guys excited?"[40] Lastly, the author of the Tumblr page entitled "Toddlers & Tiaras Confessions" wrote, "Premiere season of toddlers and tiaras tonight!!!! Yaaaaaay!!!"[41]

On the other hand, various Tumblr bloggers posted about their guilt and shame associated with "how much [they] like the show." One Tumblr user

posted a photo of *Toddlers & Tiaras* with the hashtag #toddlersandtiaras and
wrote, "My excitement level for this show should not be as high as it is, but I
can't help it."[42] Another user reblogged this photo and wrote, "I might end
up watching this. WHAT IS WRONG WITH ME I LOVE TRASHY TV I
THOUGHT TLC STOOD FOR THE LEARNING CHANNEL."[43] A twenty-two-year-old male Tumblr posted a GIF of Honey Boo-Boo dancing and
wrote, "I'm so guilty of liking her and watching this sassified show." Another Tumblr blogger posted a photoset of various pageant contestants and
wrote, "Currently watching Toddlers & Tiaras. . . . Not sure why though. It
just makes me thankful my mom never put me through that."[44] Thus, at a
base level multiple bloggers on Tumblr used their blogs and hashtags to
explain why they enjoyed watching the show and, in some cases, were simultaneously ashamed that they like watching the series.

Identification with Pageant Participants

In addition to posting about the show returning to TLC, other posts focus on
their identification with the *Toddlers & Tiaras* pageant participants. This
theme encompasses two different, yet related types of identification: identification via sameness and identification via heroism. The first subtheme, identification via sameness, is evident in posts where authors say that the pageant
participants are "just like them." One of the posts on the Tumblr blog entitled
"Toddlers & Tiaras"[45] features a GIF with a young girl in her glitz gown, and
the interviewer says about the girl, "Her favorite color is anything that spar-
kles." The author posted under the GIF, "She doesn't understand how alike
we are." Similarly, on the same page, the next post features a picture with
Makenzie stating before one of her pageants, "I'm gonna be higher than a
bird."[46] Under the photo, the blog author posted, "I feel the same way. We're
the same person." On another Tumblr entitled "Toddlers & Tiaras,"[47] there is
a post with a GIF set of six GIFs of various participants during their pageant
dances and post-show interviews. Each GIF has a funny statement from each
participant about what they would do with their prize money ("I'd own a lot
of Taco Bells" and "I would buy cheese dip"), and the Tumblr blogger's
caption states, "GIFs of children that accurately describe my life." Lastly, a
post on yet another Tumblr blog entitled "Toddlers & Tiaras"[48] also features
a set of five *Toddlers & Tiaras* GIFs with the pageant participants making
faces, dancing, and sticking out their tongues. The author's caption stated,
"All my Toddlers & Tiaras gifs that explain me."

The second subtheme, identification via heroism, is evidenced by site
authors referring to pageant participants as their heroes. This theme operates
by site authors appreciating the pageant participants and venerating their
over-the-top outfits, high-maintenance pageant preparation, glitzy outfits,
and gyrating dance routines. Although some blog authors write posts that

attack the show for the "damage it does to children," other authors admire the pageant participants' perceived beauty and fame. For example, one Honey Boo-Boo GIF with the hashtag #toddlersandtiaras said, "Y'all think you're all that and a pack of crackers, but y'all not." The blog author says, "She is my hero."[49] Moreover, a post featuring a GIF of Brock, a young boy wearing a glitter top during a dance routine, had a caption that stated, "Brock, you're my hero."[50] Whether the Tumblr authors are referring to the pageant contestants as heroes because they continue to participate in pageants, despite the attacks from anti-pageant protestors, or because they are fascinated by the glitz, glam, and pageant routines is uncertain; what I am certain of, however, is that some Tumblr bloggers continue to refer to pageant contestants as heroes and that this could be one reason that potentially motivates them to watch the series.

Adoration for Pageant Participants

The third theme that emerged during analysis is adoration for pageant participants. These posts discuss in some form or fashion how the Tumblr authors have favorite participants, love their make-up and dance routines, and overall think the participants are "so cute." For example, on one Tumblr blog entitled "Toddlers & Tiaras,"[51] there are multiple posts that discuss how and why the author likes certain participants. Under a picture of one of the participants, the author wrote, "She has such a perfect face! So cute and so natural."[52] Another post on this Tumblr writes under a picture of pageant contestant Paisely, "I wish Paisley was my daughter. She's so cute!"[53] Lastly, the quotation under a photo of pageant contestant Kayla states, "I love Kayla! She's truly gorgeous and so talented!"[54]

Additionally, the Tumblr entitled "Toddlers & Tiaras Confessions"[55] is dedicated to confessions about why viewers love the series. The owner of the blog writes, "Whether you're not ashamed of being a Toddlers & Tiaras fan, or T&T is your guilty pleasure, submit your confessions here!" Moreover, the Tumblr owner writes, "DISCLAIMER: For those who are not smart enough to understand, THESE ARE NOT MY OPINIONS. I receive messages (confessions) from people and make the art. Period." These statements would lead one to conclude that this Tumblr owner has perhaps dealt with people who voice their opposition of the series, and thus it is not surprising that the majority of posts on this Tumblr are from fans who love the show and watch it religiously. For example, one anonymous Tumblr user posted, "Emerald is such an amazing kid! She is very beautiful and stands out from the crowd! She is so talented! She will go far someday!"[56] Another post from an anonymous user stated, "Natali was my favorite from the episode! She was so cute and BEAUTIFUL!"[57] Lastly, an anonymous user posted, "Mackenzie Broach is sooooo cute. She's my favorite."[58]

Space for Opposition/Defense of Series and Participants

The fourth theme that emerged during analysis is that *Toddlers & Tiaras*
Tumblr fan sites function as a space for viewers and nonviewers alike to
debate the ethical issues and female objectification associated with the series.
For example, one Tumblr blogger posted on the "Toddlers & Tiaras"[59]
Tumblr account, "Thanks for setting back women 100 years and supporting a
psychologically damaging excuse to live through our children." In response,
the "Toddlers & Tiaras" Tumblr page author responded, "For starters, just
because I like a ******* TV show doesn't define who I am as a person.
Secondly, the children involved in this show all want to do it and never have
been forced to do it. People need to stop judging and stereotyping, it's really
ridiculous."[60] Moreover, another anonymous poster wrote on this site, "You
are disgusting. I hope you never have children and if you do, that they hate
you with all their heart."[61] In response, the page author responded, "You
need to stop ******* judging someone just because they like a show."[62]
Furthermore, a Tumblr blogger posted a photo of a pageant contestant
dressed as Julia Roberts from *Pretty Woman* with the hashtag #toddlersand-
tiaras with the caption, "The hot mess that is *Toddlers & Tiaras*. This young
pageant toddler's mom dressed her up as Julia Roberts' character from *Pretty
Woman*, which—if we need to remind you—was a prostitute."

Posts similar to these were also featured on the Tumblr entitled "Toddlers
& Tiaras Confessions."[63] One post from an anonymous Tumblr user stated,
"This is really creepy. Do you guys realizing that judging the beauty of
young children is borderline pedophilia? It's disturbing that you are okay
with the sexualization of young girls and that you encourage people to not
only watch, but to form opinions and judge these girls based on their participation in a beauty pageant."[64] The owner of the Tumbler entitled "Toddlers
& Tiaras Confessions" responded with a lengthy post in defense of the television series and beauty pageants overall:

> There is nothing sexual about child pageantry. Have you ever attended a real
> child glitz pageant? I have been to several and competed as well. These pa-
> geants are so different in person it's unreal. What's sexual? Makeup? That's
> not sexual, it's just a fun dressup. Is it the clothes? Swimsuits? Yes, some
> outfits may be a little revealing but it's nothing out of the ordinary in modern
> times. Some parents do go overboard but that's such a small percentage. Judg-
> ing it is not pedophilia. Anyways, pedophiles like little girls because they look
> like kids, not little women. So they wouldn't want all that makeup any-
> ways. . . . Majority speaking, it is not sexualizing the kids. The moms that do
> go overboard usually do not make it far, other moms or concerned people will
> step in and say something. No worries.[65]

Lastly, an anonymous poster on the "Toddlers & Tiaras Confessions"[66]
Tumblr posted about the ethical issues associated with pageantry when s/he

stated, "Pageants are such an obvious scam. No legitimate or ethical competition would charge you $300+ to prance your child around an undecorated hotel ballroom. The only winners in pageants are the pageant director's wallets."

In addition to protesting the series because of how it "damages young girls," other Tumblr bloggers use these *Toddlers & Tiaras* accounts to attack series participants, including the pageant contestants, their mothers, and even the pageant coaches. For example, posts that focus on the pageant contestants speak of their "conceited attitudes" and "bad hearts." One Tumblr user posted on "Toddlers & Tiaras Confessions"[67] about pageant contestant Alyssa Brielle and wrote, "Alyssa Brielle seems a little conceited with her 'I'm prettier than you' attitude, and that's not really good for a 4 year old." Another post on this Tumblr about pageant contestant Faithlyn stated, "Faithlyn is beautiful, but her lack of humility makes her heart ugly."[68] A post on this Tumblr about pageant contestant Mackenzie stated, "MaKenzie is only pretty when she is all dolled up. . . . Otherwise, she is hideous and a spoiled brat! Her mother needs to learn how to discipline her daughter before she ends up on 16 & pregnant!"[69] Lastly, a post on the "Toddlers & Tiaras Confessions" Tumblr discussed Alana, more popularly known as Honey Boo-Boo, and it stated, "I think Alana is so trashy. I don't know why everyone is making a big deal about her."[70]

Tumblr users also focused their disapproval of the series by talking about the pageant moms. A post about pageant participant Mia on "Toddlers & Tiaras Confessions"[71] stated, "I LOVE little Mia, she was so beautiful and cute, but I felt so sorry for her, her mother obviously did not care about her, she just cared about her winning the highest title. I almost cried when Mia went running up to her mother with her winnings and her mother barely even smiled. What a b****!" A post about pageant contestant Destiny on "Toddlers & Tiaras Confessions"[72] also focused on Destiny's mom because she let her daughter have a cigarette as a prop in one of her dance routines on the 1950s-themed pageant episode. The anonymous poster stated, "Destiny's mom is freaking stupid. Who gives a baby a cigarette? We know what [her mom] did as a teenager!" Another post on "Toddlers & Tiaras Confessions"[73] stated about pageant contestant SamiJo's mom, "I cannot stand SamiJo's mom Tricia. She acts like her child's the best ever and all 3 times she's been on she totally bombs it. And she's getting a bigger ego every time." Also, a post with the hashtags #Toddlers&Tiaras and #sexualizingyouth featured a photo of a pageant contestant and points out the maintenance that accompanies participating pageants, including the "fake eyelashes," "fake teeth," "spray tan," "caked on makeup," and "stripper outfit." The caption under the photo stated, "I was reading an article about Toddlers & Tiaras and they included this photo. I thought it was pretty sweet so I'm sharing it. It's so bizarre and disturbing that people do this to children."[74]

Lastly, posts attack the pageant coaches for their "bad attitudes" and for how terribly they treat the pageant contestants. One anonymous post on "Toddlers & Tiaras Confessions"[75] focused on pageant contestant Faithlyn and stated, "Faithlyn's coach Nikki is a useless piece of trash. Who calls a little kid ugly?!? (Especially with a face like hers!!!!!)" Similarly, another post on "Toddlers & Tiaras Confessions" focused on Faithlyn and her coach, and it stated, "Faithlyn's coach was the most hideous human being I've ever seen in my life. How do you expect your child to be polite if she grows up surrounded by a complete imbecile?"[76]

On the other end of the argument's spectrum, authors' followers posted positive comments in defense of the pageant contestants, their mothers, and the series. One Tumblr user posted on the "Toddlers & Tiaras Confessions"[77] page about pageant contestant Eden and stated, "I hated Eden until I saw an episode where she was not featured, but was at the pageant. She looked so tiny and innocent. It's easy to forget that at the end of the day they are all just children." Another anonymous Tumblr user posted a comment on the "Toddlers & Tiaras Confessions"[78] account that stated, "I saw [pageant contestant] Emerald Wulf at a pageant once and I asked her mom for an autograph and she told me to go away and Emerald laughed. . . . I lost complete respect." One of the account moderators, who visits multiple pageants in Texas, responded in defense of the pageant contestant and her mother:

> So, I'm not going to make art for this one because it is simply not true. I have met Emerald and her mom several times, 3 pageants overall. This one and the one saying she was a brat at pageants and it's all fake upset me because they are such lies. Honestly. Emerald is exactly like she is on camera but very shy and reserved at first, once you get to know her she is the sweetest. Her parents are also so kind and loving and fun and you can tell they love their daughter. Emerald has not done a pageant in almost a year. She's done like 2 or 3 since she got popular and I was at 2 of them and do not remember anything like this and I know them so well, this is such a lie. So stop sending in fake confessions because I know the truth.

Lastly, an anonymous poster on the "Toddlers & Tiaras Confessions"[79] Tumblr posted about Mackenzie, "I feel upset when people call MaKenzie a brat. She's not a bully or self-obsessed. And in real life she's very polite and really funny. T&T can portray her as a 'brat' but she's not."

Tumblr users also defend the series and pageantry overall. On the Tumblr entitled "Toddlers & Tiaras,"[80] one blogger wrote a post entitled "My first real big full glitz pageant experience" in which s/he stated:

> I can tell you, that toddlers and tiaras stuff is udder s**t. All of the younger girls were laughing, playing and just full of life. Those so-called "crazy" pageant moms are the nicest people you'll ever meet, a lot of them really go

out of their way to help each other, even help some who's competing against their child.

An anonymous post on the Tumblr entitled "Toddlers & Tiaras Confessions"[81] stated, "I am so sick of people saying that this show is the reason pedophiles exist. STOP MAKING EXCUSES FOR PEDOPHILES. Pedophiles exist because they are sick in the head. Don't blame little girls for perverts actions." Another anonymous poster on the "Toddlers & Tiaras Confessions"[82] Tumblr stated, "I hate it when people say that kids on T&T will become a slut or whore when they grow up. It's not what they're doing in pageants, it's how their parents are raising them."

Thus, a variety of themes emerged from an analysis of the Tumblr blogs that were not always evident in Facebook and Twitter conversations about *Toddlers & Tiaras*. Bloggers communicated about a wide variety of topics, with enthusiasm about the series and adoration of participants on one end of the debate spectrum and attacks of pageant moms, coaches, and participants on the other end of the spectrum. One specific attack emerged as the most dominant theme for Facebook and Twitter users—attacks on pageant mothers for "being abusive" and "subjecting their children to such terror."

Facebook Users and the "Mom-ster" Debate

Whereas bloggers on Tumblr use their sites to discuss a variety of topics pertaining to *Toddlers & Tiaras* such as blogger identification with contestants, contestant appreciation, and opposition of and defense of the series, the discourse on the official TLC Facebook page dedicated to *Toddlers & Tiaras*[83] is characterized solely by Facebook users criticizing the series for what they consider to be unethical and immoral behavior on behalf of TLC, the series producers, and the "mom-sters" in the series. Although the page has over one million "likes" (meaning subscribers) and features interactive posts about both *Toddlers & Tiaras* and *Cheer Perfection*, the majority of the comments on this Facebook page discuss users' disdain of the television series through their blame of the "mothers who force their children into these ridiculous pageants."

Describing pageant mothers as abusing their children, various Facebook users posted on the official TLC *Toddlers & Tiaras* page that pageant mothers are "abusive," "sick," and "twisted," and that their behavior is problematic because it "just encourages more weird, abusive mums" and creates "bratty, disrespectful children." For example, one commentator posted, "The mothers of these children need to seek professional help pronto! What is worse is that the TLC channel that broadcasts this garbage is promoting values that are twisted and sick."[84] Another Facebook user posted, "These mothers should be in leg irons, being escorted to jail! Not be given money to destroy the lives of children!"[85] Moreover, in a very heated and critical tirade, another Facebook user posted:

> All you parents who parade your daughters around like this and hype them up on sugar should be investigated by Child Protective Services. You should be ashamed of yourselves. These girls are beautiful the way they are. Why don't you so-called moms enter yourselves in a beauty contest. Oh, that's right, most of you are ugly, fat losers. Teach your daughters to love themselves for who they are. This is abuse.[86]

Other commentators responded to this post by arguing that pageant mothers are "harming girls irreparably by instilling the concept that exterior beauty determines their worth," that they are "sickening and psychological damaging," and that they "are living through their daughters and it's pretty sad."

While one part of the discourse showcases Facebook users blaming the pageant mothers for "forcing" their daughters into pageantry, the other part of the discourse occurring on the official TLC *Toddlers & Tiaras* Facebook page[87] showcases a villainization of the pageant mothers for creating "spoiled brats." For example, one Facebook user posted, "The parents push these kids into pageants and most of the girls hate it. The parents have the dream, but the child is in the nightmare. These children are such big brats and divas. They are disrespectful to their parents and just little terrors."[88] Another commentator responded to this post by stating, "It's all on the parents! The children are a blank canvas. The adults are to blame!"[89] An additional commentator stated she was "floored at the [contestants'] disgusting behavior," noting, "She is rude and the parents allow it! Just awful! What is it teaching her when she gets older? How will she treat others? She thinks it's perfectly okay to disrespect adults now! They need do teach them respect!"[90] Lastly, one self-identified Australian Facebook user expressed her disgust of the show when she stated:

> As an Aussie mum, I am totally horrified at the spoiled, indulgent, bad-tempered pouty little brats that these pageants produce and even more disgusting are the mothers who pout, rant and complain about their little girls being hard done by! These little ones are, for the most part, being coerced into provocatively flouncing around and making moves that would make an elephant blush! I cry inwardly every time I see a little one say she doesn't want to go on stage, that she's tired and her mother hits her with a sugar stick. To all parents of these poor little kids, remember Jonbenet Ramsey and what happened to this precious little life years ago.[91]

Thus, although there was considerable variation within the *Toddlers & Tiaras* discourses appearing on the Tumblr posts and Web pages associated with *Toddlers & Tiaras*, the majority of comments on the official TLC *Toddlers &*

Tiaras Facebook page criticize pageant mothers for "forcing" their daughters to compete in pageants and for creating "little brats."

Twitter Users and the "Abusive Parents" Discourse

Similar to Facebook users, a majority of the tweets about *Toddlers & Tiaras* criticized pageant parents for "abusing their children." Although a few tweets with the hashtag #toddlersandtiaras focused on how tweeters "couldn't stop watching the show" and a few even showcased photos of college-aged females mockingly dressed as pageant toddlers for Halloween, senior prom, and other themed parties, a majority of the tweets included commentary that criticized the series and questioned why parents "would put their child through that crap." Tweets attacked pageant moms broadly by stating that "people should be jailed" for putting their children in pageants and more specifically by attacking certain mothers from the series.

At a broad level, tweets criticized parents for subjecting their children to such "torture" and for how repulsive the series is. First, one woman tweeted, "I'm utterly #shocked that #TLC airs #ToddlersAndTiaras. It's #disgusting that #mothers do this to their #children."[92] Another tweet noted, "Am I the only one who sees a #toddlersandtiaras trailer and feels sick?"[93] Another tweet stated, "Personally, I think the show #toddlersandtiaras makes me sick to my stomach. It's just disgusting and parents should be ashamed."[94] Following the logic of critiquing pageant mothers' motives for "forcing" their daughters to participate in pageantry, another tweeter questioned, "These pageant moms are doing the worst! Why would you force your child to go through that? Spray tanning, plucking eyebrows, fake nails, huge wigs, and heavy makeup on 3–6 year olds. This is not acceptable. #toddlersandtiaras."[95] Lastly, noting her disgust with the series, another tweeter stated, "First time watching #toddlersandtiaras and I am seriously disturbed with these mothers."[96] Thus, the theme of disgust ran rampant in tweets about the series, with tweeters questioning why "mothers would do this to their children."

At a more specific level, tweeters lashed out against certain mothers from the series, noting that they "hate the scumbag moms" who "force" their children into pageantry and modify their children's bodies. One tweeter, for example, stated, "If your child cries/screams every time you make her get manicures, spray tans, and fake eyelashes, MAYBE STOP DOING IT. #toddlersandtiaras."[97] Focusing on the eyelashes and makeup modifications, one tweet stated, "Some of the moms on #ToddlersandTiaras really piss me off. Your daughter is 3, no need for spray tans, flippers, fake eyelashes, or makeup!"[98] Building upon the abhorrence of pageant mothers changing their children's physicality, one tweeter exclaimed, "Oh my god, the mom in this episode is bleaching her kids' teeth,"[99] while another tweeter stated, "Why the hell is this redneck mother shaving her 7-year-old's legs on #toddlersand-tiaras?"[100] Lastly, another tweet exclaimed, "A woman just admitted she only had her children so they could do pageants! How could you say that? #ohamerica."[101] Thus, just as Facebook users debated ethical and moral issues associated with *Toddlers & Tiaras* parents forcing their children into pageantry, Twitter users criticized pageant parents more generally and pageant mothers more specifically for their role in child pageantry.

DISCUSSION

Toddlers & Tiaras is a highly controversial reality television series because of the physical modifications that pageant toddlers and children must go through on a frequent basis, despite their frequent tantrums, cries, and fits of rage. Given the findings of past research that has explored reality television fans' online digital interactivity, this study sought to explore how *Toddlers & Tiaras* fans and critics interact on social networking sites such as Tumblr, Facebook, and Twitter.

First, findings indicate that *Toddlers & Tiaras* fans and critics are very active in their online discussions and debates about the series. Fans and critics alike "took to the message boards" to discuss their fandom and high anticipation about future seasons; debate the ethical and moral issues associated with the series more broadly and with pageantry more specifically; and outright attack and criticize pageant mothers, pageant coaches, and pageant participants. Contrary to seminal findings that characterize reality television audiences as dumbed down, passive viewers, this study shows that fans and critics of *Toddlers & Tiaras* were very active on social networking sites.

Second and more interestingly, fans and critics communicated about different topics on different social networking sites. Fans of the series dominated much of the Tumblr traffic about *Toddlers & Tiaras* as they discussed their favorite pageant contestants and favorite aspects of pageantry, even going so far as to refer to certain pageant contestants as their heroes. On the other hand, discourses about *Toddlers & Tiaras* on Facebook and Twitter solely focused on criticizing and attacking the vague "parent parents" and more specific "pageant mom-sters" who force their children to participate in pageantry. One potential explanation for the difference in discourses could be the average demographics for users on each social networking sites. Tumblr users, for example, have been described as "very dedicated to the site;" furthermore, 34 million global users state that they contribute to Tumblr on a monthly basis, and almost 50 percent of these users are between the ages of sixteen and twenty-four.[102] On the other hand, according to the most recent Pew Research Center report on social media usage, 71 percent of online adults use Facebook and 19 percent of online adults use Twitter.[103]

Thus, it appears that users on Tumblr are much younger than users on Facebook and Twitter, and this could explain some of the support for the series and attack of pageant participants. Since users on Facebook and Twitter are much older than Tumblr users, this could potentially explain why most of the conversations on Facebook and Twitter criticized the series and the pageant parents and were lengthier and slightly more articulate.

Third, pageant parents were discursively constructed in a myriad of negative ways on all three social networking sites. Users called pageant mothers "mom-sters" and a host of other negative names for enrolling their children in pageantry. In addition to demonizing pageant mothers, pageant contestants were also criticized, as some online users referred to them as "ugly brats" with bad attitudes and dim futures. This raises the question: what mechanism causes critics of the series to spend exorbitant amounts of time online criticizing the series and engaging in debates with other fans and critics than result in debate threads with over three hundred comments? This extreme dislike of the series also causes critics to engage in a form of distanced cyberbullying as they attack pageant parents and contestants and call them names. Future research should explore why critics of this series and other reality television programs engage in extensive online debates about the series and to what extent (if any) these online interactions affect series production and development.

CONCLUDING THOUGHTS

Audience digital interactivity has recently dominated media studies research, as scholars have sought to understand how media audiences interact with television series via online message boards, Web sites, and social networking sites. Joining in this conversation, this study explored how *Toddlers & Tiaras* fans and critics interact with each other on social networking sites such as Tumblr, Facebook, and Twitter. Results show that fans and critics discussed a variety of topics pertaining to *Toddlers & Tiaras* online, including their adoration for the series, their identification with pageant participants, their perceptions of ethics and morality in the series, and their blatant criticism of pageant parents and contestants. Future research should utilize ethnographic methods to uncover why reality television fans and critics spend excessive amounts of time on social networking sties criticizing reality television characters in an effort to truly uncover more nuanced dynamics of the audience-text relationship.

NOTES

1. Leandra Hinojosa Hernandez, "The Lolita Spectacle and the Aberrant Mother: Exploring the Production and Performance of Manufactured Femininity in *Toddlers & Tiaras*," in *Reality Television: Oddities of Culture*, eds. Alison F. Slade, Amber J. Narro, and Burton P. Buchanan. (Lanham: Lexington Books, 2014), 163–82.

2. Christine Tamer, "Toddlers, Tiaras, and Pedophilia? The 'Borderline Child Pornography' Embraced by the American Public," *Texas Review of Entertainment & Sports Law* 13, (2011): 85–101.

3. Liana Nobile, "The Kids Are Not Alright: An Open Call for Reforming the Protections Afforded to Reality Television's Child Participants," *UC Davis Journal of Juvenile Law & Policy* 17, (2013): 41.

4. Melissa Henson, "*Toddlers & Tiaras* and Sexualizing 3-year-olds," *CNN*, September 13, 2011, http://www.cnn.com/2011/09/12/opinion/henson-toddlers-tiaras/.

5. "The Five Biggest Controversies to Hit *Toddlers & Tiaras*," *Business Insider*, September 17, 2012, http://www.businessinsider.com/toddlers-and-tiaras-controversies-2012-9.

6. "Wednesday's Cable Ratings: 'Amish Mafia,' 'Moonshiners' Top Viewers, Demos," *The Futon Critic: The Web's Best Television Resource*, December 13, 2012, http://66.51.174. 176/ratings/2012/12/13/wednesdays-cable-ratings-amish-mafia-moonshiners-top-viewers-demos-476113/cable_20121212/.

7. Mark Andrejevic, "Watching Television Without Pity: The Productivity of Online Fans," *Television & New Media* 9, (2008): 24–46.

8. Erich M. Hayes and Norah E. Dunbar, "Do You Know Who Your Friends Are? An Analysis of Voting Patterns and Alliances on the Reality Television Show *Survivor*," in *Reality Television: Merging the Global and the Local*, ed. Amir Hetsroni. (New York: Nova Science Publishers, Inc., 2011), 7–24.

9. Ibid., 3. Steven Reiss and James Wiltz, "Why People Watch Reality TV," *Media Psychology*, 6 (2004): 363–78.

10. Hugh Dauncey, "French 'Reality Television': More Than a Matter of Taste?" *European Journal of Communication,* 11, (1996): 83-106. Zizi Papacharissi and Andrew L. Mendelson, "An Exploratory Study of Reality Appeal: Uses and Gratifications of Reality TV Shows," *Journal of Broadcasting & Electronic Media* 51 (2007): 355–70. Reiss and Wiltz, 370.

11. Carolyn Michelle, "(Re)contextualizing Audience Receptions of Reality TV," *Participations: Journal of Audience & Reception Studies* 6, (2009): 137–70. Craig Hight, "Debating Reality-TV," *Continuum: Journal of Media and Cultural Studies* 15, (2001): 389–95. Papacharissi and Mendelson, 355.

12. Hight, 390.

13. Papacharissi and Mendelson, 367. Reiss and Wiltz, 373.

14. Andrejevic, 29. Michael A. Stefanone, Derek Lackaff, and Devan Rosen, "We're All Stars Now: Reality Television, Web 2.0, and Mediated Identities," *Hypertext: Culture and Communication* (2008): 1–5.

15. Reiss and Wiltz, 373.

16. Ibid., 373–74.

17. Robin L. Nabi, Erica N. Biely, Sara J. Morgan, and Carmen R. Stitt, "Reality-Based Television Programming and the Psychology of Its Appeal," *Media Psychology,* 5, (2003): 303–30.

18. Papacharissi and Mendelson, 366.

19. Michelle, 139. Jane Roscoe, Harriette Marshall, and Kate Gleeson, "The Television Audience: A Reconsideration of the Taken-for-Granted Terms 'Active,' 'Social,' and 'Critical,'" *European Journal of Communication,* 10, (1995): 87–108.

20. Stefanone, Lackaff, and Rosen, 1.

21. Ibid., 1.

22. Will Brooker, "Overflow and Audience," in *The Audience Studies Reader*, eds. Will Brooker and Deborah Jermyn. (London: Routledge, 2002), 322–33. Su Holmes, "But This Time You Choose!: Approaching the 'Interactive' Audience in Reality TV," *International Journal of Cultural Studies*, 7, (2004): 213–31.

23. Holmes, 218.
24. Ibid.
25. S. Elizabeth Bird, "Are We All Producers Now? Convergence and Media Audience Practices," *Cultural Studies*, 25, (2011): 502–16.
26. Axel Bruns, "Towards Produsage: Futures for User-Led Content Production," Retrieved from http://snurb.info/files/12132812018_towards_produsage_0.pdf.
27. Bird, 3–5. Holmes, 216.
28. Andrejevic, 43.
29. Victor Costello and Barbara Moore, "Cultural Outlaws: An Examination of Audience Activity and Online Television Fandom," *Television & New Media*, 8, (2007): 124–43.
30. Ibid.
31. Papacharissi and Mendelson, 359. Costello and Moore, 128.
32. Costello and Moore, 128.
33. Hernandez, 177.
34. Rebecca M. Curnalia, "Frugal Reality TV during the Great Recession: A Qualitative Content Analysis of TLC's *Extreme Couponing*," in *Reality Television: Oddities of Culture*, eds. Alison F. Slade, Amber J. Narro, and Burton P. Buchanan. (Lanham: Lexington Books, 2014), 101–21.
35. Costello and Moore, 128. James Lull, *Inside Family Viewing: Ethnographic Research on Television's Audiences* (New York: Routledge, 1990).
36. Curnalia, 105.
37. Ibid.
38. http://toddlersntiaras.tumblr.com/.
39. Ibid.
40. Ibid.
41. http://toddlersandtiarasconfessions.tumblr.com/.
42. http://jarrodt83.tumblr.com/post/27921579752/my-excitement-level-for-this-show-should-not-be-as.
43. http://bellafrankensteinthe23rd.tumblr.com/.
44. http://trickster-chick.tumblr.com/post/27026708905/currentley-watching-toddlers-tiaras-not-sure-why.
45. Ibid.
46. Ibid.
47. http://toddlers-a-n-d-tiaras.tumblr.com/.
48. http://toddlersandtiarasgirl.tumblr.com/.
49. http://wicked-youth.tumblr.com/post/97134244394/damnnngina-the-honey-boo-boo-child-strikes.
50. http://dontgetcomfortable.tumblr.com/post/7341189501/brock-youre-my-hero.
51. http://toddlersandtiarasgirl.tumblr.com/.
52. Ibid.
53. Ibid.
54. Ibid.
55. http://toddlersandtiarasconfessions.tumblr.com/.
56. Ibid.
57. Ibid.
58. Ibid.
59. http://toddlersntiaras.tumblr.com/.
60. Ibid.
61. Ibid.
62. Ibid.
63. http://toddlersandtiarasconfessions.tumblr.com/.
64. Ibid.
65. Ibid.
66. Ibid.
67. Ibid.
68. Ibid.

69. Ibid.
70. Ibid.
71. Ibid.
72. Ibid.
73. Ibid.
74. http://ad-busting.tumblr.com/post/35018463702/i-was-reading-an-article-about-toddlers-tiaras.
75. http://toddlersandtiarasconfessions.tumblr.com/.
76. Ibid.
77. Ibid.
78. Ibid.
79. Ibid.
80. http://toddlersandtiarasgirl.tumblr.com/.
81. http://toddlersandtiarasconfessions.tumblr.com/.
82. Ibid.
83. https://www.facebook.com/ToddlersandTiaras.
84. Ibid.
85. Ibid.
86. Ibid.
87. Ibid.
88. Ibid.
89. Ibid.
90. Ibid.
91. Ibid.
92. https://twitter.com/hourie.
93. https://twitter.com/xpijx.
94. https://twitter.com/AaronLanghorne.
95. https://twitter.com/Miss_Twantwa.
96. https://twitter.com/ChanellOliver.
97. https://twitter.com/Hmer.
98. https://twitter.com/Brooke_Wood1.
99. https://twitter.com/zoe_lewis_.
100. https://twitter.com/Nataskiia.
101. https://twitter.com/Britt9Goodwin.
102. Hillary Heino, "Social Media Demographics—Instagram, Tumblr, and Pinterest," *Agile Impact*, March 13, 2014. http://agileimpact.org/social-media-demographics-instagram-tumblr-and-pinterest/.
103. Pew Research Center, "Social Networking Use." Retrieved from http://www.pewresearch.org/data-trend/media-and-technology/social-networking-use/.

REFERENCES

Andrejevic, Mark. "Watching Television Without Pity: The Productivity of Online Fans." *Television & New Media*, 9, (2008): 24–46.
Bird, S. Elizabeth. "Are We All Produsers Now? Convergence and Media Audience Practices." *Cultural Studies*, 25, (2011): 502–16.
Brooker, Will. "Overflow and Audience." In The Audience Studies Reader, edited by Will Brooker and Deborah Jermyn, 322–33. London: Routledge, 2002.
Bruns, Axel. "Towards Produsage: Futures for User-Led Content Production." Retrieved from http://snurb.info/files/12132812018_towards_produsage_0.pdf.
Buchanan, Burton P. "Portrayals of Masculinity in the Discovery Channel's Deadliest Catch." In *Reality Television: Oddities of Culture*, edited by Alison F. Slade, Amber J. Narro, and Burton P. Buchanan, 1–20. Lanham: Lexington Books, 2014.
Costello, Victor, and Barbara Moore. "Cultural Outlaws: An Examination of Audience Activity and Online Television Fandom." *Television & New Media*, 8, (2007): 124–43.

Curnalia, Rebecca M. "Frugal Reality TV During the Great Recession: A Qualitative Content Analysis of TLC's Extreme Couponing." In *Reality Television: Oddities of Culture*, edited by Alison F. Slade, Amber J. Narro, and Burton P. Buchanan, 101–21. Lanham: Lexington Books, 2014.

Dauncey, Hugh. "French 'Reality Television': More Than a Matter of Taste?" *European Journal of Communication*, 11, (1996): 83–106.

"The Five Biggest Controversies to Hit Toddlers & Tiaras." *Business Insider*, September 17, 2012. Retrieved from http://www.businessinsider.com/toddlers-and-tiaras-controversies-2012-9.

Hayes, Erich M., and Norah E. Dunbar. "Do You Know Who Your Friends Are? An Analysis of Voting Patterns and Alliances on the Reality Television Show Survivor." In *Reality Television: Merging the Global and the Local*, edited by Amir Hetsroni, 7–24. New York: Nova Science Publishers, Inc., 2011.

Heino, Hillary. "Social Media Demographics—Instagram, Tumblr, and Pinterest." *Agile Impact*, March 13, 2014. Retrieved from http://agileimpact.org/social-media-demographics-instagram-tumblr-and-pinterest/.

Henson, Melissa. "Toddlers & Tiaras and Sexualizing 3-year-olds." CNN, September 13, 2011. Retrieved from http://www.cnn.com/2011/09/12/opinion/henson-toddlers-tiaras/.

Hernandez, Leandra Hinojosa. "The Lolita Spectacle & the Aberrant Mother: Exploring the Production and Performance of Manufactured Femininity in Toddlers & Tiaras." In *Reality Television: Oddities of Culture*, edited by Alison F. Slade, Amber J. Narro, and Burton P. Buchanan, 163–82. Lanham: Lexington Books, 2014.

Hight, Craig. "Debating Reality-TV." *Continuum: Journal of Media and Cultural Studies*, 15, (2001): 389–95.

Holmes, Su. "But This Time You Choose! Approaching the 'Interactive' Audience in Reality TV." *International Journal of Cultural Studies*, 7, (2004): 213–31.

Lull, James. *Inside Family Viewing: Ethnographic Research on Television's Audiences*. New York: Routledge, 1990.

Michelle, Carolyn. "(Re)contextualizing Audience Receptions of Reality TV." *Participations: Journal of Audience & Reception Studies*, 6, (2009): 137–70.

Nabi, Robin L. and Erica N. Biely, Sara J. Morgan, and Carmen R. Stitt. "Reality-Based Television Programming and the Psychology of Its Appeal." *Media Psychology*, 5, (2003): 303–30.

Nobile, Liana. "The Kids Are Not Alright: An Open Call for Reforming the Protections Afforded to Reality Television's Child Participants." *UC Davis Journal of Juvenile Law & Policy*, 17, (2013): 41.

Papacharissi, Zizi and Andrew L. Mendelson. "An Exploratory Study of Reality Appeal: Uses and Gratifications of Reality TV Shows." *Journal of Broadcasting & Electronic Media*, 51, (2007): 355–70.

Pew Research Center. "Social Networking Use." Retrieved from http://www.pewresearch.org/data-trend/media-and-technology/social-networking-use/.

Reiss, Steven and James Wiltz. "Why People Watch Reality TV." *Media Psychology*, 6, (2004): 363–78.

Roscoe, Jane, and Harriette Marshall and Kate Gleeson. "The Television Audience: A Reconsideration of the Taken-for-Granted Terms 'Active,' 'Social,' and 'Critical.'" *European Journal of Communication*, 10, (1995): 87–108.

Stefanone, Michael A., and Derek Lackaff and Devan Rosen. "We're All Stars Now: Reality Television, Web 2.0, and Mediated Identities." *Hypertext: Culture and Communication* (2008): 1–5.

Tamer, Christine. "Toddlers, Tiaras, and Pedophilia? The 'Borderline Child Pornography' Embraced by the American Public." *Texas Review of Entertainment & Sports Law*, 13, (2011): 85–101.

"Wednesday's Cable Ratings: 'Amish Mafia,' 'Moonshiners' Top Viewers, Demos." *The Futon Critic: The Web's Best Television Resource*, December 13, 2012. Retrieved from http://66.51.174.176/ratings/2012/12/13/wednesdays-cable-ratings-amish-mafia-moonshiners-top-viewers-demos-476113/cable_20121212/.

Chapter Ten

Zombie Fans, Second Screen, and Television Audiences

Redefining Parasociality as Technoprosociality in AMC's #Talking Dead

Sabrina K. Pasztor and Jenny Ungbha Korn

INTRODUCTION: THE EVOLUTION OF TALK SHOWS, FROM PHIL DONAHUE TO CHRIS HARDWICK

Television programming, content accessibility and audience viewership have radically shifted through recent decades.[1] Traditional models of talk-show audience engagement[2] focused on maximizing audience interaction through prize giveaways, in-studio comments and caller questions, à la Phil Donahue and Oprah Winfrey. The advent of online social media technology (e.g., Twitter, Facebook, online fandom communities, audience polling, emails) has transformed talk-show television from the viewpoints of production and consumption and has significant implications for contemporary fandom, digital participatory culture, and the media industry. Interaction is no longer confined to in-studio and phone-in audience members; collaboration now includes fan participation through Twitter, Facebook, blogs, emails, and other Web sites. Using the Chris Hardwick-hosted, AMC cable network program *Talking Dead* (2011–present) as a successful (and understudied) case of a modern talk show, this research calls attention to how television in combination with online social media practices now yields a contemporary model for television engagement, redefining *parasociality*[3] into a new construct the authors define as *technoprosociality*.

While scholars have focused on AMC's *The Walking Dead* (2010–present), research on its companion talk show, *Talking Dead*, has not yet been published. It is, therefore, a site not only ripe for further analysis, but relevant as a pioneer modality in which new ways of audience interaction, production and consumption represent an evolution of the American television industry. This chapter will examine the relationship among online social media and contemporary fandom, social capital, audience engagement, modalities of community-building, the use of social media technology, and media business models.[4] The analysis will redefine the construct of *parasociality* and its outmoded role in the evolution of the new economic model into one employing *technoprosociality*, which for purposes of this study is defined as the integration of social media technology to maximize audience engagement and interpersonal relationship development between celebrities and fans.

Audiences today reflect a more inclusive, ethnically, racially, sexually, and geographically diverse, technologically savvy, and demanding population. When viewers of television's "first screen" engage in online social media through the "second screen"[5] of laptops and mobile phones, they often also participate in online social media practices to consume, critique, share, and even influence television content.[6] Although the authors do not ascribe to the overly reductionist theory of technological determinism advanced by Croyteau and Hoynes,[7] in this chapter it is likewise argued that newer technology—in this case, social media—in companionship with traditional forms of human communication via phone lines and face-to-face question-and-answers provided by Hardwick and his *Talking Dead* guests—are not neutral, as one "cannot disregard the social and cultural conditions that technology has produced."[8] The clever and profitable use of technology, specifically social media (Facebook, Twitter, and online polls), enables technoprosociality to happen: the social distance between Hardwick, his guests, studio audience and virtual audience gets minimized which contributes to audience engagement. *Talking Dead* employs technoprosociality as the foundation of its programmatic model, and should be thoughtfully considered by broadcast and cable networks not only for its sheer popularity but the ability to be replicated as a lucrative profitability framework.

The contemporary act of television viewing as part of a *digital participatory culture*, then, along with the impact of social media and Internet technology on changing perceptions of family, agency, interaction, and community, brings inherent tensions to light. Is contemporary television viewing a solo experience, community-based, or both and how do social media influence television viewing practices? In what ways do contemporary talk show television programs employ social media mechanisms, measuring online consumer interactions through status updates, mentions, retweets, Skype videos, and the like to build fandom?[9] What are the implications for future media industry business models? This chapter seeks to provide insight into the answers to these questions through a rhetorical criticism[10] of the only talk show in which asking whether zombies can gain weight or get pregnant is perfectly normal, uploading an online picture of oneself to be remade as a zombie or "deaded" becomes a unifying community experience, debating if killing a "walker" via hatchet, sword, or gun is preferred is standard discourse, and shedding tears over the loss of fictional characters in the companionship of a studio and digital audience are reflective of contemporary audience behaviors.

BACKGROUND: THE COMPANION HORROR TALK SHOW

Television's horror genre has a pedigreed past. The classic *Twilight Zone* (1959–64), *Tales from the Crypt* (1989–96), *Buffy the Vampire Slayer* (1997–2003), and *American Horror Story* (2011–present) represent a few of the complex, compelling horror programming shows that have fans on the edge of their seats. *The Walking Dead* (2010–present), the latest horror genre entry from the AMC cable network channel, continues this tradition. Concluding its fourth season in 2014, *The Walking Dead* has become a top cable television program for AMC, particularly for the coveted demographic of adults aged eighteen through forty-nine, with an average of 13 million viewers during its fourth season.[11] *The Walking Dead* began with 5 million zombie fans and has grown now to over 16 million "walker stalker" viewers consistently watching the program during the coveted cable spot of 9 p.m., Eastern time on Sunday nights.[12] Told in fifty-one, sixty-minute episodes, the show portrays the post-apocalyptic world after a viral outbreak has turned the majority of the human population into what most Americans consider zombies, but *The Walking Dead* introduced a new term for the living dead: walkers. Battling against the walking dead, the survivors, comprised of multicultural and multiracial individuals with various skills, ambitions and behaviors from the Atlanta region primarily, fight regularly for resources and safety.

The Walking Dead opened its second season by departing from its horror show predecessors by modernizing the classic talk show format and audience engagement processes and debuting a companion talk show, *Talking Dead*, in October of 2011. The title and host of the series reflected a different tone to traditional talk shows, which were originally developed as first-person, issue-oriented confessionals "where average people discuss a personal or personalized political issue (obsessive compulsive disorder, teenage pregnancy, or racism) with the host, expert, and the audience."[13] Traditional talk show programs periodically incorporated segments on personal, financial, or home makeovers and relationship advice, but the host maintained a serious, oftentimes somber tone and stoic demeanor similar to those of a news anchor. Furthermore, audiences had limited and heavily controlled levels of interactivity in the context of their viewing experience, heavily reinforcing parasociality: the "one-way flow" of programming meant studio and at-home viewing audiences participated in formats restricted by the telephone and their physical presence in the studio.[14] Over-the-phone questions from program viewers, in-person questions from the studio audience, audience clapping, booing, or hissing to show support or displeasure about a topic or invited guest, and exuberance about receiving a take-away gift or prize from the host showcased maximum audience engagement. Fans, however, remained physically and, in many cases, emotionally distant from both the talk show host and guest. *Talking Dead* has permanently changed this construct by renegotiating the traditional forms of parasociality through its unique use of technology to spur audience engagement into technoprosociality.

METHOD: RHETORICAL ANALYSIS OF EPISODES

Situated within this backdrop, this chapter employs rhetorical criticism to analyze episodes across the first three seasons (2011, 2012, and 2013) of *Talking Dead*, discussing the narrative arc, segment features, interactive elements of audience participation, and host and guest rhetorical discourse as a systematic process to illuminate and analyze persuasive techniques and "products of human activity" observed in host-audience interaction.[15] The study focuses heavily on examples of social media mechanisms that measure online consumer interactions via second-screen, centering around the technoprosocial elements (Internet, social media) and the subtle communication tactics employed by Hardwick to encourage guest, audience, and fan engagement. It concludes with a brief discussion of this new and expanding programmatic model's implications for future media industry business models.

This chapter limits the episode sample to eighteen, thirty- and sixty-minute television episodes across three seasons, the first premiering on October 16, 2011, and the last concluding on March 30, 2014. Episodes were accessed online without fee via *netmov.tv*.

Six episodes were selected from each season to provide adequate opportunity for thematic redundancies and accurate representation of the show's progression through the seasons. In addition to analyzing the first and last episode of every season, selected because of their positionality as opening and concluding a season (when one would expect higher levels of viewership given the success of the program), the study also examined the top four episodes with the highest viewership (see Table 10.1), with the presumption that higher levels of audience viewership would correlate with greater frequency of audiences engaged in second-screen behavior and vice versa. The first season of *Talking Dead* was comprised of thirteen thirty-minute episodes. During the middle of its second season, *Talking Dead* lengthened its episodes from thirty minutes to sixty minutes, a doubling in length that corresponded with a doubling in viewership. The study examined key inter-actional elements of *Talking Dead*, including the use of social media Twitter hashtags, audience emails, Facebook posts and pre-recorded Skype videos with fan questions; contrasting these second-screen modalities to traditional modes of fan participation via phone calls, audience questions and interviews with cast members.

Table 10.1.

Season 1	Ep. 1	Ep. 2	Ep. 5	Ep. 9	Ep. 12	Ep. 13
2011–12	1.16 Million	1.11 Million	1.16 Million	1.12 Million	1.33 Million	4.30 Million
Season 2	Ep. 1	Ep. 10	Ep. 11	Ep. 12	Ep. 15	Ep. 16
2012–13	2.14 Million	4.01 Million	3.50 Million	3.75 Million	4.49 Million	5.16 Million
Season 3	Ep. 1	Ep. 8	Ep. 9	Ep. 14	Ep. 15	Ep. 16
2013–14	5.10 Million	6.02 Million	5.85 Million	5.42 Million	5.29 Million	7.34 Million

"Episode" abbreviated to "Ep." in this table.

COMMUNITY AUDIENCES: SECOND SCREEN'S EXPANSION OF *PARASOCIALITY* INTO *TECHNOPROSOCIALITY*

Capitalizing on parasocial-to-technoprosocial relationships is the cornerstone of *Talking Dead*'s successful business model. The first horror companion talk show to leverage both in-studio and online audiences, *Talking Dead* is adept in constructing technoprosociality through physical elements and so-cial media. After describing how *Talking Dead*'s online practices promote parasocial relationships, this section focuses on specific strategies for culti-vating parasociality, namely online exchanges, second screen usage, and friendly familiarity.

Parasociality vs. Technoprosociality

Online social media are changing the meaning of "community audiences," creating geographically unbounded interactions around television shows. The 1950s image of the white "nuclear family" gathered around the television set to watch programs as a communal activity has morphed into one that is more inclusive, with an ethnically, racially, and sexually diverse audience. While parasociality has been used to describe interpersonal relationships in which the television audience member knew more about the celebrity than the ce-lebrity did about a fan,[16] the process through which this one-sidedness occurs has changed since the advent of online social media. The celebrity-fan relationship has become multisided rather than one-sided.

While it is likely that audience knowledge of *The Walking Dead* is deeper than the characters' knowledge of their audience, the gap between the two has radically decreased due to online social media sites kept by *The Walking Dead* cast, producers, and crew. When an update from an actress on *The Walking Dead* is shared, it reaches millions of fans, news that may or may not be related to *The Walking Dead*'s plot. When an actress shares about her activism, fans learn about the cause as well. As Fraser and Brown suggest, "one powerful representation of value congruence and behavioral adaptation is the practice of impersonation, which is primarily a communication phenomenon. People adopt the values and behaviors of celebrities with whom they develop perceived relationships, whether interpersonal or through media."[17] Because online social media invites replies, the actress also receives direct and indirect reactions to her posts. Fans may choose to like, favorite, retweet or share posts; they may also opt to write publicly to the actress. These technological affordances built into online social media for registering one's opinion operate to let *The Walking Dead* crew, producers, and cast know what their audiences are thinking about story lines and other items unrelated to *The Walking Dead*. This further enhances the reach of technoprosociality.

Technoprosocial Online Exchanges

While the parasocial process happens in an ad hoc manner across the discourse patterns of millions of individuals, *Talking Dead* is the first contemporary talk show to organize such interactivity into consistent, repeated features of its programming. *Talking Dead* has capitalized on the experience of "belonging" that television imparts to viewers by extending such social surrogacy to other mechanisms for interaction, namely Facebook, Twitter, Skype, telephone, and its own Web sites.[18] For example, in every episode, the audience is reminded to "Tweet us at #AMCTalkingDead, e-mail us, or call us at 1-855-DEADLIVE," with the "Dead Live" component of the phone number in a drawn-out breath that became synonymous with *Talking Dead*.

From the onset, *Talking Dead*'s emphasis has been on fan interaction through the use of the "second screen" provided by the Internet via cell phones, tablets and laptops, encouraging *The Walking Dead* fans to engage with other fans in real-time.[19] On-screen polls, video games, contests, and applications allow fans to create zombie alter-egos, do phone-ins and post Tweets, Facebook comments, and questions. By its third season, *Talking Dead* presented a unique social media element involving the use of pre-recorded and live Skype videos, during which various audience members expressed a comment or asked a question. When asked via Skype, "What real-world celebrity would likely be you in *The Walking Dead*?" cast member Andrew Lincoln who plays primary protagonist Rick Grimes responded with, "William Shatner," simultaneously giving a verbal "nod" to the audience's reverence of popular culture icon James T. Kirk from Star Trek, and recognition of host Hardwick's nerd persona and the audiences' engagement with technology.

Talking Dead considers its audience in both its physical and online forms. The show utilizes a dedicated Twitter feed, an established phone number, a public Facebook page, email on TalkingDead.com, and even a Skype account (*AMCTalkingDead*) soliciting twenty-second video messages containing audience questions. All contests for *Talking Dead* rely upon Internet interaction from the audience to win. For example, the show's Web site offered walk-on roles as zombies and attendance at the upcoming season premiere to winners of various contests, but the submission had to be conducted via Facebook. In its second season, *Talking Dead* asked fans to create videos to prove they were the ultimate fan, which *The Walking Dead*'s executive producer judged, and uploaded those videos to the *Talking Dead* Facebook page. Fans could "dead" themselves, which required the audience to download a mobile application from the TalkingDead.com Web site that converted fan pictures into distorted, grotesque, zombie self-portraits, one of which was selected as the "Dead Fan of the Week."

Every week, audience polls were set up at the beginning of the *Talking Dead* episode, and the results were shown at the episode's conclusion. The poll questions were drawn from that night's episode, like "If you got pregnant during the zombie apocalypse, where would you want to give birth?" which referenced the birth of Carl Grimes's sister Judith in the prison. Voting in these polls was conducted entirely online and only through TalkingDead.com, not Twitter, Facebook, or email. Fans were also encouraged to purchase related merchandise from a different Web site separate from TalkingDead.com entitled ShopTheTalkingDead.com.

With *Talking Dead* providing additional Internet-mediated avenues for community audiences to learn about their beloved *The Walking Dead* characters, *Talking Dead* expands parasocial relationships to online audiences that are more diverse in terms of age, ethnicity, and other socioeconomic characteristics than those typically associated with solo, noncommunal television viewing. *Talking Dead* reduces the social distance between *The Walking Dead* celebrities and their fans by providing direct access for the latter to the former on its weekly show through in-person and digital interactions. Audience members witness first-hand as actors and entertainers reveal themselves to be fans through their own online engagement: several, like Joe Manganiello, Aisha Tyler, Yvette Nicole Brown, Retta, and Nathan Fillion note their having read *The Walking Dead* graphic novels and are able to discuss in great detail the nuanced differences between the novels and television program, as well as indicate their participation in the Twittersphere and show blogs. The lines among host, guests, actors, cast, entertainers, and audience subsequently become blurred in *Talking Dead*, a process that mimics events as they unfold on *The Walking Dead* and the audience questions whether living versus surviving is truly possible.

Second Screen

From an industry standpoint, *Talking Dead* is among the first shows to target second-screen users and incorporate computer-mediated communication into its weekly programming. It has effectively tapped into an expanding demographic of American television fans that rely upon the primary device of television while utilizing a secondary device: 44 percent of Americans use a second screen half of the time that they watch television.[20] Second-screen usage has significant implications for television fandom, as online social media are changing the meaning and interactions of "community" and "sociable, computer-mediated group viewing experiences . . . [as part of] distributed, shared television viewing."[21] Story Sync, a custom Web-viewing application designed explicitly for AMC in 2012, allows audiences to "watch an interactive presentation that plays in real-time during the initial broadcast of an episode."[22] Fans can answer a Story Sync question about one of the show's characters (for example, predict which character will kill the most zombies), discuss the current status of characters (e.g., will Michonne move past Andrea's death?), and discover in-depth information about special effects, tools, and gadgets employed by the show (if a fan is curious about the axe Rick wields, Story Sync directs the viewer to information provided by the show's set designers).

Perhaps the most unique identifying characteristic of *Talking Dead* is its management of second screen modalities to gain audience participation. Facebook and Twitter supplied questions and comments for the entire show: on average, of the twenty to twenty-five fan questions addressed in each episode, thirteen to fifteen were sent in through Facebook, five to seven through Twitter hashtags/the show's Twitter site, and the remaining were call-ins. While *Talking Dead*'s host relied upon online submissions of content for each show, Hardwick also spontaneously created Twitter hashtags to encourage the online audience to join in the camaraderie as a form of rhetorical persuasion. For example, in the third season, the host rewarded audience members that participated in the question-and-answer section with a spoonful of chocolate pudding and playfully shouted to the audience to get #pudding going in reference to that night's episode in which Carl Grimes voraciously consumed the rare find of a gigantic can of chocolate pudding. When *Talking Dead* returned from a commercial break, the host announced, "We're back, and I am pleased to announce that #pudding is trending in the United States!"

That the mere mention of #pudding on *Talking Dead* was enough to send that hashtag soaring into a national trend displayed the interactive power that the show had with its digital audience.

Programmatic content that combines interactive technology with community-building represents an innovative economic media model for television fandom. The act of television viewing is now a part of digital participatory culture. Today, when viewers of television's first screen engage in online social media through computers and mobile phones, they concurrently participate in "second screen" online social media practices to consume, critique, share, and even influence television content.[23] This further correlates with increased viewership, as a 2013 Nielsen study indicated: among eighteen- to thirty-four-year-olds, "an 8.5 percent increase in Twitter volume corresponds to a 1 percent increase in TV ratings for premiere episodes, and a 4.2 percent increase in Twitter volume corresponds with a 1 percent increase in ratings for midseason episodes."[24] Among thirty-five- to forty-nine-year-olds, a 14 percent increase in Twitter volume equates to a 1 percent ratings increase.[25] Consequently, the development of companion talk shows that incorporate social media trends reflect not only television industry programmatic changes but directly contribute to viewership and improved ratings.

Friendly Familiarity

Shattering the confining format of traditional talk shows, *Talking Dead* capitalized on the needs of fans and audience members to interact via technology and sociability by selecting a personable and self-proclaimed culture nerd to host the program. Chris Hardwick is a forty-three-year-old contemporary Renaissance man with a University of California, Los Angeles, degree in philosophy who references famous Star Wars lines, attends Comic-Con (San Diego, 2014) and holds verbal jousts with his guests and the audience easily while sporting contemporary *GQ* suits. His experience as an actor, host of @Midnight, Nerdist.com author and podcaster, die-hard Tweeter, DJ, and social media fan make him an ideal representation of the digital and millennial generation's "thinking man" and a credible negotiator of guest and audience participation. Hardwick himself strives to create a more lighthearted, personal, comedic, empathic, persuasive, and sarcastic at times, atmosphere in which the audience and celebrity as fan takes center stage in a variety of modalities.

A standard episode of *Talking Dead* began with the camera on a full body image of Hardwick, who typically appeared in a dark but dapper suit with a more colorful button-down shirt and tie, stylish narrow-leg pants and business shoed and presented the modern-day persona of a "hip nerd." Up-to-speed on clothing styles, with an obvious nod to the importance of appearance, Hardwick dressed like an urban, tech-loving, witty pop culture icon.

The term "nerd" coined during post-war modernism was a representation of hegemonic ideas capable of displaying *consumer-viable* characteristics of a "geek-chic, technosexual" being.[26] Hardwick's image deliberately portrayed a cool, tech-savvy, knowledgeable, intellectually stimulating "boy next door" with a subtle sexuality that was most obvious in his rapid-fire speech, periodic emotional outbursts, and ruthless questioning of his guests and the audience.

Typically, Hardwick briefly introduced the guests on that evening's show by adding a short biography to the person's name: "Executive Producer Robert Kirkman joins us" (season 1, episode 1); "Comedian Aisha Tyler is here with us tonight, and we'll also be joined by Michael Rooker who plays Merle" (season 1, episode 5); "Welcome back Executive Director Scott Gimple and Nathan Fillion from *Castle* on ABC" (season 3, episode 1). Though brief, these linguistic identifiers established the credentials and highlighted the variety of guests, including entertainers who are fans, the show's creative talent/cast and producers/directors/screenwriters who are regularly encouraged by Hardwick to share advanced knowledge of upcoming episodes. For example, Hardwick indicated that he was so anxious to find out the next story line for season 3 of *The Walking Dead* that he texted Executive Producer Robert Kirkman to see if he would be willing to shed light on the upcoming season. As expected, Kirkman did not reveal any spoilers, but Hardwick's Internet-mediated action to text created an additional element of intrigue, interest, and affiliation with his audience who also want to know, "Now what?"

Besides celebrity fans, guests on *Talking Dead* were drawn from the cast and producers of *The Walking Dead*. At least one of the guests was always affiliated with the show as an actor, producer, writer, or director. In fact, *Talking Dead* touted its direct access to *The Walking Dead* as part of its appeal, opportunities that were afforded solely due to *Talking Dead* through AMC. *Talking Dead* interviews included the actors and actresses playing Rick Grimes, Daryl Dixon, Michonne, Maggie Greene, Glenn Rhee, Hershel Greene, Merle Dixon, the Governor, Lori Grimes, Andrea, Carol Peletier, Shane Walsh, Tyreese, Sasha, and Gareth, among others. *Talking Dead*'s celebrity fans were an eclectic mix of comedians (e.g., Aisha Tyler, Patton Oswalt, Fred Arminsen, Sarah Silverman, and Keegan-Michael Key), musicians and mixed martial artists (e.g., Marilyn Manson and CM Punk), and actors from other shows and films (e.g., Jon Heder from several films including *Napoleon Dynamite*, Joe Manganiello from HBO's *True Blood*, Eliza Dushku from the CW's *Buffy the Vampire Slayer* and *Angel*, and Yvette Nicole Brown from NBC's *Community*).

After introducing the guest panel, Chris Hardwick engaged the guests by asking a series of questions directly related to the current episode. He referenced pre-written notecards with questions from Facebook and Twitter, questions that were submitted earlier from the online audience. In season 1, episode 13, Executive Producer Glenn Mazzarra was asked what it was like to burn a barn full of zombies, providing Mazzarra an opportunity to explain in-depth how that scene was orchestrated and his personal feelings regarding its relation to the larger narrative. Norman Reedus, the actor who plays Southern-redneck-turned-compassionate-gentleman-and-leader Daryl Dixon, responded to Hardwick's question regarding his possible death and fan reaction with: "A lot of people were like, 'If you kill Daryl Dixon, we'll be really upset.'" Reedus then noted that "fans have been really, really gracious about me bringing this role into life" (season 2, episode 16), emphasizing the communal (prosocial) ties between audience, fans, and guests. Upon the return of the show from its mid-season break (season 3, episode 9), Hardwick said to actor Danai Gurira who played fierce Michonne, "I missed you guys so much—it's been a long couple of months," and then asked her and Executive Director Greg Nicotero to discuss her evolving vulnerability and interaction with her adopted family of Rick and son Carl. The actor was encouraged to dissect, interpret, and provide a nuanced perspective of her character's strengths and weaknesses and predict potential outcomes from this personal growth, an introspective exercise that brought the audience closer to both Hardwick as "pseudo-therapist" and actor as "one of their own."

Choosing a host who emits friendliness is important in creating a show that operates implicitly on maximizing interpersonal engagement. Hardwick did not treat *Talking Dead* fans with formality; he chatted with them in the same manner as one would with a friend. The audience was allowed to believe that *The Walking Dead* guests were just another version of individuals that they would invite into their own living room for sociality and conversation, giving host, guests, and audience a high degree of relatedness, or "the extent to which a person feels connected to the person around him or her" which correlates with greater engagement in prosocial behavior.[27] Friendly familiarity as set and maintained by the host of today's talk show is vital for sustaining this interpersonal dynamic and is a hallmark feature of technoprosociality.

In addition to the host, the design of the set for *Talking Dead* played an important role in establishing social interaction among the host, guests, and audience. The set was designed in a neutral palate of colors (cream, taupe), designed to reassure and even calm the audience. The furniture consisted of a three-cushioned couch in beige, where guests sat each week, with three smaller pillows that occasionally changed from season to season. In front of the couch was a long coffee table upon which an aesthetically unique object related to *The Walking Dead* was placed: a clay model of the hands of the walking dead, reaching up for air, rising from the earth. To the right of the couch directly facing the audience was a single beige chair in which Hardwick sat. The walls behind the couch were a dull brick material reflective of an urban loft; the wall to the left of the audience was lined with bookshelves containing tchotchkes related to *The Walking Dead*, and during the first season, a mini-bar/cocktail shelf with several types of alcoholic beverages and drinking glasses, reminiscent of a local "speak-easy."

The right wall behind the couch contained a gigantic poster of a *The Walking Dead* scene, again reflective of a topic or feeling appropriate to the given season that changed with the advent of each season or mid-season premiere, a targeted move designed to mimic not only the narrative plotline but also give the audience a sense of visual reassurance each week around a common topic. Season 1's poster was a "deaded" picture of *The Walking Dead*'s Executive Director Greg Nicotero, zombie jaw wide open, eyes staring maliciously at its intended prey (the denizens of Atlanta and, indirectly, the audience).

By its second season, *Talking Dead* had a change in decoration, with the poster reflecting the visage of Rick Grimes with a prison backdrop, as that is where many of Grimes's family and fellow survivors were holed up attempting to create a new and protected world. During the third season, the poster shifted to a representation of the "Terminus" sign, a designated and allegedly save haven to which Grimes's band fled after the complete destruction of the prison, and later shifted to Rick and son Carl in the same season, as the travelers were now separated and there was a focus now maintaining relationships, finding other members of the group who have been lost and isolation that necessitated an even greater reliance upon formal and informal family structures. Completing the 180-degree audience view, a microphone stand was placed to Hardwick's left, slightly off-stage but close enough so that an audience member called from the studio during the show could easily join the host and his guests on stage to ask a question.

The set was expressly designed to evoke the feeling of intimacy. Darker, muted lighting, comfortable yet stylish furniture, the mini-bar, a coffee table of distressed wood, the proximity of the audience to Hardwick's seating area, the unframed poster, the camera angle focused not on the audience's faces, but only the tops of their heads in the first row (visually reducing the separation of audience-to-host-to guests), all worked in tandem to draw in the audience and create a sense of community. "We are through, people," Hardwick declares as the second season of *Talking Dead* ends, implying a sense that the audience in companionship with the host and his guests over the sixteen episodes, have jointly survived the apocalyptic landscape. The set design definitively and strategically echoes that sentiment. The choice of *Talking Dead*'s set of a living room is designed to extend and mirror the viewer's living room. Combining friendly familiarity with physical set design is meant to encourage social interaction among the viewer who longs to know more about every aspect *The Walking Dead*.

IMPLICATIONS FOR THE MEDIA INDUSTRY

In response to *Talking Dead*'s revamping of traditional talk show formats, advertisers have had to change their marketing tactics. Advertisers have entered into what has been referred as the "matrix era" of television viewing, "where media users select, sample and interconnect a diverse ensemble of technologies, texts and experiences."[28] To reach multitasking viewers, more than 80 percent of whom use smartphone and tablet devices while viewing television,[29] Story Sync showcases its own set of ads during *The Walking Dead*'s commercial breaks. *Talking Dead* optimizes the elements of matrixed viewing by placing the audience at the center of viewing experience, capitalizing on their use of diverse technology platforms and sociability, and building the network business around that synergistic sweet spot.

Moreover, companion talk shows that air after many contemporary television programs are now strategically building upon the success of *Talking Dead*. A&E's *Bates Motel* (2013–present) now has a companion talk show named *After Hours* (2013–present), and AMC's iconic serial drama *Breaking Bad* (2008–13) gave birth to *Talking Bad* (2013, one season), also hosted by Hardwick. Companion talk shows, especially those that exploit second-screen culture, have been developed as a unique way to retain audiences. Programmatic content that combines interactive technology with community-building represents an innovative economic media model for television fandom.

Corporate Sponsors

Talking Dead partnered with corporate sponsors to promote companies through specific program segments and visible logo placement. For example, iTunes was mentioned every week as a source for downloading episodes and music from *The Walking Dead* and *Talking Dead*. On episode 9 of season 3, *Talking Dead* informed viewers that the perfect Valentine's Day card would be a zombie electronic card that either features Nicotero's "zombified" face or could be customized for any fan's face, which was available at iTunes as a mobile phone application. Subway purchased the right to the name of *Talking Dead*'s segment on insider information about how scenes were produced. When Subway sponsored the segment, the name was Subway's Fresh Buzz, which referenced Subway's slogan to "eat fresh." Subway sponsored "Fresh Buzz" segments, linguistically linking their product attributes (fresh sandwiches and salads) to the concept of copious energy, excitement, and stimuli focused around online audience participation and conversation (buzz). When Subway did not sponsor the segment, its name was shortened simply to Buzz.

In a nod to Oprah whose talk show made audience gift-giving an expected tradition, Hardwick also provided the audiences each season with various giveaways. Season 2, episode 15 shows a promotion for Activision's *The Walking Dead* Survival Instinct Video Game, featuring brothers Merle and Daryl Dixon in voice-overs; studio audience members each received a copy. In season 3, the audience received an official *The Walking Dead* Survival Kit (bug-out bag), available on ShoptheWalkingDead.com. The Lincoln Motor Company and Hyundai, whose cars are regularly driven by characters in *The Walking Dead*, sponsor *Talking Dead* show segments ("In Memoriam" or "Fan of the Week"), and regular segments appear in two to three episodes each season sponsored by X-Box or Nintendo that show a "super fan" playing the video game and commenting on the experience.

The use of advertising in promoting brands during programmatic content and the targeted media planning and buying process tied to audience viewership and ratings is not a new phenomenon. Nor is the sudden popularity of a cult program like *The Walking Dead* which is continuing its epic climb up the charts, landing at number eighteen from number thirty-three in the Total 12+ rankings, a significant jump from its number=forty-six standing just in 2011.[30] What *Talking Dead* does uniquely and nearly perfectly that reinforces its status as a groundbreaking model is two-fold. First, the maximization of distributed technology (social media in all its forms) is used to make the television viewing experience both hyper-collective yet intimate by circumventing barriers to sociable interaction.[31] Audiences can access Hardwick, his guests, and each other through the matrixed second screen access points (Facebook, Twitter, show Web site, blog, the Facebook game, board game, call-ins, Skype audience questions, Fan of the Week, and "Dead Yourself" applications), which support the interactive and social experience of a digitally-oriented fan base.

Second, the strategic selection of host and guests move them away from being unattainable celebrities and into the realm of diehard fans much like the audience itself. While the sheer variety of guests is impressive, what distinguishes *Talking Dead* is these individuals' presence as fans rather than "unattainable" celebrity. Audiences can perceive themselves to be closer to the guests through the common denominator of allegiance to *The Walking Dead*. *Talking Dead* creates a sanctioned space in which fans and guests are one-in-the-same, and harkens back to a time in the past when television audiences could spend collated time dissecting the minutiae of the latest episode (narrative, character evolution, plot twists, and cliffhangers); in this case, Hardwick has welcomed each fan in to his "party" in his "living room," and fans and celebrities/guests happily comingle across once-insurmountable barriers that traditionally separated fans from celebrities.

CONCLUSION

Talking Dead is an example of how online social media has transformed audience interaction from a model of parasociality to one of technoprosociality in which the perceived distance between celebrity-host-audience is significantly diminished. In this chapter, the authors are the first to employ rhetorical criticism to examine its narrative arc, online social media linkages and audience connections that successfully redefine parasociality by introducing technology-mediated interaction through technoprosociality. The show owes its success to strategic use of Internet-mediated sociality that engages individuals and broadens participation, blurring the lines between celebrities-as-fans and host-audiences. This new talk show format combines all the best aspects current digital technology and social mediums have to offer with audience engagement, and easily relatable host and guest interpersonal dynamics, and an intimate studio setting. As a result, it is highly unique and profitable paradigm from an audience viewership and advertising industry perspective. It therefore behooves media producers intent on developing unique programmatic content to pay close attention to *Talking Dead* as the technoprosocial pioneer in this new industry model.

NOTES

1. Mukesh Nathan, Chris Harrison, Svetlana Yarosh, Loren Terveen, Larry Stead, and Brian Amento, "CollaboraTV: Making television viewing social again," in *Proceedings of the First International Conference on Designing Interactive User Experiences for TV and Video,* (ACM, 2008), 85–94. Harry Castleman and Walter Podrazik, *Watching TV: Six decades of American television* (Syracuse, NY: Syracuse University Press, 2003).

2. Patrick Bateman, Peter Gray, and Brian Butler, "Community commitment: How affect, obligation, and necessity drive online behaviors," in *ICIS 2006 Proceedings,* (ICIS, 2006), 63.

3. Donald Horton and R. Richard Wohl, "Mass communication and para-social interaction: Observations on intimacy at a distance," *Psychiatry* 19, no. 3 (1956): 215–29.

4. James Coleman, "Social capital in the creation of human capital," *American Journal of Sociology* (1988): S95–S120. Louise Barkhuus, "Television on the internet: New practices, new viewers," in *CHI '09 Extended Abstracts on Human Factors in Computing Systems,* (ACM, 2009), 2479–88. Nicole Ellison, "Social network sites: Definition, history, and scholarship," *Journal of Computer-Mediated Communication* 13, no. 1 (2007): 210–30. Henry Jenkins, *Convergence culture: Where old and new media collide* (New York: New York University Press, 2006). Michael Curtin and Jane Shattuc, *The American Television Industry* (United Kingdom: Palgrave Macmillan, 2009).

5. Patrick Lochrie and Paul Coulton, "Mobile phones as second screen for TV, enabling inter-audience interaction," in *Proceedings of the 8th International Conference on Advances in Computer Entertainment Technology* (2011), 73.

6. Lelia Green, *Technoculture: From alphabet to cybersex* (Sydney: Allen & Unwin, 2002).

7. David R. Croteau and William D. Hoynes, *Media society: Industries, images, and audiences* (Thousand Oaks, CA: Sage, 2013).

8. Jo Brown, Amanda Broderick, and Nick Lee, "Word of mouth communication within online communities: Conceptualizing the online social network," *Journal of Interactive Marketing* 21, no. 3 (2007): 2–20.

9. Sara Bibel, "The Walking Dead Season 4 Finale Delivers 15.7 Million Viewers & 10.2 Million Adults 18–49, A Season Finale Record," *TV by the Numbers*, 2014, http:// tvbythenumbers.zap2it.com/2014/03/31/the-walking-dead-season-4-finale-delivers-15-7-million-viewers-10-2-million-adults-18-49/249361 .

10. Victoria O'Donnell, *Television Criticism* (Thousand Oaks, CA: Sage, 2013).

11. Nielsen, "Tops of 2013: TV and Social Media," *Nielsen*, 2013, http://www.nielsen.com/ us/en/newswire/2013/tops-of-2013-tv-and-social-media.html .

12. Ibid.

13. Curtin and Shattuck, *The American Television Industry*, 149.

14. Carrie Heeter, "Interactivity in the context of designed experiences," *Journal of Interactive Advertising* 1, no. 1 (2000): 3–14.

15. O'Donnell, *Television Criticism*.

16. Horton and Wohl, *Mass Communication and Parasocial Interaction*.

17. Benson P. Fraser and William J. Brown, "Media, celebrities, and social influence: Identification with Elvis Presley," *Mass Communication and Society* 5, no. 2 (2009): 183–206.

18. Nicolas Ducheneaut, Robert J. Moore, Lora Oehlberg, James D. Thornton, and Eric Nickell, "Social TV: Designing for distributed, sociable television viewing," *International Journal of Human-Computer Interaction* 24, no. 2 (2008): 136–54.

19. Cory Janssen, "Second Screen," *Techopedia*, 2014, http://www.techopedia.com/ definition/29212/second-screen .

20. Shea Bennett, "Social TV: The value of the second screen," *Mediabistro*, 2013, http:// www.mediabistro.com/alltwitter/value-second-screen_b51567 .

21. Ducheneaut, Moore, Oehlberg, Thornton, and Nickell, *International Journal of Human-Computer Interaction*, 137.

22. Bryan Bishop, "How a second-screen app made 'The Walking Dead' come alive," *The Verge*, 2014, http://www.theverge.com/entertainment/2014/2/13/5406498/how-a-second-screen-app-made-the-walking-dead-come-alive .

23. Curtin and Shattuck, *The American Television Industry*, 149.

24. Nielsen, "New Study Confirms Correlation between Twitter and TV Ratings," *Nielsen*, 2013, http://www.nielsen.com/us/en/insights/news/2013/new-study-confirms-correlation-between-twitter-and-tv-ratings.html.

25. Ibid.

26. Christine Quail, "Nerds, Geeks, and the Hip/Square Dialectic in Contemporary Television," *Television New Media* 12 (2011): 460–84.

27. Louisa Pavey, Tobias Greitemeyer, and Paul Sparks, "Highlighting relatedness promotes prosocial motives and behaviors," *Personality and Social Psychology Bulletin* 37 (2011): 905–19.

28. Mark Lochrie and Paul Coulton, "Mobile phones as second screen for TV, enabling inter-audience interaction." In *Proceedings of the 8th International Conference on Advances in Computer Entertainment Technology*, (ACM 2011), 73.

29. Nielsen, *Tops of 2013*, 2013.

30. Ross Walton, "HBO, AMC, and A&E Rise in 'Must Keep TV' Rankings," *Solutions Research Group Consultants*, 2013, http://www.srgnet.com/2013/07/08/hbo-amc-and-ae-rise-in-must-keep-tv-rankings.

31. Jaye L. Derrick, Shira Gabriel, and Kurt Hugenberg, "Social surrogacy: How favored television programs provide the experience of belonging," *Journal of Experimental Social Psychology* 45 (2009): 352–62.

REFERENCES

Barkhuus, Louise. "Television on the internet: New practices, new viewers." In CHI'09 Extended Abstracts on Human Factors in Computing Systems, 2479–88. ACM, 2009.

Bateman, Patrick, Peter Gray, and Brian Butler. "Community commitment: How affect, obligation, and necessity drive online behaviors." ICIS 2006 Proceedings, 63. 2006.

Bennett, Shea. "Social TV: The value of the second screen." Mediabistro. 2013. http://www. mediabistro.com/alltwitter/value-second-screen_b51567 .

Bibel, Sara. "The Walking Dead season 4 finale delivers 15.7 million viewers & 10.2 million adults 18–49, a season finale record." TV by the Numbers. 2014. http://tvbythenumbers. zap2it.com/2014/03/31/the-walking-dead-season-4-finale-delivers-15-7-million-viewers-10-2-million-adults-18-49/249361.

Bishop, Bryan. "How a second-screen app made 'The Walking Dead' come alive." The Verge. 2014. http://www.theverge.com/entertainment/2014/2/13/5406498/how-a-second-screen-app-made-the-walking-dead-come-alive.

Brown, Jo, Amanda Broderick, and Nick Lee. "Word of mouth communication within online communities: Conceptualizing the online social network." *Journal of Interactive Marketing* 21, no. 3 (2007): 2–20.

Castleman, Harry, and Walter Podrazik. *Watching TV: Six Decades of American Television.* Syracuse, NY: Syracuse University Press, 2003.

Coleman, James. "Social capital in the creation of human capital." *American Journal of Sociology* (1988): S95–S120.

Croteau, David R., and William D. Hoynes. *Media/Society: Industries, Images, and Audiences.* Thousand Oaks, CA: Sage, 2013.

Curtin, Michael, and Jane Shattuc. *The American Television Industry.* United Kingdom: Palgrave Macmillan, 2009.

Derrick, Jaye L., Shira Gabriel, and Kurt Hugenberg. "Social surrogacy: How favored television programs provide the experience of belonging." *Journal of Experimental Social Psychology* 45 (2009): 352–62.

Ducheneaut, Nicolas, Robert J. Moore, Lora Oehlberg, James D. Thornton, and Eric Nickell. "Social TV: Designing for distributed, sociable television viewing." *International Journal of Human-Computer Interaction* 24, no. 2 (2008): 136–54.

Ellison, Nicole. "Social network sites: Definition, history, and scholarship." *Journal of Computer-Mediated Communication* 13, no. 1 (2007): 210–30.

Fraser, Benson P., and William J. Brown. "Media, celebrities, and social influence: Identification with Elvis Presley." *Mass Communication and Society* 5, no. 2 (2009): 183–206.

Green, Lelia. *Technoculture: From Alphabet to Cybersex.* Sydney: Allen & Unwin, 2002.

Heeter, Carrie. "Interactivity in the context of designed experiences." *Journal of Interactive Advertising* 1, no. 1 (2000): 3–14.

Horton, Donald, and R. Richard Wohl. "Mass communication and para-social interaction: Observations on intimacy at a distance." *Psychiatry* 19, no. 3 (1956): 215–29.

Janssen, Cory. "Second Screen." Techopedia. 2014. http://www.techopedia.com/definition/29212/second-screen .

Jenkins, Henry. *Convergence Culture: Where Old and New Media Collide.* New York: New York University Press, 2006.

Lochrie, Mark, and Paul Coulton. "Mobile phones as second screen for TV, enabling inter-audience interaction." In Proceedings of the 8th International Conference on Advances in Computer Entertainment Technology, p. 73. ACM, 2011.

Nathan, Mukesh, Chris Harrison, Svetlana Yarosh, Loren Terveen, Larry Stead, and Brian Amento. "CollaboraTV: Making television viewing social again." In *Proceedings of the 1st International Conference on Designing Interactive User Experiences for TV and Video*, pp. 85–94. ACM, 2008.

Nielsen. "Tops of 2013: TV and social media." Nielsen. 2013. http://www.nielsen.com/us/en/newswire/2013/tops-of-2013-tv-and-social-media.html .

O'Donnell, Victoria. *Television Criticism.* New York: Sage, 2013.

Pavey, Louisa, Tobias Greitemeyer, and Paul Sparks. "Highlighting relatedness promotes prosocial motives and behaviors." *Personality and Social Psychology Bulletin* 37 (2011): 905–19.

Quail, Christine. "Nerds, geeks, and the hip/square dialectic in contemporary television." *Television New Media* 12 (2011): 460–84.

Walton, Ross. "HBO, AMC, and A&E rise in 'Must Keep TV' rankings." Solutions Research Group Consultants. 2013. http://www.srgnet.com/2013/07/08/hbo-amc-and-ae-rise-in-must-keep-tv-rankings .

Chapter Eleven

Memes, Tweets, and Props

How Fans Cope When Shows Go Off the Air

Alane Presswood and Steven Granelli

Social networking sites have permanently altered the face of fandom. Fan activism, formerly limited to picketing studios and launching postcard campaigns, can now reach exponentially larger numbers with only a fraction of the effort previously required through Facebook, Twitter, online fan forums, and Reddit. For example, seven years after it was unceremoniously cancelled, *Veronica Mars* made the leap from cult favorite teen detective television serial to big-screen fan service homage. The movie and its creator, director Rob Thomas, had long been denied by studio Warner Brothers due to a conviction that there wouldn't be enough fan interest to warrant a film project. To combat that conception, Thomas turned to crowdfunding Web site Kickstarter—where the *Veronica Mars* movie proposal swiftly broke all of the site's previously held records. The project raised $5.7 million through 91,585 backers, far exceeding the original project goal of $2 million. The project is still listed in the top five best-funded efforts in Kickstarter's history. The *Veronica Mars* movie project needed only eleven hours to reach its original goal of $2 million, a feat that would not have been physically possible without the reach of the Internet. Such fan efforts are now a regular feature of mainstream news, which heralds these passionate efforts as touching and dedicated—a far cry from the fan ridicule typically portrayed in popular culture (a la the 1980s favorite *Revenge of the Nerds*).

What happens, then, when television shows stop airing and the beloved presence of these characters—and the ensuing viewer-character relationship—is abruptly ended? This chapter will examine how fans cope with the cessation of a favored show and the loss of a significant, yet fictional presence in their lives—and how those behaviors consequentially affect networks and producers.

The fandoms selected for examination in this piece were chosen via three major criteria. First, production on the show must have ceased—either before a creator's chosen time period or at a natural time of conclusion. Second, our selected shows needed to have accrued a dedicated and active fanbase on social media that has taken action in attempts to either rescue or perpetuate their favored show (action that commonly includes reaching out to networks or studios to demonstrate that enough popular interest warranted a continuation/comeback, or launching awareness campaigns on Twitter/Facebook). To that end, each show selected for inclusion also needed a continuously active fanbase—preferably a fanbase that encompassed both viewers who watched at home, and professionals involved with the show during its airing (actors, director, writers, and so on). Based on these criteria, four series were selected for analyses: *Breaking Bad*, *Community*, *Firefly*, and *Buffy the Vampire Slayer*.

Consistently ranked by fans and critics as one of the best shows ever aired on television, *Breaking Bad* aired on AMC from January 2008 until September 2013. Vince Gilligan's tense serial drama followed the exploits of Walter White (Bryan Cranston), a former high school chemistry teacher who, after being diagnosed with cancer, begins to manufacture meth to pay for his treatment and provide for his family. Five seasons chronicle his transformation into a drug kingpin and murderer, along with the struggles of his family and his associates. Over the course of its run, the show accrued ten primetime Emmy awards.

Laden with pop culture references and tongue-in-cheek humor, Dan Harmon's *Community* ran for five seasons before NBC pulled the plug in May 2014. Yahoo picked the show up for an additional sixth season, aired online in June 2014. Featuring an ensemble cast and a fictional community college, the series was nominated for an "Outstanding Writing in a Comedy Series" Emmy award in 2012, and was on several critical lists of the best shows on television in 2010, 2011, and 2012.

Joss Whedon's space western *Firefly* aired in 2002; just eleven of the fourteen produced episodes were run before Fox network cancelled the series. The show follows Captain Mal (Nathan Fillion) and his companions on the starship *Serenity*, after coming out on the losing side of an interplanetary civil war in the year 2517. In 2003, the series won an Emmy for Outstanding Special Visual Effects. Despite the low viewership and early cancellation, strong DVD sales and fan campaigns led to the production of the 2005 film continuation *Serenity*.

Joss Whedon's first television show and breakaway hit (featuring petite, blonde Sarah Michelle Gellar in the feisty titular role), *Buffy the Vampire Slayer* (henceforth referred to as *Buffy*) ran from 1997 through 2003. The

Emmy-winning series has been widely influential not only on later generations of television shows (particularly regarding the "supernatural hero trying to balance everyday life" themes) but also in academia—it's regularly the subject of study and even has its own conference. After a seventh and final season, the series was continued in comic book format. An animated series was discussed, but never materialized. *Buffy* inspired a spinoff series after its third season, *Angel the Series*; the *Buffy* fandom frequently co-exists with the spin-off fandom.

After selecting the four fandoms to be included, we conducted a content analysis on the major social media outlets fans used to express their emotions after production ceased. For a period of four months from April 2014 through July 2014, social media usage relating to the four fandoms under analysis was monitored and collected using Facebook, Twitter, Reddit, and assorted show-specific message boards. In total, twenty-one separate show-specific sites were monitored for data collection, and eight hashtag specific searches were conducted on Twitter covering the time period between April 2014 and July 2014. We discovered that fans express mourning in six significant ways: through creating humor, sharing the series universe (the imagined surrounding world in which the source material exists), advocating for series renewal, providing space for criticism, expanding the universe, and inhabiting the universe.

UNDERSTANDING THE IMPLICATIONS OF CONTEMPORARY FANDOM

In many ways, it has become significantly easier to be a fan in the twenty-first century. The major movie blockbusters of spring and summer 2014 included *Captain America: The Winter Soldier*; *X-Men: Days of Future Past*; the fourth *Transformers* installment; *Godzilla*; *The Amazing Spider-Man II*; *Guardians of the Galaxy*; and *Teenage Mutant Ninja Turtles*. Fans of the Marvel Comics universe and 1980s action toys are no longer lurking in comic book stores; they're lined up around the block for midnight movies on multi-million dollar release nights with the rest of media-consuming America. And while shows are still cancelled and fans still mourn, comic serialization, webisodes, Netflix binging, and a proliferation of discussion opportunities all prolong the life of the vanished universe and ease the fan's sadness at its absence. Fan mourning post-series-finale—which once was itself mocked as a sign of a pathetic inability to cope with real life—is now chronicled with admiration in major news outlets. Has fandom, then, been normalized at last?

The myriad ways in which we have defined fandom in scholarship may bring us closer to answering that question. Finding a consensus on the definition of a "fan," even among scholarship on popular culture, can be difficult. To the lay public, terming a viewer a "fan" may bring to mind compulsive behaviors, immature attitudes and habits (such as collecting action figures and costumes), an inability to successfully interact and socialize with others, and unhealthy obsessions with possessing fictional characters and/or inhabiting fictional universes—possibly leading to threatening actions such as stalking in real life. In his influential 1992 work *Textual Poachers*, Henry Jenkins is quick to remind readers that the word itself—"fan"—stems from the Latin "'fanaticus' . . . which quickly assumed the negative connotation 'of persons inspired by orgiastic rites and enthusiastic frenzy.'"[1]

Coping tactics and rhetoric are studied exhaustively with regard to many stressful events in "real life": trouble at school, terminal illness in the family, and difficult workplaces are just a few avenues where coping communication has been studied. Booth-Butterfield, Booth-Butterfield, and Wanzer define coping as "a transactional process in which a threatening, challenging, or difficult personal situation leads to a negative or positive emotional response."[2] Social media outlets facilitate the transactional aspects of coping during periods of loss and mourning after a show ceases production.

Of course, what counts as "real life" (and therefore, what merits implementing coping strategies) is hardly a standard question from person to person. Rebecca Bley found that individual fans alternately described "real life" as "'that thing I try desperately to avoid,' 'the boring bits,' and 'all the boring, lonely, pathetic hours I'm not online.'"[3] The average American watches seven hours of television per day—nearly as much time as they spend in the workplace, and far more than the three-and-a-half minutes the average parent spends meaningfully conversing with their child(ren) per day. The characters and places portrayed in the shows taken in by viewers become a legitimate part of their daily routine; viewers follow the emotional highs and lows of these characters, track their whereabouts and their relationships, and monitor their progress, both literally and figuratively. The relationships formed with these fictional characters and locations, then, holds very real weight in the lives of fans, and the loss thereof results in a collection of identifiable mourning behaviors as they attempt to adjust to filling the hours previously spent engaging with the objects of fandom.

HUMOR AS A COPING MECHANISM

Hey @NBC: if I promise to laugh at one Jay Leno joke, will you promise to air @nbcCommunity soon? I'm basically offering you my soul here.[4]

Fans retain deep emotional connections to these shows after they cease airing. Humorous outlets, especially memes, are used to display these emotions while simultaneously keeping vulnerability to a minimum. Sometimes the jokes are aimed at those outside the fandom, sometimes the humor is pointed at fans themselves, and sometimes the humor masks anger at the network responsible for the loss of the show.

When used to mask or subversively express anger at networks for canceling a favored show, fans display identical use of humor as a coping mechanism as do individuals who feel they have been the victim of an interpersonal transgression, or a "nontraumatic social transaction that individuals nonetheless perceive to be morally wrong and personally harmful."[5] Further, we cope with these perceived wrongs by formulating an emotional response deemed appropriate for situation-specific contexts.[6] While non-fans or casual viewers dismiss the end of show production as a non-event, or an event in due course, strong fans view cancellation or ceased production as a personal affront, and they transact a more extreme emotional response accordingly.

To exemplify humorous postings as an expression of resentment toward a network for canceling a show, a Fireflyfans.net user posted the following message to a board in May 2014, formatted as a sarcastic e-card featuring the image of a man praying: "Dear God, Please make the jerks who canceled (sic) *Firefly* spend an eternity watching Jersey Shore reruns. Amen."[7] In just one line of text and a single image, the poster conveys his anger at Fox, his disbelief that an inferior show (in his opinion) had a longer run, and his sadness over the loss of *Firefly*.

Similarly, a Reddit user posted a meme featuring the juxtaposition of a shot from the music video for the song "What Does the Fox Say" (by the band Ylvis) with a promo shot of the *Firefly* cast, overlaid with one word: cancelled. Reddit, which bills itself as "the front page of the Internet," is a news aggregator Web site—content is contributed and commented on by users. Every post on Reddit, whether it be the original post in a discussion thread or a reply, is presented with two arrows left-justified in relation to the title of the thread or post content. Those two arrows, one pointing up and one down, allow users of Reddit to register their approval or disapproval of that thread or post based on topicality, quality of content, and general popularity. Redditors refer to the practice of clicking the "up" arrow as "upvoting," and clicking the "down" arrow as "downvoting." When a post or thread is upvoted, in increases the visibility of that content, as the most upvoted content remains closest to the top of the page. The juxtaposition of "What Does the Fox Say" with the Firefly cast was upvoted by 1,830 Reddit users, testifying to the fan base's collective anger and need for constructive expression.

Community fans demonstrated nearly identical humorous blaming behaviors that showcased their anger (this time directed at NBC rather than Fox). A Redditor using the pseudonym FunkyHenryGale posted a pie chart under the title "NBC's Programming Priorities," demonstrating that the network cares about only two things: first, the reality show *The Voice*, and second, "making it 1997 again through science or magic."[8] Again, the combination of humor and vitriol allows fans to express their resentment in an acceptable format, inspiring fellow fans to partake in conversation inspired by witty comments or jabs at a common enemy (NBC). These two posts stand out as some of the more obviously angry examples of humorous content created by fans. The expression and sharing of such feelings of anger is a vital step in allowing fans to overcome bitterness and celebrate the object of their fandom without resentment, transforming their emotional response to a perceived personal affront into a more active emotion.

But humor is not used exclusively as a way to showcase anger or resentment; as a coping mechanism, it is actually one of the most positive methods fans may deploy, and repeated use even leads to reevaluation of the situation in a more positive light.[9] Melanie Booth-Butterfield, Steven Booth-Butterfield, and Melissa Wanzer highlight that humorous communication techniques have demonstrable physical and mental rehabilitative effects, ranging from increased interpersonal attraction to a stronger immune system.[10] And because the literature on coping widely considers it to be an interactive process, the proliferation of message boards available to fans enables broader access to both posting and viewing humorous messages. Consider this response Nathan Fillion gave after being asked how he would want Fox to apologize for canceling *Firefly*; rather than evading or answering aggressively, Fillion said "I'm gonna need a well-appointed tropical island to which every year a cruise ship would arrive with *Firefly* fans. We would have a life-sized replica of the ship, on which we can play out scenes of the show!"[11] By redirecting that question away from blaming Fox, and toward a celebration of the fans' enjoyment, Fillion effectively encouraged followers to be the master of their own emotions. "Laughing off" pain is not always a valuable coping tactic. However, in the case of perceived network transgression against fans, humorous coping strategies encourage healthier emotional responses and the shedding of anger and resentment.

Similarly, humorous responses and interactions allow fans to acknowledge the aspects of their behavior that non-fans might deem problematic, namely obsessive behaviors or explicit displays of emotionality. Stephanie Kelly-Romano and Victoria Westgate explain that cartoons serve multiple functions in allowing their creators to move past a personal or public crisis—most notably, creative/humorous images help individuals contextualize the problem and examine their own behaviors and actions before versus after the conflict (conflict, in this case, being the loss of a favored show).[12] Creating and sharing photo-cartoons and image-based memes allow fans to subtly apologize for those less-socially acceptable behaviors; at this stage, fans acknowledge that their deeply emotional responses are perceived as unusual by a majority of non-fans. They accept that outlook, but simultaneously cannot completely overcome those displays: commiserating in image-based format provides an integral middle-ground. For example, a user on the *Buffy* subReddit posted a humorous photo-collage of Joss Whedon featuring the quotation "I don't think of [my fans] as fanatical so much as extraordinarily tasteful. They have great taste, they're very smart, and they're better than other people."[13] This post, upvoted 524 times, humorously exemplifies how self-aware fans may both mock and embrace their own fannish behaviors as they progress away from anger at the loss of a show and toward a state of fond acceptance.

SHARING THE UNIVERSE

Thank you, Community fans, for 5 seasons of incredible, heartwarming support. So deeply proud to have been a part of this. #darkesttimeline[14]

When fans have the opportunity to come together with the cast and/or crew of their favorite shows, opportunities arise to share the universe with like-minded individuals and extend the life of the fandom. Additionally, such opportunities provide fans with added validation in the form of proof that the actors and crew who once worked so intimately on the object of their fandom still have a vested interest in the well-being of the show's legacy.

One of the primary ways fans feel connected to the actors and crewmembers who worked on their favorite shows are through AMAs hosted on the Web site Reddit. An AMA ("Ask Me Anything") is an interactive forum in which a public persona sets up a specific time window for Internet users to ask questions and receive answers in a live chat forum. The subReddit (any smaller page on the larger Reddit Web site devoted to a specific purpose or topic) for *Breaking Bad* alone claims twenty-seven separate AMAs from cast and crew, posted over more than two years. These forums give fans the chance to ask actors their opinions on particular scenes or plot theories, give their thanks for bringing life to the show, and bring up queries on the actor's personal life or career history. The actor, in turn, may choose not to answer particularly invasive questions—but because the fans are able to interface with celebrities in real time, from their own home, it brings a sense of intimacy to fan-actor relationships that would have been unimaginable two decades ago.

Twitter also allows fans previously unprecedented levels of intimacy with the creators of shows. Immediately prior to the *Breaking Bad* series finale in September 2013, the main cast members tweeted out their heartfelt thanks to the fan community for keeping the show alive, as well as touching behind-the-scenes photos of the friendships that had developed over the airing of the show. Bryan Cranston, the actor behind lead character Walter White, told his fans, "Well, this is it. The last episode ever of *Breaking Bad*. Thank you for sharing this ride with me. Without you we never would have lasted."[15] By inviting the public to utilize the hashtag #GoodbyeBreakingBad, the cast and crew communicated to fans that they were part of this intimate, bittersweet farewell. Between both the east and west coast airings, the series finale garnered 1.24 million tweets from fans and cast members alike.

Twitter was also a galvanizing force when NBC canceled *Community* after airing its fifth season. The responses from the fan *Community* all expressed despair and anger at NBC—but how they did so varied. Some were pithy: "NBC cancelled *Community*. Help me." Some were long, violent, and specific: "FUCK YOU @nbc WHY WOULD YOU CANCEL @nbcCom-munity WHEN WE ONLY NEEDED ONE MORE SEASON FUCK YOU FUCK YOU EXECUTIVES AND FUCK YOUR CHANNEL DIE."[16] *Community* fans were luckier than many fan communities; their Twitter outreach and other visibility campaigns helped convince Yahoo to pick up the show for a sixth season, to be aired online. Used in this respect, tweets become a vehicle for fans to express sadness or joy, or merely to identify as being part of the conversation—sharing in the broader emotions sweeping through the fandom universe.

Fans also use actor sightings in public to heighten their sense of intimacy with these public personas. On the *Buffy* subReddit, a user posted a picture of married couple Alyson Hannigan and Alexis Denisof—two actors featured frequently in director Whedon's work—dressed up for Halloween with their two children. Comments on the photograph reflected both a personal admiration for the couple's actual personas ("One of the only celebrity couples worth following," "I love them and want them to adopt me") as well as a projection of their characters onto their real selves ("I love that the two brains of the Scoobies got together").[17] Social media both enables fans to get instant updates on their favorite stars' whereabouts, statements, and preferences, and allows them to bond and discuss those updates with other likeminded fans. This method of sharing the universe of their fandom existed through tabloids and newspapers long before the advent of social media and the Internet, but the Internet enables continuous updates and real-time discussion, as well as unimaginably heightened reach.

PROLONGING HOPE FOR RENEWAL

Despite these shows being off the airwaves for a decade or more, fans continue to display their fervent hope for renewal in posts. This may take the form of speculation over cryptic interview sound-bites with show-runners like Joss Whedon, discussion of what story lines would be covered, what format would be best for renewal, how to raise money/awareness, and so on. Sometimes these posts are serious in their attempts to bring the shows back to life in their original format; just as often, the poster admits that they know there is probably no real hope of resurgence, but the act of carrying on hope itself has become ritualistic and emblematic of a true fan.

Twitter hashtags are a primary way in which fans express their fervent hope for renewal. Shared conversations on Twitter allow fan communities to consolidate their voices under the same banner; indeed, we have already discussed how Twitter serves as an expedient way for fans to share their universe with each other and with the cast and crew of the show in question. However, tweets can also stand as messages of support during loss or extended mourning, and are frequently based in advocacy in attempts to save a show or character. Vince Gilligan, the creator of *Breaking Bad*, credits the movement his show gained on social media with keeping it alive past season two—without the buzz created over Twitter (#savewalterwhite and #breakingbad) and Facebook and the binge-watching capabilities afforded by Netflix and Amazon, the show likely would have been cancelled early. As Gilligan states in an interview with *Wired*, "My gut tells me that it's very possible we wouldn't have made it to sixty-two episodes without . . . these technologies and this cultural creation of binge-watching."[18] Such testimonies from directors and show runners only reify fan convictions that holding out hope for more episodes of a favored show is now a telltale sign of diehard fandom.

Community has already been discussed here as an example of a successful Twitter campaign, but it bears repeating as an excellent exemplar of the sort of armchair advocacy that fans demonstrate on social media in the hope of renewing their favorite shows. The hashtag #sixseasonsandamovie began as an in-universe joke in *Community*—the phrase was initially introduced by Abed, a character in the show who closely represents a typical social conception of a fan, as his prediction for how long a favored superhero show would last. Show creator Dan Harmon flashed the phrase on-screen as the final visual during the season 3 finale, which many viewers thought might become the series finale. After that, the phrase was adopted as the fan community's rallying cry.

Of course, not all renewal efforts are as successful as that of the *Community* fans. *Firefly* was cancelled in 2002; more than a decade later, fans continue to hold out hope that they will see their favored space captain ride again. Hope grows particularly vigorous after a large-scale success, like *Community*'s; one Reddit user posted humorously that after he heard that Yahoo was saving *Community*, all he could think was "Save us, Ask-Jeeves!"[19] The post implicitly acknowledges the unlikelihood of this with its choice of a now-defunct search engine, and the respondents to the post admit that it is increasingly unlikely that *Firefly* will ever return. However, while they acknowledge that it is most likely "too late" to resurrect their beloved show, the fans cannot give up hope that a similar renewal will come their way.

When the hope of renewal is lost, fans can still serve as advocates for their show, creating awareness to inspire new generations of fans to take part in their universe. On the discussion forum *Firefly*fans.net, a user rallied to gain support for a grassroots Internet campaign to convince NASA to name a shuttle Serenity, after the spaceship featured on *Firefly*. Other posters requested the forum users' knowledge of guerilla marketing tactics—dropping flyers or posting stickers that might covertly pique interest in the show, temporarily boosting DVD sales or Netflix viewings. Through such tactics, fans create their own illusion of movement and continuity while also recruiting new fans to join in their campaigns for recognition and respect.

PROVIDING A SPACE FOR CRITICISM

> [*Breaking Bad*] isn't only about what Gilligan and the writers intend. It's about what the audience invents as well, given our cultural context and the long history of symbolic meanings in Western Culture. [20]

One of the most intriguing functions of the fan communities studied here is the perceived ability to offer a heightened level of insight and critique. These fan communities utilize social media as not only the gathering place for sharing critique, but also to mutually reinforce their positionality as intimately engaged and enlightened fans, as established during the discussion of how fans share their universe. While it is commonplace for base-level evaluation of material in terms of liking or disliking to be shared on Reddit on a per-episode or even a per-minute basis, Reddit users in the subReddit dedicated to *Breaking Bad* have created a thread entitled "The Official *Breaking Bad* Critical Analysis Thread,"[21] which has amassed more than eight hundred responses. This thread has housed criticism and praise on all aspects of the show, from casting, cinematography, writing, and intricate analyses of lighting choices that reflect common themes throughout the series. Those Reddit users posting in this thread provide links to annotated screenshots, reference interviews with creator Vince Gilligan to support theories, and cite outside academic material on identity management when debating the duality of Walter White and Heisenberg.

Fans of *Community* on Reddit have taken critical analysis a step further, as they have created a separate "study room," in line with the central theme of the show, to house critique. "Study Room F" is home to 4,174 members, and their stated aim is to "let *Community* fans share and discuss their views on what's done well and what's lacking in terms of character development, plot arc, and tone of episodes and seasons of *Community*."[22] The discussions span from analysis of a single episode to debating the legacy of the show once it has concluded. The cancellation and subsequent renewal reinvigorated discussions in this subReddit that were dormant since the conclusion of the fifth season. Posted in the sidebar of Study Room F is the following message which communicates the tone expected by all members posting, it reads, "Remember, we are all Human Beings here, and we are all on the same side: the side of people who want to seriously critique the show. That critique doesn't always have to be positive or negative!"[23]

These critical spaces also give fans a place to air critics' problems with the series and discuss their merits (or lack thereof). A user asked the *Buffy* subReddit in July 2014 what some of the popular criticisms of the show were while it was airing, as many fans on the forum only came to know *Buffy* through Netflix or DVDs years after it went off the air. The ensuing thirty-eight-response discussion ranged from thematic arcs in particular seasons, to specific character arcs, to the continuity present within the show's magical subplots. These spaces are not just a place for fans to air their own grievances (including those who daringly state their opposition to popular romantic pairings), but a space for the fandom to come together and create a critical communal consensus. Knowing the opinion and viewpoints of the group is one hallmark of a "true" fan, as opposed to a casual or part-time viewer; these critical spaces further the formation of and education on those opinions.

EXPANDING THE UNIVERSE

This is the best update you've ever released! I can't believe you actually made it, all that work for an incredible result. I'm playing it again and it's like a new game. Congratulations to you, to all the fans, because we're a great community who will still live a long time after the show :-)[24]

These shared spaces inhabited by so many devoted fans allow for a blossoming of creativity that is inspired from the source material. The collected users take aspects of the show, whether they are big or small, and bring them to life. The expansion of the universe in which the show resides provides the opportunity for a participative fandom. Spaces are created that allow the viewers to have the same experiences as the characters in the show, and in doing so the fans use the source material and their own creativity to build worlds that are far more in depth than the source material intended. The world created by *Community* is amazing in its seemingly mundane setting, yet fosters so many opportunities for fans to expand on small details within the show.

One such example has been identified previously in Paul Booth's examination of the extension of the show-within-a-show, *Inspector Spacetime*. Within the *Community* universe, characters Abed and Troy are devoted fans of *Inspector Spacetime*. The show-within-a-show bears a close resemblance to *Doctor Who* and the devoted fans of that show, which serve as a blueprint as well as comedic fodder. Booth posits that *Inspector Spacetime* serves as an example of hyper-mediated fandom that marries fans of *Community, Doctor Who,* and *Inspector Spacetime.*[25]

Dedicated message boards, tumblr posts, and fan-produced shorts all provided fans of *Community* with an opportunity to engage *Inspector Spacetime* on the same levels as characters Abed and Troy. Booth's work provides an insight into the creation of *Inspector Spacetime* as an entity derived from *Community,* and explores the tension that exists when that entity becomes its own commodity.

While the message board dedicated to *Inspector Spacetime* was largely inactive during the period of cancellation and renewal for *Community,* fans of the show have created another extension of the universe that has remained a consistent presence. Reddit users have created an interactive downloadable video game based on events in the episode "Digital Estate Planning," where all study group members must cooperatively play a video game titled "Hawkthorne," and conquering the game would result in character Pierce Hawkthorne receiving his inheritance. The bulk of the episode is shown as the game being played live, and the various characters, background animations, goals, and obstacles provided base material for Reddit users to create a playable model of the game. To play the game, one can download the latest version at the subReddit devoted to Hawkthorne,[26] where users also share tips, post questions, and leave praise for the game. The Hawkthorne subReddit houses 12,879 group members, with devoted players posting updates daily.

Fan fiction is another way fans have vastly expanded the universe of the show they love, and is particularly important if fans are dissatisfied with the way the series ended. Fan fiction provides an outlet for fans to explore avenues not appropriate for primetime television, such as character's sexual relations (either in actual canonized pairings, or in "slash" pairings that counter official canon). Popular Web site fanfiction.net contains over 32,000 *Buffy* and *Angel*-centric fan fiction submissions, ranging back to 1998;[27] viewers have the opportunity to review and respond to these stories, further immersing themselves in the fictionalized universe even decades after the original show has ceased airing.

INHABITING THE UNIVERSE

For some fans, discussion of the series and interaction with actors is not enough to satiate their need to experience the universe they love so much; they desire to fully inhabit it. These are the fans who become so attached to a show that after it ceases production, they have difficulty accepting the fact that it will no longer be a new, continuously evolving part of their weekly routine. These fans seek any means by which they can break down the barrier between the everyday world they live and work in and the world of their favored show. The increasingly popular practice of dressing up as your favorite character, known as "cosplaying," is a primary way for fans to marry their desire to inhabit the universe that houses the shows they are fans of, and their real life surroundings. One fan considered *Firefly* such an integral part of his experience that he themed his wedding around it, asking the subReddit users for advice on constructing the perfect outfit.[28] In the *Buffy* subReddit, fans have maintained an informal, ongoing, low-key competition regarding who can best recreate the vampire face make-up employed on both *Angel* and *Buffy*. Other users assess the validity of these get-ups and offer critiques on how they may be altered or improve for future wear.

Similarly, fans often use each other as resources for acquiring famous props from the series (such as a replica of the scythe Buffy wielded in the series finale) or to show off their fandom-inspired artwork (handknitted hats inspired by *Firefly*, or character portraits from every series). Creating something from within the universe allows fans to use their hands to fully immerse themselves in the show world for a period of time; crafting something wearable, such as the aforementioned knitted hat further allows them to physically embody the show's universe or characters. *Breaking Bad* fans have actually created an entire subReddit for their fan art—r/breakingbadart—on which fans share both the works they themselves have created and commercially available items.

In further efforts to inhabit the universe of the show, fans frequently extend and perpetuate the assumption that the actors who appeared in the show actually exist as the characters. To further the illusion that their show lives on in some format, fans will point out to each other when actors appear in other shows—but referenced by their character name, not their actual name. By treating each show as its own universe, the fans can use actors who have appeared in multiple shows as bridges between the worlds. When Whedon actors Alyson Hannigan and Alexis Denisof appeared together in a movie, a fan excitedly posted to the *Buffy* subReddit that "Willow and Wesley are in marriage counseling!" (referring to the characters respectively portrayed by each in *Buffy*).[29] Similarly, fans on *Firefly*fans.net excitedly point out when "Shepherd Book" (Ron Glass) or "Jayne" (Adam Baldwin) appear in other shows, allowing fans as a community to imagine that while *Firefly* is over, their favorite characters still exist, having adventures in other universes.

To further exemplify this, the long career of actor Bryan Cranston and his appearances are so well-established that fans of *Breaking Bad* imagine Walter White as an enduring character with a pre-*Breaking Bad* backstory. Cranston's experience on Malcolm in the Middle has inspired a subReddit titled "Breaking In The Middle,"[30] where the fans create mash-ups between the two shows and imagine Walter White and Hal as interchangeable characters. By continuing White's lifespan as a character, fans can forget their sadness at the show's conclusion and focus on further expanding the universe, creating ties to enmesh the two entities as much as possible. Creating linkages infinitely increases the amount of opportunity for humor, fan fiction, external referencing, and criticism.

CONCLUSION

Jenkins clarified that gossip and discourse are always critically important to fans, even in the midst of a show's successful run, because they have "practical value in perpetuating fan culture and because they offer new ways of thinking about the programs."[31] This practical value increases for fans who are striving to compensate for the loss of a show's active presence in their lives; the communal nature of coping demands bonding with like-minded individuals to expedite the journey to wholeness. We have discussed six categories that fan coping strategies fall into after a favored show has ceased production: creating humor, sharing the series universe, advocating for renewal, providing space for criticism, expanding the universe, and inhabiting the universe.

Fans who create humorous outlets for handling their distress return more swiftly to balanced emotional states and handle their anger more productively; additionally, one humorous outpost is likely to spawn others. Producers and networks could effectively harness the power of humor in moving fans past their anger at cancellations by encouraging actors to treat the cancellation with gentle humor, following Nathan Fillion's example with *Firefly*. Humorously redirecting fans toward spin-offs—such as the upcoming *Breaking Bad* spin-off, *Better Call Saul*—might also encourage fans to channel their emotions in a healthier direction.

Sharing the universe of the series is one of the primary ways in which fans feel bonded to one another. Including discussion with actors and crew members on social media and fan-to-fan discussions, these practices extend the life of the universe and re-convince fans that the men and women who worked on their favored show shared the intimate connection that they (the fans) feel toward the object of their fandom. When actors and crew members actively share the series universe with fans, but producers and network executives do not, it strengthens feelings of ill will toward those executives; to weaken inclinations toward attacking networks, executives might consider involving themselves in social media conversations, making themselves a part of the shared-universe bond.

Fans are often adamant about their hope for the renewal of their show, regardless of the circumstances in which it stopped airing. Renewal efforts have drastically diverged from the postcard campaigns of yesteryear; the reach of social media results in massive support, both emotionally and often, financially. Additionally, carrying on hoping for the renewal of your show—even if more than a decade has passed since it last aired—is considered one of the marks of a true fan, and fans who admit defeat are frequently denigrated by their peers.

The spaces provided by social media give fans a space to offer criticism as well have response to their views. The validation of that criticism with others in a shared space reinforces the positionality of the enlightened and advanced fan, able to move beyond mere viewing and provide an informed reading of the source material.

The extension of the universe beyond the source material is a key component of how fans cope with the cessation of new material. By expanding on aspects of shows that are provided limited screen time, fans pay tribute to the creativity of content creators by seemingly continuing their vision. The creation of playable games and web series based on show material are indicative of the dedication of fans and the desire to continue the experience of engaging with source material beyond the boundaries of broadcast.

The ability for fans to become part of the universe through cosplay and creating bridges between their favorite shows allow them to redefine the boundaries of original material. By inhabiting characters, bringing them to life within the fan's real-world surroundings, and expanding the backstories of the characters to incorporate the additional roles of the actors in their favorite shows, fans have found their inroads to interact with the universe that has helped shaped their identity.

Discussing fan's reactions to the loss of entertainment programming as mourning might strike some readers as disrespectful to the grief felt by those who have lost a real, physical presence in their lives. But to devoted fans, the characters and places portrayed in favored shows are sometimes more real than their flesh-and-blood surroundings; finding and immersing themselves in a fictional universe becomes an escape from a daily world that is characterized by boredom, fear, or isolation.[32] Just as fandom culture is becoming more accepted by the broader sphere of popular culture, more research is needed in this area to discover exactly how fans define reality, as well as how strong emotional ties to a fictional universe impact behavior in the "real world."

Lastly, this chapter focused on the behaviors fans display when they are mourning the loss of a show. While it was not within the scope of this inquiry to determine, the researchers strongly suspect that further work would reveal that these behaviors are cycled through as fans move within the cycle of grief. More work in this direction could reveal whether this is in fact correct, and if so, how these behaviors accord with such a cycle.

NOTES

1. Henry Jenkins. *Textual Poachers* (New York: Routledge, 1992), 12.

2. Melanie Booth-Butterfield, Steven Booth-Butterfield, and Melissa Wanzer. "Funny Students Cope Better: Patterns of Humor Enactment and Coping Effectiveness." *Communication Quarterly* 55, no. 3 (2007): 299.

3. Rebecca Bley, "RL on LJ: Fandom and the Presentation of Self in Online Life," in *Buffy and Angel Conquer the Internet*, ed. Mary Kirby-Diaz (Jefferson, NC: McFarland and Company, 2009), 43.

4. "Performing Unspeakable Acts to Get Community on Air," Reddit, last modified October 24, 2012. http://www.reddit.com/r/Community/comments/1207ta/performing_unspeakable_acts_to_get_Community_on/.

5. Michael McCollough et al. "Rumination, Fear, and Cortisol: An In Vivo Study of Interpersonal Transgressions," *Health Psychology* 26, (2007): 127.

6. Sara LaBelle, Melanie Booth-Butterfield, and Keith Weber, "Humorous Communication and Its Effectiveness in Coping with Interpersonal Transgressions," *Communication Research Reports* 30, no. 3 (2013): 223.

7. "Firefly and Serenity Guerilla Marketing," FireflyFans.net, last modified May 9, 2014, http://www.Fireflyfans.net/mthread.aspx?tid=56507.

8. "NBC's Programming Priorities," Reddit, last modified May 2014. http://www.reddit.com/r/community/comments/255h8n/nbcs_programming_priorities.

9. LaBelle, Booth-Butterfield, and Weber, "Humorous Communication and Its Effectiveness," 227.

10. Booth-Butterfield, Booth-Butterfield, and Wanzer, "Funny Students Cope Better," 299.

11. "Nathan's Answer to the Question, 'If FOX Was Ever to Apologize for Canceling This Show, How Would You Want Them to Do It?'" Reddit, last modified January 2013. http://www.reddit.com/r/Firefly/comments/174e13/nathans_answer_to_the_question_if_fox_was_ever_to/.

12. Stephanie Kelley-Romano and Victoria Westgate. "Drawing Disaster: The Crisis Cartoons of Hurricane Katrina," *Texas Speech Communication Journal* 31, no. 1 (2007): 13.

13. "Joss Whedon Fans According to Joss Whedon," Reddit, last modified February 2013, http://www.reddit.com/r/buffy/comments/19b406/joss_whedon_fans_according_to_joss_whedon.

14. Alison Brie. Twitter post. May 9, 2014, 9:49 pm. https://twitter.com/alisonbrie.

15. Brian Anthony Hernandez, "'Breaking Bad' Finale Didn't Quite Break Twitter or Facebook," Mashable, September 30, 2013, http://mashable.com/2013/09/30/breaking-bad-finale-twitter-facebook/.

16. Matthew Bramlett, "Community Gets Canceled and Fans Are Absolutely Devastated," *The Wrap*, May 9, 2014, http://www.thewrap.com/Community-canceled-twitter-reactions/.

17. "Alyson Hannigan and Alexis Denisof's Adorable Halloween Costumes," Reddit. Last modified November 1, 2013, http://www.reddit.com/r/Buffy/comments/1poqxa/alyson_hannigan_and_alexis_denisofs_adorable/.

18. Angela Watercutter, "Breaking Bad Creator Vince Gilligan on Why Binge-Watching Saved His Show," *Wired*, June 6, 2013, http://www.wired.com/2013/06/breaking-bad-season-5-dvd/.

19. "All I Can Think of with Yahoo Saving Community," Reddit. Last updated July 2014. http://www.reddit.com/r/Firefly/comments/29kyff/all_i_can_think_of_with_yahoo_saving_Community.

20. Theplott, comment on "The Official Breaking Bad Critical Analysis Thread," Reddit, comment posted September 9, 2013, http://www.reddit.com/r/breakingbad/comments/m341p/the_official_breaking_bad_critical_analysis_thread/?sort=confidence.

21. "The Official Breaking Bad Critical Analysis Thread," Reddit, last modified October2013. http://www.reddit.com/r/breakingbad/comments/1m341p/the_official_breaking_bad_critical_analysis_thread.

22. "Study Room F," Reddit, last modified July 28, 2014, http://www.reddit.com/r/studyroomf/.

23. Ibid.
24. "Journey to the Center of Hawkthorne v.0.9.0 Has Been Released (Two Year Anniversary Edition)," Reddit, last modified July 31, 2014, http://www.reddit.com/r/hawkthorne/comments/261y0r/journey_to_the_center_of_hawkthorne_v090_has_been/chol5bu.
25. Paul Booth, "Reifying the Fan: Inspector Spacetime as Fan Practice," *Popular Communication* 11 no. 2 (2-13): 148–49.
26. "Hawkthorne." Reddit, last modified July 29, 2014, http://www.reddit.com/r/hawkthorne.
27. "Buffy: The Vampire Slayer," Fanfiction.net, accessed August 1, 2014. https://www.fanfiction.net/tv/Buffy-The-Vampire-Slayer.
28. "Getting Married (Costume Theme Wedding) and I Want To Dress as Mal from Shindig," Reddit, last modified July 9, 2014, http://www.reddit.com/r/Firefly/comments/2a7c40/getting_married_costume_theme_wedding_and_i_want.
29. "Willow and Wesley Are in Marriage Counseling," Reddit, last modified July 5, 2014, http://www.reddit.com/r/Buffy/comments/29mlu7/willow_and_wesley_are_in_marriage_counseling.
30. "Breaking in the Middle" Reddit, last modified July 14, 2014, http://www.reddit.com/r/Breakinginthemiddle.
31. Jenkins, *Textual Poachers*, 82.
32. Bley, "RL on LJ," 44.

REFERENCES

"All I Can Think of with Yahoo Saving *Community* . . ." Reddit. Last modified July 2014. http://www.reddit.com/r/*Firefly*/comments/29kyff/all_i_can_think_of_with_yahoo_saving_Community.
"Alyson Hannigan and Alexis Denisof's Adorable Halloween Costumes." Reddit. Last modified November 1, 2013. http://www.reddit.com/r/*Buffy*/comments/1poqxa/alyson_hannigan_and_alexis_denisofs_adorable.
Bley, Rebecca. "RL on LJ: Fandom and the Presentation of Self in Online Life." In *Buffy and Angel Conquer the Internet*, edited by Mary Kirby-Diaz, 43-61. Jefferson, NC: McFarland and Company, 2009.
Booth, P. "Reifying the Fan: *Inspector Spacetime* as Fan Practice." *Popular Communication* 11, no. 2 (2013): 146–59.
Booth-Butterfield, Melanie, Steven Booth-Butterfield, and Melissa Wanzer. "Funny Students Cope Better: Patterns of Humor Enactment and Coping Effectiveness." *Communication Quarterly* 55, no. 3 (2007): 299–315.
Bramlett, Matthew. "'*Community*' Gets Canceled and Fans Are Absolutely Devastated." *The Wrap*, May 9, 2014. http://www.thewrap.com/*Community*-canceled-twitter-reactions.
"Breaking in the Middle." Reddit. Last modified July 14, 2014. http://www.reddit.com/r/Breakinginthemiddle.
Brie, Alison. Twitter post. May 9, 2014, 9:49 pm. https://twitter.com/alisonbrie.
"*Buffy*: The Vampire Slayer." FanFiction.net. Accessed August 1, 2014. https://www.fanfiction.net/tv/*Buffy*-The-Vampire-Slayer.
"*Firefly* and *Serenity* Guerilla Marketing." Fireflyfans.net. Last modified May 9, 2014. http://www.*Firefly*fans.net/mthread.aspx?tid=56507.
"Getting Married (Costume Theme Wedding) and I Want to Dress as Mal from Shindig." Reddit. Last modified July 9, 2014. http://www.reddit.com/r/*Firefly*/comments/2a7c40/getting_married_costume_theme_wedding_and_i_want.
"Hawkthorne." Reddit. Last updated July 29, 2014. http://www.reddit.com/r/hawkthorne.
Hernandez, Brian Anthony. "*Breaking Bad* Finale Didn't Quite Break Twitter or Facebook." *Mashable*, September 30, 2013. http://mashable.com/2013/09/30/breaking-bad-finale-twitter-facebook.
Jenkins, Henry. *Textual Poachers.* New York: Routledge, 1992.

"Joss Whedon Fans According to Joss Whedon." Reddit. Last modified February 2013.http://www.reddit.com/r/buffy/comments/19b406/joss_whedon_fansaccording_to_joss_whedon.

"Journey to the Center of Hawkthorne v.0.9.0 Has Been Released (Two Year Anniversary Edition)." Reddit. Last updated July 31, 2014.http://www.reddit.com/r/hawkthorne/comments/261y0r/journey_to_the_center_of_hawkthorne_v090_has_been/chol5bu.

Kelley-Romano, Stephanie, and Victoria Westgate. "Drawing Disaster: The Crisis Cartoons of Hurricane Katrina." *Texas Speech Communication Journal* 31, no. 1 (2007): 1–15.

LaBelle, Sara, Melanie Booth-Butterfield, and Keith Weber. "Humorous Communication and Its Effectiveness in Coping with Interpersonal Transgressions." *Communication Research Reports* 30, no. 3 (2013): 221–29.

McCollough, M., Orsulak, P., Brandon, A., and Akers, L. "Rumination, Fear, and Cortisol: An In Vivo Study of Interpersonal Transgressions." *Health Psychology* 26, (2007): 126–32.

"Nathan's Answer to the Question, 'If FOX Was Ever to Apologize for Canceling This Show, How Would You Want Them to Do It?'" Last modified January 2013. http://www.reddit.com/r/*Firefly*/comments/174e13/nathans_answer_to_the_question_if_fox_was_ever_to.

"NBC's Programming Priorities." Reddit. Last modified May 2014.http://www.reddit.com/r/community/comments/255h8n/nbcs_programming_priorities.

"The Official *Breaking Bad* Critical Analysis Thread." Reddit. Last modified October 2013.http://www.reddit.com/r/breakingbad/comments/1m341p/the_official_breaking_bad_critical_analysis_thread.

"Performing Unspeakable Acts to Get *Community* On Air." Reddit. Last modified October 24, 2012.http://www.reddit.com/r/*Community*/comments/1207ta/performing_unspeakable_acts_to_get_*Community*_on.

"Study Room F." Reddit. Last modified July 28, 2014. http://www.reddit.com/r/studyroomf.

Theplott. Comment on "The Official *Breaking Bad* Critical Analysis Thread." Reddit. Comment posted on September 9, 2013.http://www.reddit.com/r/breakingbad/comments/1m341p/the_official_breaking_bad_critical_analysis_thread/?sort=confidence.

Watercutter, Angela. "*Breaking Bad* Creator Vince Gilligan on Why Binge-Watching Saved His Show." *Wired,* June 6, 2013.http://www.wired.com/2013/06/breaking-bad-season-5-dvd.

Whedon, Joss. Interview by Adam Tanswell. *Cult.* BBC. September 2005.

"Willow and Wesley Are in Marriage Counseling." Reddit. Last modified July 5, 2014.http://www.reddit.com/r/*Buffy*/comments/29mlu7/willow_and_wesley_are_in_marriage_counseling.

Chapter Twelve

So Are the Days of Our Tweets

An Examination of Twitter Use by American Daytime Serials and Their Fans

Marsha Ducey

There's something about the dedication and commitment that soap fans have for their shows that really infused the online fan experience with an intensity that many other Internet fan groups lack. It comes, I think, from the fact that, when the Web became a big part of soap fans' lives, many fans had already been engaged with these soap story worlds for years—in many cases, fans' involvement predated the Internet by decades. The Web, which permits for a really wide range of discussions and actions that can be micro-interventions or can go on for months or years, almost seems like it was specifically built as a platform for soap fans, who have decades' worth of information and insight to discuss.
—Abigail De Kosnik, co-editor of *Survival of the Soap Opera: Transformations for a New Media Era*, talking with media scholar Henry Jenkins [1]

Daytime serials, more commonly known as soap operas, have been part of American popular culture for more than eighty years and are among the longest running broadcasts. Named by journalists for the product they once advertised, cleaning supplies, soap operas are dramatic programs with ongoing stories and characters.[2] "They never begin and they never end."[3] In June 2014, more than 12 million combined viewers tuned into the four remaining daytime serials to see the antics of their favorite characters.[4] Soaps have several distinct characteristics. American daytime serials have been broadcast Monday through Friday for decades, first on radio and then, in the 1950s, moving to television and making what used to be considered the three major television networks, ABC, CBS, and NBC. The pace of soap operas is deliberately slow because they were originally designed for housewives who would be working while listening to or watching the stories.[5] Flashbacks and repetition are two elements commonly used to help those following the program catch up on what they may have missed. Soap operas have multiple characters, including "sensitive" male characters and powerful female characters and tend to center on a home or "some other place that functions as a home."[6] At the heart of the stories told on soap operas are love and family. Traditionally, stories of this focus were thought to be "women's stories."[7] Indeed, women viewers are still the majority of soap opera viewers. Male viewers make up only 20 to 25 percent of soap opera viewers.[8] Soap operas are known for being emotional and dramatic, and this association, in addition to the fact that the majority fan base has been and continues to be women and the programs air during the daytime, has contributed to a cultural disrespect for soap fans. "Followers of soap opera have often been depicted as unproductive and powerless, as housewives enraptured and captured by the weepy melodrama supplied by daytime dramas during the same hours of the day that more powerful members of American society (adult men and women employed outside the home) are at their most productive."[9] Soaps are considered by some to be "the absolute bottom of the television hierarchy, lumped with game shows and professional wrestling in terms of their perceived moral worth."[10] But media scholar Robert Allen argues that soap operas are "demonstrably one of the most narratively complex genres of television drama whose enjoyment requires considerable knowledge by its viewers" and that its fans "cut across social and demographic categories."[11] Fans of the genre flock by the millions to catch the latest goings on in the fictional towns of Genoa City, Los Angeles, Port Charles, and Salem.

The literature includes a plethora of research on soap operas and their fans, starting with communication research pioneer Herta Herzog's 1940s work, "What Do We Really Know About Daytime Serial Listeners?"[12] However, while the literature includes studies of soap fans and the Internet,[13] scant research exists on soap opera fans and social media. This chapter examines the history of soap operas and soap opera fans and includes an exploration of how some soap opera fans are using the social media site Twitter to express themselves, and conversely, how soap operas are using the social medium to engage with viewers. Beyond merely expressing their opinions about what is happening in their "stories," this research argues that some fans use Twitter in an attempt to drive story lines of "their" shows, to make their voices heard by the television network, and to show their own derivative works based on their favorite shows. In turn, soap operas use Twitter to promote shows and show-related events, to catch followers up on what has happened that day and to interact with fans in multiple ways.

A HISTORY OF SOAP OPERAS

American daytime serials started on Chicago radio in 1930 at a time when radio broadcasters were trying to lure manufacturers of household products to advertise to female listeners.[14] Irna Philips wrote and acted in the first radio soap opera, a fifteen-minute show called *Painted Dreams*, a family drama. (Philips scripted many early soap operas, including *Guiding Light*, which would later move on to television, and co-created *As the World Turns* and *Days of Our Lives* for television.[15]) By 1937, soap operas filled daytime radio and attracted sponsors like Procter and Gamble, maker of household products like detergents and creator of its own radio soap opera company. "In 1948 the ten highest-rated daytime programs were all soap operas, and of the top thirty daytime shows all but five were soaps."[16]

The first soap opera on TV, *The First Hundred Years*, debuted in 1950 and was produced by—you guessed it—Procter and Gamble. In 1956, Phillips's *As the World Turns* would be the first thirty-minute soap, and in the 1960s, soap operas would be exclusively on television, where they were not only popular, but less expensive to produce than primetime programming.[17] Soap operas, which were considered "women's fiction,"[18] attracted a key demographic for advertisers: women eighteen to forty-nine years old. And that meant network soaps competed for viewers by experimenting with stories (for example, bringing in younger characters and social issues and new settings).[19] In 1970, CBS offered eight soap operas, and ABC and NBC five soaps each, for a total of eighteen daytime soap operas.[20] In the early 1970s, 20 million viewers watched American soap operas.[21] As more women entered the workforce in the 1970s, soap opera magazines began being published to fill women in on what was happening in their soaps. *Soap Opera Digest* started in 1975, the same year that *Another World* became the first one-hour soap opera.[22] By 1982, the number of soap magazines was ten.[23] In the 1980s, despite the fact that ratings for soap operas overall were declining because of women working outside the home and options offered by cable television and the VCR led viewers elsewhere, *General Hospital* would earn the highest rating ever for a daytime soap opera when Luke and Laura married as more than 30 million viewers watched the nuptials.[24] The 1980s also saw the emergence of primetime soap juggernaut *Dallas*, which was broadcast in fifty-seven countries.[25]

Since the 1980s, number of viewers for soap operas has decreased significantly. In 1990, 78 million viewers tuned into daytime soaps, compared to 17 million in 2009.[26] Some say the soaps never recovered from the viewers who left soaps to watch the O. J. Simpson murder trial in 1995.[27] Others say the loss in viewership in a simple matter of more women working outside the home. Women made up 31 percent of the labor force in 1952 and nearly 47 percent in 2010.[28] A cable channel devoted to soap operas, SOAPnet, debuted in 2000. The channel re-ran episodes in primetime, giving viewers who worked during the day an option to catch up on their shows. Soap opera ratings continued to decline, and in 2009, *Guiding Light*, the soap opera Irna Phillips started on radio, broadcast its last episode after seventy-two years on radio and TV. It was the second-longest show on television, after *Meet the Press*.[29] *As the World Turns* was canceled and ended its fifty-four-year run in 2010. The outlook for soaps was bleak. "We are seeing the end of a genre, I think," a former television network executive told the *LA Times*.[30] Two more soap operas—*All My Children* (AMC) and *One Life to Live* (OLTL)—were canceled by their network in 2011 and lifestyles shows that were less expensive to produce replaced them. AMC and OLTL aired on ABC for forty-one and forty-three years, respectively. ABC, which owned both shows, licensed the shows to the company Prospect Park in 2011, and after solving financial and labor union problems, forty new thirty-minute episodes were available on iTunes and Hulu in 2013.[31] The shows also aired for a short time on OWN, Oprah Winfrey's network. *One Life to Live* and *All My Children*'s brief new runs ended because of a lawsuit with ABC over characters and actors and financial problems.[32] This left four soap operas airing daily on ABC, CBS, and NBC combined.

SOAPnet disappeared on New Year's Eve 2013. The popularity of DVRs meant that potential viewers could easily record shows and had impacted SOAPnet's viewership.[33] TVGN, a network owned by CBS, continues to repeat CBS soap operas *The Bold & the Beautiful* and *The Young & the Restless*. ABC and NBC offer episodes of their soaps for viewing on their Web sites.

CURRENT STATE OF AMERICAN SOAP OPERAS

At the time this chapter was written, only four soap operas remained on network television: *General Hospital* (GH) on ABC, *The Young & the Restless* (Y&R) on CBS, *The Bold and the Beautiful* (B&B) on CBS, and *Days of Our Lives* (DOOL) on NBC. Increased viewership in the 2013–14 season led to renewed optimism. The number of viewers on all four soaps had increased 10 to 14 percent.[34] "Soap fans can rest assured—this genre has not staged its last dramatic shocker yet."[35] Experts have speculated that the increase might be from viewers of the canceled soaps switching to those still on the air,[36] increased use of and research of DVRs to record soap opera episodes,[37] "new storylines and characters targeting the key female 18–49 demographic,"[38] and Nielsen ratings incorporating social media, recording episodes and watching episodes online into their ratings system.[39] While ratings for soap operas are a fraction of what they were during the height of the genre, they, at the moment, no longer appear to be in decline.

Fandom

As previously noted, the majority of soap opera viewers are women, who make up 75 to 80 percent of the audience. While the demographic advertisers seek is women eighteen to forty-nine, the reality is the median age of the soap opera viewer—and television viewers as a whole—is climbing.[40] "The median age of the average daytime soap viewer currently stands around the mid-fifties."[41] For television overall, the fastest growing age bracket of viewers is fifty-five and older.[42]

Merriam-Webster's online dictionary defines "fan" as "an enthusiastic devotee (as of a sport or a performing art) usually as a spectator" or "an ardent admirer or enthusiast (as of a celebrity or a pursuit)." Scholars argue fandom is more complicated than that. Historically, being a fan has been portrayed as abnormal, particularly when it involves fandom of media or celebrity, because it is not a "real" event like sports.[43] For decades, scholars have been studying fans of everything from Star Trek to soccer (and indeed, soap operas as well). Herzog found that fans of radio soap operas were listening to soap operas for "emotional release," "wishful thinking" (wishing life like that), and "valuable advice."[44] Some argue that fans are nothing but "cultural dopes who passively consume the texts of popular culture."[45] Grossberg rejects this argument, instead asserting that audiences are active and working to make texts meaningful to their lives.[46] Grossberg states that fans construct their identities through fandom and by seeing differences between themselves and others, and that being a fan can create a sense of empowerment.[47] "By participating in fandom, fans construct coherent identities for themselves."[48]

Fans make investments, not only of their time, but with their money.[49] They buy soap opera magazines, attend soap opera conventions, and go to appearances by soap opera actors. Because of the long-running nature of soap operas, fans are often still there watching when writers are long gone.[50] Soap opera fans have long claimed "ownership" of their characters, stories, and communities.[51] C. Lee Harrington refers to it as "moral ownership."[52] Even in the 1940s, fans were vocal about their stories and their characters.[53] The Internet has given fans a new outlet to express their opinions and feelings about soap operas. Before the Internet, many fans were limited to writing letters or writing to soap opera magazines like *Soap Opera Digest*, to get their voices heard.[54] Now, with blogs, Facebook, and Twitter, fans can connect with other fans and express their opinions to shows instantaneously. The Internet also allows fans to join forces. When ABC cancelled *All My Children* and *One Life to Live*, thousands of fans protested and created Web sites in an attempt to save their shows; one fan even created a song, "We Won't Let Our Shows Die," that she posted to YouTube.[55] "Nothing is ever dead when you're a soap opera fan" Coleman Bell, founder of Fans United Against ABC, told a reporter.[56]

SOCIAL MEDIA AND TWITTER USE

Women use social media significantly more than men. The Pew Research Center's Internet and American Life Project found in 2013 that 73 percent of adults who are online use social networking sites, with 78 percent of women saying they use social networking sites compared to 69 percent of men.[57] While younger survey respondents were more likely to use social media, 65 percent of those ages fifty to sixty-four and nearly half—46 percent—of those sixty-five and older said they did. Pew found that the social media platform used most often was Facebook, which is used by 71 percent of online adults. Overall, 18 percent of online adults were using Twitter in September 2013, and 19 percent of moms were, according to Pew. Younger people are more likely to use Twitter. Pew found that 31 percent of online adults ages eighteen to twenty-nine and 19 percent of online adults ages thirty to forty-nine used Twitter.[58] Only 9 percent of online adults ages fifty to sixty-four and 5 percent of adults sixty-five and older reported using Twitter.

Each remaining soap opera has at least one official Twitter account. While the shows are watched by millions, the Twitter accounts are followed by thousands. Followers for the official soap opera Twitter accounts number between a low of 30,600 and 64,500 for the two *Bold and Beautiful* accounts to a high of 167,000 for *General Hospital*'s sole Twitter account. *Days of Our Lives* has two Twitter accounts, one of which previously included a phone number for fans to provide comments but has changed instead to an email address for comments. It was the only official soap Twitter account to offer a way to reach the soap to comment by phone or email. The *Days* account also has a list of fifty-five fan accounts listed as "Fan Campaigns" for characters, actors, and story lines that other Twitter users can follow. *The Young & the Restless* has 122,000 followers for what it refers to as it's "the OFFICIAL The Young and the Restless Twitter"[59] and 117,000 for its @YRinsider account, which focuses on "latest news, exclusive interviews and more!"[60]

What follows is a look at the themes that emerged when analyzing a sample of Twitter messages or "Tweets" from fans and the official soap opera accounts. Twitter is a social media platform launched in 2006 that allows people to send text messages or "Tweets" of up to 140 characters. Users may also send videos or pictures, and messages may be sent from a computer or mobile device. Tweets are public and can be seen by anyone who follows the user or Tweeter *unless* the Tweeter makes his or her account private. Twitter claims that 500 million Tweets a day are sent out, and it has 271 million monthly users, with 77 percent of accounts being outside of the United States.[61] Some Twitter users may demonstrate that they like or agree with a Tweet by doing something called "favoriting," which is clicking on the favorite link on the Tweet. Another way Twitter users can show their agreement or interest in a Tweet is by clicking on the "retweet" link on the Tweet, which will then send that Tweet on to other Twitter users.

A quick search using the Twitter search engine Topsy reveals that thousands of Tweets about soap operas are sent out daily. For example, a search of the hashtag #Days revealed almost 13,000 Tweets had been sent out using that hashtag in a five-day period. To make this exploratory analysis manageable, only a sample of Tweets from soap opera fans over a three-month period was analyzed. Likewise, only a three-month sample of Tweets from soap opera accounts was analyzed. Using qualitative content analysis, the researcher grouped Tweets into categories and looked for emerging themes.

THEMES IN FAN TWEETS

Three themes emerged when analyzing a sample of tweets and Twitter feeds from soap opera fans: the sense of identification that scholar Grossberg has described in fandom;[62] self-expression in both words and art; and advocacy, which fits into both what Grossberg has argued about fans' sense of empowerment[63] and what scholar Harrington has called the fans' sense of "moral ownership" of their soaps and their characters.[64] Twitter provides a space where fans can identify as part of a group. Some fans identify themselves as soap fans in general with their Twitter names or their Twitter bios. For example, @SassySoapFan describes herself on her Twitter bio as "a devoted soap fan & proud SWAN & I'm NOT giving up on OLTL & AMC!!—watched GH since it first aired!"[65] That Twitter bio not only represents @SassySoapFan's identification as a fan, but also her sense of advocacy and empowerment. OLTL and AMC are two soap operas canceled by ABC that had a brief post-cancelation life in 2013 through Hulu on the Internet. @SassySoapFan is expressing her hope that the shows will return again someday. Another Twitter user, Luke W., who goes by the Twitter handle @Soap_Dude, expresses his love for soap operas in his Twitter feed bio, "Spreading the joy of soaps to the masses! #GH #YR #DAYS #BB also love #Dallas, #Supernatural, #Revenge & #Nashville! Adrift in Soapland." [66] The Twitter bio for Marie (@MarieSoapFan) states "♥ Love Y&R, B&B, GH ♥ Miss ATWT, GL and OLTL."[67]

Other fans are more targeted in their fandom for particular soap operas, characters, or story lines. Examples of this include Y&RDeeVa (@YR_Deeva),[68] @stickyvitch (a fan of characters Victoria and Stitch on

YR),[69] Team Austin (@TeamMuldoon, referring to the character Austin on *DOOL*),[70] Chad & Abigail (@abbychadfans, referring to two characters on *DOOL*),[71] and Morgans & McAvoy (@JasamUnited, referring to characters on GH).[72] In some cases, these Twitter users are not only identifying themselves as fans, they are also advocating for a particular couple, actor or story line on the soap opera with their choice of Twitter name or handle. Some of these "advocacy" or campaign sites, like the previously mentioned @JasamUnited, which states it is a "Fanpage dedicated to GH's #JaSam and YR's Dylan McAvoy, plus their portrayers Steve Burton, Kelly Monaco and Billy Miller!"[73] have more than one thousand Twitter users following them. The site "Forbidden Love," which bills itself as a "wonderful online community for fans of NBC's Days of our Lives who love EJ DiMera (James Scott) & Sami Brady DiMera (Ali Sweeney) aka EJami,"[74] has 1,900 followers. An example of a community Tweet follows: "Hope all our #EJami friends had fun in Boston with @IamJamesScott and crew! Secure travels back home, everyone!"[75] The Tweet had five favorites, four retweets, and two comments.

Other Twitter soap fans stick to putting their opinions in their Tweets, not the Twitter bios, as Jason B. (@JasonXHowe) does: "Just watched Friday's #DAYS. #Snore. #DOOL Can Kristen & Daniel please kill each other so we can all move on?"[76] Fan creativity extends beyond their words to videos showing their favorite characters or story lines, memes (pictures with words) and faux Twitter accounts. Texts from Salem (@TextsFromSalem) is a Twitter account followed by more than two thousand Twitter users and has a bio that reads, "Texts, Chats, and More from Salem's Finest #Days."[77] It purports to be based in Salem, United States, the fictional town of *Days of Our Lives*. Tweets from the person or persons behind this account, which is followed by more than 1,500 Twitter users, include an image of the actress who plays Theresa in a digitally altered "Clue" card that lists her as a suspect in a murder. Her weapon is listed as a fire poker, which is what she hit her father-in-law with on the show. Other images include digitally created text messages to and from characters, a memorial page for a murdered character, and faux bumper stickers for characters' cars. Twitter accounts like Texts from Salem in particular show that some fans take a great deal of pride and invest themselves in the fictional stories told on soap operas. They are *their* stories, and they want to share them with others in their community. They know that "Texts from Salem" will mean something to fellow fans, community members who know what Salem is and why the image of Theresa Donovan in a "Clue" game card is funny. It's an inside joke told in a public forum.

Although this chapter focuses on daytime soap operas, at the time this chapter was being written a perfect example of soap opera fans advocating saving "their soap" from cancellation was taking place on Twitter, but this time the effort was for a nighttime soap opera, the resurrected *Dallas* on

TNT. After three seasons, TNT cancelled *Dallas* in October 2014. Using the hashtag #SaveDallas, fans pushed for another network to pick up *Dallas* and for TNT to reconsider its decision. In a thirty-day period from mid-October to mid-November, more than 838,000 tweets with the hashtag #SaveDallas had been sent, according to the Twitter search engine Topsy. The social media effort was the focus of many blogs, and even mainstream media outlets like Fox and Yahoo did stories on it.[78]

Perhaps adding to a feeling of "ownership" of "their stories" by fans is the fact that the Twittersphere allows for access to soap opera actors in a more direct and instantaneous way that writing letters or posting to an online forum does. Some official soap opera Twitter accounts provide the Twitter handles of their stars. Sometimes a soap opera actor or actress will Tweet a message to his or her followers on Twitter, allowing for a kind of instantaneous "direct" contact that wasn't available twenty years ago. When actor James Scott, who played E. J. DiMera on *Days of Our Lives*, received a bottle of wine and cake for his birthday from a fan community on Twitter, Scott Tweeted a picture of himself with the gifts and the message, "Thanks @FL_EJami_fans for the vino and cake!"[79] That tweet received fifty-two retweets and 216 favorites. Besides interacting on Twitter in informal ways, soap actors also take part in more formal Twitter chats that are scheduled and advertised by the soap operas. This allows fans access to the actors, albeit it virtually.

Tweets from Soap Opera Accounts

This exploratory study also revealed three main themes in official soap opera accounts. They are to promote shows and show-related events, to catch followers up on what has happened that day, and to interact with fans in multiple ways, including "favoriting" Tweets, retweeting, and responding to Tweets. An example of a common promotion Tweet is this @YRinsider Tweet: "Get ready for an #EXPLOSIVE week on #YR! Video."[80] One example of catching fans up—and soliciting feedback—is this Tweet from *The Bold and the Beautiful*'s @BandB_CBS account: "An unexpected kiss potentially changes the game plan for a couple in love. Who should Brooke be with? ow.ly/zRAIO ."[81] This tweet garnered eleven favorites, ten retweets, and eighteen comments. Twitter can be used to gauge viewer opinion quickly. Compare the response to the previous B&B Brooke tweet with the following GH tweet. When GH Tweeted, "Who's Loving #DrunkOlivia?"[82] it received 206 favorites, thirty-four retweets, and seventy-five comments.

The official soap opera accounts also interact with fans by "favoriting" tweets written by fans, retweeting fan tweets, and responding to tweets. This both lets the fan know his or her tweet has been seen and shares the comments with the thousands following the official soap opera Twitter feed. For example, the official Twitter account for *Days* added a comment and a photo to Twitter user @cb11814c5e5e401's tweet and then retweeted it: "Creep Town, USA. RT: @cb11814c5e5e401 @nbcdays I don't have a picture, but can Liam's smile be any creepier???"[83] This retweet was then favorited thirty-one times, retweeted ten times, and garnered more than twenty-five comments. Compare that to the original tweet by @cb11814c5e5e401, which had zero favorites, retweets, or comments.

Official Twitter accounts for soap operas also promote opportunities to meet actors in person and to interact with actors through social media. Y&R, for example, tweeted a photo of actors Jess Walton and Tristan Rogers along with the text, "Get ready to tune into #YR and tweet along with @JessWaltonYR and @tristanrogers! Join now bit.ly/PRSvdH."[84] Soap operas promote events in which actors can be meet in person and retweet soap opera-related tweets from the actors.

CONCLUSION

The future of soap operas is unclear, but it is certainly looking brighter than it did three years ago. Ratings are on the rise, although they are far from where they were twenty years ago. The viewership is aging out of the core demographic that advertisers seek, but that is true of primetime as well as daytime. Advertisers and soaps may be forced to adjust. Soap fans have been studied for decades, and although the ways they interact with their favorite shows, the shows' actors and each other may change, one thing has not: Soap fans are passionate about this never-ending drama of love and family. Twitter and other forms of social media offer new avenue for soap fans to connect with their shows and with each other. Although, as noted earlier, Twitter tends to skew toward younger users, the fact that 65 percent of those ages fifty to sixty-four and nearly half—46 percent—of those sixty-five and older said they use social media offers soap operas new ways to connect with fans and potential new audience members alike.[85] Some soap opera fans use Twitter as a space in which they can share opinions about soap operas and be part of the soap fan community, much like twenty years ago soap fans used online bulletin boards and decades before that fans utilized letters and soap magazines. The exploratory study presented here has limitations, foremost among them the fact that is only a small sample in a moment of time. Will social media provide not only a space for fans to gather, talk about, create derivative works related to and take ownership of their shows, but also an avenue to gathering more viewers to a dramatic form now more than eighty years old? As soap opera fans have been doing for years, you'll have to tune in tomorrow to find what happens next.

NOTES

1. Henry Jenkins, "The Survival of the Soap Opera (Part Four)," *Confessions of an Aca-Fan*, last modified December 10, 2010, http://henryjenkins.org/2010/12/the_survival_of_soap_opera_par_3.html.

2. Robert C. Allen, "Soap Opera," *The Museum of Broadcast Communications Encyclopedia of Television*, accessed July 31, 2014, http://www.museum.tv/eotv/soapopera.htm.

3. Muriel G. Cantor and Suzanne Pingree, *The Soap Opera* (Beverly Hills, CA: Sage, 1983), 22.

4. Sara Bibel, "Soap Opera Ratings: All Soaps Up in Total Viewers and Women 18–49 Viewers," *TV by the Numbers*, last modified June 14, 2014, http://tvbythenumbers.zap2it.com/2014/06/14/soap-opera-ratings-all-soaps-up-in-total-viewers-women-18-49-viewers/272600/.

5. Cantor and Pingree, *The Soap Opera*, 23.

6. Mary Ellen Brown, "The Politics of Soaps: Pleasure and Female Empowerment," *Australian Journal of Cultural Studies* 4, no. 2 (1987): 4.

7. Cantor and Pingree, *The Soap Opera*, 28.

8. Jenkins, "The Survival of the Soap Opera (Part Four)," *Confessions of an Aca-Fan* Also, Roger Newcomb, "Daytime Ratings, Male Viewers & 18–49," *We Love Soaps*, last modified December 2, 2008, http://www.welovesoaps.net/2008/12/daytime-ratings-male-viewers-18-49.html. This percentage is consistent with what C. Lee Harrington and Denise D. Bielby found for the 1990s and 1980s in "Soap Fans: Pursuing Pleasure and Making Meaning in Everyday Life, edited by C. Lee Harrington and Denise D. Bielby (Philadelphia: Temple University Press, 1995), 15.

9. Sam Ford, Abigail De Kosnik and C. Lee Harrington, "Introduction: The Crisis of Daytime Drama and What It Means for the Future of Television," in *Survival of the Soap Opera: Transformations for a New Media Era*, edited by Sam Ford, Abigail De Kosnik, and C. Lee Harrington (Jackson, MS: University Press of Mississippi, 2011), 7.

10. C. Lee Harrington and Denise D. Bielby, introduction to "Soap Fans: Pursuing Pleasure and Making Meaning in Everyday Life, ed. C. Lee Harrington and Denise D. Bielby (Philadelphia: Temple University Press, 1995), 5.

11. Robert C. Allen, "Soap Opera," http://www.museum.tv./eotv/soapopera.htm.

12. Shearon A. Lowery and Melvin L. DeFleur. *Milestones in Mass Communication Research: Media Effects Third Edition* (White Plains, NY, Longman Publishers, 1995).

13. Nancy K. Baym. *Tune In, Log On: Soaps, Fandom, and Online Community* (Thousand Oaks, CA: Sage, 2000). Other examples include Christine Scodari, "'No Politics Here': Age and Gender in Soap Opera 'Cyberfandom,'" *Women's Studies in Communication* 21, no. 2 (1998).

14. Robert C. Allen, "Soap Opera," http://www.museum.tv/eotv/soapopera.htm.

15. The Paley Center for Media, "Irna Phillips: Radio and Television Creator, Writer," *She Made It: Women Creating Television and Radio*, accessed July 31, 2014, http://www.shemadeit.org/meet/biography.aspx?m=47.

16. Robert C. Allen, "Soap Opera," http://www.museum.tv/eotv/soapopera.htm.

17. Robert C. Allen, "Soap Opera," http://www.museum.tv/eotv/soapopera.htm.

18. Cantor and Pingree, *The Soap Opera*, 28.

19. Robert C. Allen, "Soap Opera," http://www.museum.tv/eotv/soapopera.htm.

20. Brian Steinberg, "Who Dropped the Soap on Daytime TV?" *Advertising Age*, August 9, 2010.

21. Robert C. Allen, "Soap Opera," http://www.museum.tv/eotv/soapopera.htm.

22. Brian Steinberg, "Who Dropped the Soap on Daytime TV?" *Advertising Age*, August 9, 2010.

23. Robert C. Allen, "Soap Opera," http://www.museum.tv/eotv/soapopera.htm.

24. Buck Wolf, "Luke and Laura: Still the Ultimate TV Wedding," *ABCNews.com*, last updated November 14, 2006, http://abcnews.go.com/Entertainment/WolfFiles/story?id=236498; Brian Steinberg, "Who Dropped the Soap on Daytime TV?" *Advertising Age*, August 9, 2010; Robert C. Allen, "Soap Opera," http://www.museum.tv/eotv/soapopera.htm.

25. Robert C. Allen, "Soap Opera," http://www.museum.tv/eotv/soapopera.htm.

26. C. Lee Harrington, "Perspective: Scholars Barbara Irwin and Mary Cassata on the State of U.S. Soap Operas," in *Survival of the Soap Opera: Transformations for a New Media Era*, Edited by Sam Ford, Abigail De Kosnik, and C. Lee Harrington (Jackson, MS: University Press of Mississippi, 2011), 23.

27. C. Lee Harrington, "Perspective: Scholars Barbara Irwin and Mary Cassata on the State of U.S. Soap Operas," 23.

28. Brian Steinberg, "Daytime TV's New Entries Push Soaps Toward Drain," *Advertising Age*, last modified August 9, 2010, http://adage.com/article/media/tv-soap-operas-losing-viewers-marketing-dollars/145291/.

29. James Hibberd, "'Guiding Light' Ending After 72 Years," *Hollywood Reporter*, last modified April 1, 2009, http://www.hollywoodreporter.com/news/guiding-light-72-years-81740.

30. Brian Steinberg, "Daytime TV's New Entries Push Soaps Toward Drain," *Advertising Age*.

31. Meg James, "Soap Opera 'All My Children' Washed Up?" *Los Angeles Times*, last updated November 13, 2013, http://articles.latimes.com/2013/nov/13/entertainment/la-et-ct-soap-opera-all-my-children-dead-20131113.

32. Meg James, "Soap Opera 'All My Children' Washed Up?" Also Leslie Goldberg, "'One Life to Live,' 'All My Children' Cut to Two Episodes a Week," *Hollywood Reporter*, last updated May 16, 2013, http://www.hollywoodreporter.com/live-feed/one-life-live-all-my-524025.

33. Meg James, "Disney's SOAPnet Channel Ends After 14-Year Run," *Los Angeles Times*, last modified December 31, 2013, http://www.latimes.com/entertainment/envelope/cotown/la-et-ct-disneys-soapnet-channel-ends-14-year-run--20131231-story.html.

34. Tim Kenneally, "Daytime Ratings Resurgence: If Soaps Are 'Dead,' Why Are Their Audiences Growing?" *The Wrap*, last modified December 12, 2013, http://www.thewrap.com/young-and-the-restless-bold-and-the-beautiful-general-hospital-days-of-our-lives-ratings-nbc-cbs-abc/.

35. Tim Kenneally, "Daytime Ratings Resurgence: If Soaps Are 'Dead,' Why Are Their Audiences Growing?" *The Wrap*.

36. Tim Kenneally, "Daytime Ratings Resurgence: If Soaps Are 'Dead,' Why Are Their Audiences Growing?" *The Wrap*.

37. Tim Kenneally, "Daytime Ratings Resurgence: If Soaps Are 'Dead,' Why Are Their Audiences Growing?" *The Wrap*.

38. Catherine Boyle, "Twitter Aids Soap Opera's Return from the Dead," *CNBC.com*, last updated December 31, 2013, http://www.cnbc.com/id/101288736#.

39. Catherine Boyle, "Twitter Aids Soap Opera's Return from the Dead," *CNBC.com*.

40. C. Lee Harrington and Denise Brothers, "Constructing the Older Audience: Age and Aging in Soaps," in *Survival of the Soap Opera: Transformations for a New Media Era*, edited by Sam Ford, Abigail De Kosnik, and C. Lee Harrington (Jackson, MS: University Press of Mississippi, 2011), 300.

41. Lucas Kavner, "Soap Operas: Can They Survive?" *Huffington Post*, last updated June 8, 2011, http://www.huffingtonpost.com/2011/04/08/soap-operas-all-my-children_n_845089.html.

42. Michael Schneider, "55-plus Age Bracket Is Fastest Growing Demo," *Variety*, last updated August 28, 2008, http://variety.com/2008/scene/news/nielsen-finds-audiences-are-aging-1117991318/.

43. C. Lee Harrington and Demise D. Beilby, *Soap Fans: Pursuing Pleasure and Making Meaning in Everyday Life*, 2–4.

44. Shearon A. Lowrey and Melvin L. DeFleur, *Milestones in Mass Communication Research: Media Effects Third Edition* (White Plains, New York: Longman Publishers, 1995) 107–8.

45. Lawrence Grossberg, "Is There a Fan in the House? The Affective Sensibility of Fandom," in *Adoring Audience: Fan Culture and Popular Media*, edited by Lisa A. Lewis (Florence, Kentucky: Routledge, 1992), 52.

46. Lawrence Grossberg, "Is There a Fan in the House? The Affective Sensibility of Fandom," in *Adoring Audience: Fan Culture and Popular Media.*

47. Lawrence Grossberg, "Is There a Fan in the House? The Affective Sensibility of Fandom," in *Adoring Audience: Fan Culture and Popular Media.*

48. Lisa A. Lewis, Introduction to *Adoring Audience: Fan Culture and Popular Media,* edited by Lisa A. Lewis (Florence, Kentucky: Routledge, 1992), 3.

49. C. Lee Harrington and Demise D. Beilby, *Soap Fans: Pursuing Pleasure and Making Meaning in Everyday Life.*

50. C. Lee Harrington, "The Moral Economy of Soap Opera Fandom," *Spreadable Media,* accessed August 1, 2014, http://spreadablemedia.org/essays/harrington/#.U96aCpRdWh0.

51. C. Lee Harrington, "The Moral Economy of Soap Opera Fandom," *Spreadable Media.*

52. C. Lee Harrington and Demise D. Beilby, *Soap Fans: Pursuing Pleasure and Making Meaning in Everyday Life,* 155.

53. C. Lee Harrington, "The Moral Economy of Soap Opera Fandom," *Spreadable Media.*

54. C. Lee Harrington, "Perspective: Scholars Barbara Irwin and Mary Cassata on the State of U.S. Soap Operas," 25.

55. To see the video of "We Won't Let Our Shows Die," go to https://www.youtube.com/watch?v=HGDiYq6gflg.

56. Sam Schechner, "As Venerable Soap Operas Die Off, Fans Fight for One More Life to Live," *Wall Street Journal,* last modified June 18, 2011, http://online.wsj.com/news/articles/SB10001424052702304186404576389693935512286?tesla=y&mg=reno64-wsj.

57. Pew Research Center's Internet and American Life Project, *Social Networking Fact Sheet,* accessed May 9, 2014, from http://www.pewinternet.org/fact-sheets/social-networking-fact-sheet/.

58. Maeve Duggan and Aaron Smith, *Social Media Update 2013,* Pew Research Center's Internet and American Life Project, released December 30, 2013, 5. Accessed at http://www.pewinternet.org/files/2013/12/PIP_Social-Networking-2013.pdf.

59. *Young and Restless,* Twitter feed, accessed Nov. 14, 2014, from https://twitter.com/YandR_CBS.

60. YR Insider, Twitter feed, accessed Nov. 14, 2014, from https://twitter.com/YRInsider.

61. Twitter, "About Twitter: Company," *Twitter.com,* accessed August 1, 2014, from https://about.twitter.com/company.

62. Lawrence Grossberg, "Is There a Fan in the House? The Affective Sensibility of Fandom," in *Adoring Audience: Fan Culture and Popular Media,* edited by Lisa A. Lewis (Florence, Kentucky: Routledge, 1992), 60–65.

63. Lawrence Grossberg, "Is There a Fan in the House? The Affective Sensibility of Fandom," in *Adoring Audience: Fan Culture and Popular Media.*

64. C. Lee Harrington and Demise D. Beilby, *Soap Fans: Pursuing Pleasure and Making Meaning in Everyday Life,* 155.

65. Frankie Hoornaert, Twitter feed, accessed November 12, 2014, from https://twitter.com/SassySoapFan.

66. Luke W., Twitter feed, accessed November 12, 2014, from https://twitter.com/@Soap_Dude.

67. Marie, Twitter bio, accessed November 12, 2014, from https://twitter.com/MarieSoapFan.

68. Y&RDeeVa, Twitter feed, accessed November 12, 2014, from https://twitter.com/YR_Deeva.

69. Stickyvitch, Twitter feed, accessed November 12, 2014, from https://twitter.com/stickyvitch.

70. Team Austin, Twitter feed, accessed November 12, 2014, from https://twitter.com/TeamMuldoon.

71. Chad and Abigail, Twitter feed, accessed November 12, 2014, from https://twitter.com/abbychadfans.

72. Morgans and McAvoy, Twitter feed, accessed November 12, 2014, from https://twitter.com/JasamUnited.

73. Morgans and McAvoy, Twitter bio, accessed November 12, 2014, from https://twitter.com/JasamUnited.

74. Forbidden Love, Twitter feed, accessed November 12, 2014 from https://twitter.com/FL_EJami_Fans.

75. Forbidden Love, Twitter post, August 3, 2014, 7:28 p.m., https://twitter.com/FL_EJami_Fans.

76. Jason B., Twitter post, August 3, 2014, 8:41 p.m., https://twitter.com/JasonXHowe.

77. Texts from Salem, Twitter bio, accessed November 12, 2014, from https://twitter.com/textsfromsalem.

78. Mike Norris, "Social Media Campaign Underway to Save 'Dallas' TV series," myfoxdfw.com, October 6, 2014, http://www.myfoxdfw.com/story/26715667/social-media-campaign-underway-to-save-dallas-tv-series; Kimberly Potts, "'Dallas' Showrunner on Future of Series: No Netflix, but Optimistic the Show will Return," tv.yahoo.com, October 12, 2014, https://tv.yahoo.com/blogs/yahoo-tv/the-future-of-dallas-netflix-173121376.html.

79. James Scott, Twitter post, January 14, 2014, https://twitter.com/IamJamesScott.

80. *Young & The Restless*, Twitter post, August 3, 2014, https://twitter.com/YRInsider.

81. *Bold & The Beautiful*, Twitter post, August 1, 2014, https://twitter.com/BandB_CBS.

82. *General Hospital*, Twitter post, July 29, 2014, https://twitter.com/GeneralHospital.

83. *Days of Our Lives*, Twitter post, April 11, 2014, https://twitter.com/nbcdays.

84. *Young and Restless*, Twitter post, April 9, 2014, 12:28 p.m., https://twitter.com/YandR_CBS.

85. Pew Research Center's Internet and American Life Project, *Social Networking Fact Sheet*, accessed May 9, 2014, from http://www.pewinternet.org/fact-sheets/social-networking-fact-sheet/.

REFERENCES

Allen, Robert C. *Soap Opera*, in *The Museum of Broadcast Communications Encyclopedia of Television*, ed. Horace Newcomb, accessed July 31, 2014, http://www.museum.tv/eotv/soapopera.htm.

Babbie, Earl. *The Practice of Social Research*. Belmont, California: Wadsworth/Thompson Learning, 2001.

Baym, Nancy K. *Tune In, Log On: Soaps, Fandom, and Online Community*. Thousand Oaks, CA: Sage, 2000.

Bibel, Sara. "Soap Opera Ratings: All Soaps Up in Total Viewers and Women 18–49 Viewers." *TV by the Numbers*. Last modified June, 14, 2014. http://tvbythenumbers.zap2it.com/2014/06/14/soap-opera-ratings-all-soaps-up-in-total-viewers-women-18-49-viewers/272600/.

Boyle, Catherine. "Twitter Aids Soap Opera's Return from the Dead," *CNBC.com*, last updated December 31, 2013, http://www.cnbc.com/id/101288736#.

Brown, Mary Ellen. "The Politics of Soaps: Pleasure and Female Empowerment." *Australian Journal of Cultural Studies* 4, no. 2 (1987): 4.

Brown, Mary Ellen. *Soap Opera and Women's Talk: The Pleasure of Resistance*. Thousand Oaks, CA: Sage, 1994.

Cantor, Muriel G. and Pingree, Suanne. *The Soap Opera*. Beverly Hills, CA: Sage, 1983.

Duggan, Maeve and Smith, Aaron. "Social Media Update 2013." *Pew Research Center's Internet and American Life Project*. Released December 30, 2013. Accessed at http://www.pewinternet.org/files/2013/12/PIP_Social-Networking-2013.pdf.

Ford, Sam, De Kosnik, Abigail, and Harrington, C. Lee. "Introduction: The Crisis of Daytime Drama and What It Means for the Future of Television." In *Survival of the Soap Opera: Transformations for a New Media Era* edited by Sam Ford, Abigail DeKosnik, and C. Lee Harrington, 3–21. Jackson, MS: University Press of Mississippi, 2011.

Goldberg, Leslie. "'One Life to Live,' 'All My Children' Cut to Two Episodes a Week." *Hollywood Reporter*. Last updated May 16, 2013. http://www.hollywoodreporter.com/live-feed/one-life-live-all-my-524025.

Grossberg, Lawrence. "Is There a Fan in the House?: The Affective Sensibility of Fandom." In *Adoring Audience: Fan Culture and Popular Media* edited by Lisa A. Lewis, 50–65. Florence, KY: Routledge, 1992.

Harrington, C. Lee. "The Moral Economy of Soap Opera Fandom." *Spreadable Media*. Accessed August 1, 2014. http://spreadablemedia.org/essays/harrington/#.U96aCpRdWh0.

———. "Perspective: Scholars Barbara Irwin and Mary Cassata on the State of U.S. Soap Operas." In *Survival of the Soap Opera: Transformations for a New Media Era* edited by Sam Ford, Abigail De Kosnik, and C. Lee Harrington, 22–28. Jackson, MS: University Press of Mississippi, 2011.

Harrington, C. Lee and Bielby, Denise B. *Soap Fans: Pursuing Pleasure and Making Meaning in Everyday Life*. Philadelphia: Temple University Press, 1995.

Harrington, C. Lee and Brothers, Denise. "Constructing the Older Audience: Age and Aging in Soaps." In *Survival of the Soap Opera: Transformations for a New Media Era* edited by Sam Ford, Abigail De Kosnik, and C. Lee Harrington, 300–14. Jackson, MS: University Press of Mississippi, 2011.

Hibberd, James. "'Guiding Light' Ending After 72 Years." *Hollywood Reporter*. Last modified April 1, 2009. http://www.hollywoodreporter.com/news/guiding-light-72-years-81740.

James, Meg. "Soap Opera 'All My Children' Washed Up?" *Los Angeles Times*. Last updated November 13, 2013. http://articles.latimes.com/2013/nov/13/entertainment/la-et-ct-soap-opera-all-my-children-dead-20131113.

———. "Disney's SOAPnet Channel Ends After 14-Year Run." *Los Angeles Times*. Last modified December 31, 2013. http://www.latimes.com/entertainment/envelope/cotown/la-et-ct-disneys-soapnet-channel-ends-14-year-run--20131231-story.html.

Jenkins, Henry. "The Survival of the Soap Opera (Part Four)." *Confessions of an Aca-Fan*. Last modified December 10, 2010. http://henryjenkins.org/2010/12/the_survival_of_soap_opera_par_3.html.

Kavner, Lucas. "Soap Operas: Can They Survive?" *Huffington Post*. Last updated June 8, 2011. http://www.huffingtonpost.com/2011/04/08/soap-operas-all-my-children_n_845089.html.

Kenneally, Tim. "Daytime Ratings Resurgence: If Soaps Are 'Dead,' Why Are Their Audiences Growing?" *The Wrap*. Last modified December 12, 2013. http://www.thewrap.com/young-and-the-restless-bold-and-the-beautiful-general-hospital-days-of-our-lives-ratings-nbc-cbs-abc/.

Lewis, Lisa A. Introduction to *Adoring Audience: Fan Culture and Popular Media*. Edited by Lisa A. Lewis (Florence, KY: Routledge, 1992), 1–6.

Newcomb, Roger. "Daytime Ratings, Male Viewers & 18–49." *We Love Soaps*. Last updated December 2, 2008. http://www.welovesoaps.net/2008/12/daytime-ratings-male-viewers-18-49.html.

Norris, Mike. "Social Media Campaign Underway to Save 'Dallas' TV Series." *Myfoxdfw.com*. Last updated October 6, 2014. http://www.myfoxdfw.com/story/26715667/social-media-campaign-underway-to-save-dallas-tv-series.

The Paley Center for Media. "Irna Phillips: Radio and Television Creator, Writer." *She Made It: Women Creating Television and Radio*, accessed July 31, 2014. http://www.shemadeit.org/meet/biography.aspx?m=47.

Pew Research Center's Internet and American Life Project. *Social Networking Fact Sheet*. Accessed May 9, 2014, from http://www.pewinternet.org/fact-sheets/social-networking-fact-sheet/.

Potts, Kimberly. "'Dallas' Showrunner on Future of Series: No Netflix, but Optimistic the Show will Return." *tv.yahoo.com*. Last updated October 12, 2014. https://tv.yahoo.com/blogs/yahoo-tv/the-future-of-dallas-netflix-173121376.html.

Schechner, Sam. "As Venerable Soap Operas Die Off, Fans Fight for One More Life to Live." *Wall Street Journal*. Last modified June 18, 2011. http://online.wsj.com/news/articles/SB10001424052702304186404576389693935512286?tesla=y&mg=reno64-wsj.

Schneider, Michael. "55-plus Age Bracket is Fastest Growing Demo." *Variety*. Last updated August 28, 2008. http://variety.com/2008/scene/news/nielsen-finds-audiences-are-aging-1117991318/.

Scodari, Christine. "'No Politics Here': Age and Gender in Soap Opera 'Cyberfandom.'" *Women's Studies in Communication* 21, no. 2 (1998).

Steinberg, Brian. "Daytime TV's New Entries Push Soaps Toward Drain." *Advertising Age.* Last modified August 9, 2010. http://adage.com/article/media/tv-soap-operas-losing-viewers-marketing-dollars/145291/.

————. "Who Dropped the Soap on Daytime TV?" *Advertising Age*, August 9, 2010.

Twitter. "About Twitter: Company." *Twitter.com.* Accessed August 1, 2014, from https://about.twitter.com/company.

Wolf, Buck. "Luke and Laura: Still the Ultimate TV Wedding." *ABCNews.com.* Last updated November 14, 2006, http://abcnews.go.com/Entertainment/WolfFiles/story?id=236498.

Chapter Thirteen

Army Wives Connect

Lifetime Viewers' Everyday Lives and Fandom Converge in Online Communities

Darcey Morris

Lifetime Television emerged in 1984 with the joining of Viacom's Cable Health and Hearst/ABC's Daytime TV—both cable networks attracted a female audience, a group advertisers sought fervently, and joined to further capitalize on delivering this segment of the market.[1] At its inception, Lifetime defined itself by aligning with women, declaring that it best represented and understood their unique interests and needs. Throughout the past twenty-five years, the cable network's ratings provided support to these claims, but in the post-network era of increased competition for the cable audience, it struggles to maintain its resonance with women across demographic boundaries and control its image. From her tenure as Lifetime's president and chief executive officer in 2007 until her resignation in 2010, Andrea Wong worked to redefine the cable network, endeavoring to keep current fans and viewers while attracting a younger, hipper female audience. Integral to this rebranding attempt was the hit show, *Army Wives*, which brought Lifetime record high ratings and drew attention to the network's new brand identity. Coordinated with the introduction of a revamped Web site and online presence at the newly dubbed MyLifetime.com, the show's launch provides the perfect case study of an established cable network's attempts to reinvent itself in a convergent, new media environment. Wong spoke frankly about its value to the network's efforts in this endeavor during a 2009 interview. When asked what her chief priority had been as CEO of Lifetime, she responded:

The biggest goal for me has been to really bring an energy, relevance and
contemporary-ness to the brand and our programming—and I think we've
really set ourselves on a course to do that. *Army Wives* was the first real calling
card for where Lifetime is going.[2]

Army Wives helped reposition the network toward a younger audience and
served as an embodiment of the reinvigorated network's new slogan—"My
Story is on Lifetime," which also debuted in 2007 along with a revamped
Web site.

Army Wives, although fiction, sheds light on the unique experience of
women who are married to men serving in the U.S. military today. With two
wars in the last decade, servicemen and women are under great strain, and
while the show spoke to those sacrifices, it focused on the sacrifices military
families make as unsung heroes. *Army Wives* attracted a wide range of female viewers, but understandably many are real Army wives and watch
because they appreciate and enjoy seeing their side of the story told. While
Lifetime gave voice to these women by creating the program, they also
provided a space online for fans of the show, many of them military wives, to
connect and talk about their experiences and views. Many networks provide
Internet message boards for fans, yet participants in the *Army Wives* discussions used the space to do more than just talk about the show. These fans also
used the forum to provide support to one another, offer advice and even
provide helpful information about specific bases or areas to families transferring locations. Catherine Johnson writes, "Brands only gain meaning and
value through what people do with and say about them."[3] Building off this
understanding of branding, The Lifetime brand identity was strengthened and
reinforced by the online behavior of *Army Wives* fans, especially given that
these brand boosting activities took place on the official Lifetime network
Web site.

First, as Lifetime identifies as "Television for Women," the first section
in this chapter will be a brief discussion of female media consumers that
includes how they use media texts, what aspects of a program resonates with
them and the unique ways that female fans interact with a television text
online. The next portion will provide an overview of *Army Wives*, its promotional strategies and how executives marketed the program to attract female
viewers who are military wives. Finally, this study will conduct an analysis
of the *Army Wives* message boards[4] to identify any possible themes that real
Army wives focused on in their online discussions—offering advice and
support to other military wives and criticizing how true to life the show's
depiction of military life is. The study was conducted after reading every
comment and discussion posted within Lifetime's *Army Wives* Internet community during the first three seasons of the series. This study is focused on
message board participation and responses in this time frame because it

coincided with the network's rebranding efforts in 2007–09. Although fans discussed the show and its relevance to their own lives as military wives in different online arenas, this research is focused specifically on Lifetime and how the network strategically targets such viewers' attention, behavior, and fandom.

FEMALE AUDIENCES

One of the pioneers of feminist media research, Janice Radway,[5] conducted a research study on romance novels and their female fans. Radway noticed that in talking about their enjoyment of this pastime, romance readers focused on the act of reading rather than the characters, plot, or signification of the novels themselves. She concluded that the women sought escape from their everyday lives and the accompanying burdens. "Not only is it a relaxing release from the tension produced by daily problems and responsibilities, but it creates a time or a space within which a woman can be entirely on her own, preoccupied with her personal needs, desires and pleasures."[6] Escapism is most often cited by women as a reason they consume a particular media product. Former Lifetime executive Susanne Daniels believes the network meets this need for their viewers when she said, "It's a television brand where women can find programming that will allow them to escape."[7] The network even utilizes the concept in a marketing campaign featuring a jet meant to draw attention to "how the network provides an 'escape' for its viewers."[8]

Another landmark study on female audiences is Ien Ang's *Watching Dallas*.[9] The research involved a collection of written responses from thirty-nine women and girls who wrote to explain why they either loved or hated the popular television show, *Dallas*. Ang stated that whether an individual found the show good or bad corresponded with whether they found it realistic (good) or unrealistic (bad). *Dallas* is a primetime soap opera, like *Army Wives*, which is a popular format for female audiences because it primarily focuses on personal relationships and feelings, thus lending itself to the meaning-making practices female audiences employ. Ang used the phrase "tragic structure of feeling" to describe the show's oscillation between feelings of happiness and feelings of melancholy. She wrote that the fans of the show "can 'lose' themselves in *Dallas* because the programming symbolizes a structure of feeling which connects up with one of the ways in which they encounter life."[10] She further concluded that female viewers enjoy shows and characters that they can identify with in some real way.

Wood[11] studied female viewers of talk shows and morning programs and interrogates the term "referential viewing" which has often been used to describe how women relate to media texts. Referential viewing, as she defined it, is the way female audiences correlate their life experiences with those played out in television texts. Wood explained that women choose and form attachments to media texts according to how it relates to their own personal experiences with life. For her study, she watched talk shows with subjects and recorded both the media text being watched and their remarks as they viewed the programs—showing side-by-side comparisons which effectively illustrate the comparative form of viewing. The viewers shared personal stories as they remarked on the show's text, and Wood argued this is part of the viewing pleasure and that it causes a blurring between the show and the subject's own life. This explains the popularity of *Army Wives* with real women married to men who work in the military—women who have lives very similar to that portrayed in the television program.

Female fans utilize the Internet to connect with other fans, develop connections and build rich communities for a multitude of purposes. Nancy K. Baym[12] studied soap opera fans who participated in online communities and asserted they were one of the first groups to use the Internet in this way. Baym's research investigated the pleasure such fans found in the ability to discuss the show with other fans, learn how fellow viewers interpret the show and its characters and to form interpersonal relationships through these activities. It follows that as female audiences enjoy programs that focus on interpersonal relationships and characters they can identify with, they would extend these feelings to their online experience with the show. For many fans of soap opera, Baym discovered talking about the show online offered even more pleasure than watching the show itself. Ross[13] found similar results in an audience study of online fans of *Buffy the Vampire Slayer*, and *Xena: Warrior Princess*. She wrote, "for these fans, it was going online to discuss the characters and the stories that sustained their interest and passion—a form of interacting with the text past its origin that truly requires the presence of a group (i.e., social audience)."[14] As with soap fans, cult fans also interact with the media text and enjoy collective meaning making. The key for both groups is having a sense of community. Further in her book, Ross also explained how the network and its sponsors create spaces for online communities in a way for fans to bond with other fans, but also for fans to feel bonded to the network and advertisers. Networks house message boards on program Web sites so they can be a part of (and manipulate) the fan community. These concepts are key to Lifetime's restructuring and relaunch of myLifetime.com—the idea are inherent in the name which implies the Web site is a place for women to congregate and be themselves.

MEDIA INDUSTRY STUDIES

The theoretical and methodological approaches employed in this research project draw heavily from the field of media industry studies. In analyzing television texts and paratexts (such as branding and promotional campaigns), as this chapter does, the understanding of television used acknowledges it is simultaneously an object made up of industrial, economic cultural, historical processes, and complex interrelationships between media companies, audiences, and texts. The emphasis that media industry studies places on the overlapping mechanisms of structural roles, industry goals and policies, production practices and the financial interests of conglomerates frames the analysis for this project. Media industry studies, a relatively new field of study to academia, developed in response to the rapidly changing media environment of the post-network era.[15] Media industry studies integrates themes and ideas from both cultural studies and political economy including: an emphasis on structural power and individual agency, an assertion of the importance of cultural discourses and value judgments, the tension between the understanding that the media is both ideological and economic in nature, and the belief that media industries have a tremendous impact on viewers' cultural meaning-making processes.

This chapter examines the relationship between audiences and the Lifetime television network within the larger context of industrial strategies, branding practices and promotional efforts. Media industry studies scholars argue for a holistic investigation of all the parts that contribute to the media industries. This echoes du Gay et al.'s theory of the circuit of culture, which claimed that researchers must consider all the intricate dynamics and interrelationships between the various sites because looking at just the media text or network promotions would not be enough to fully understanding the complexities at work.[16] According to Havens, Lotz, and Tinic,[17] media industry studies means an emphasis on the negotiation between corporate practices and consumer agency along with the recognition that industrial strategies are at once cultural and culture-producing. To secure high ratings and viewer loyalty, Lifetime targeted military spouses as an audience for *Army Wives*, even going so far as to encourage the show's fans to use the web forum as a meeting place to discuss their personal experiences and feelings. Although such behavior may be beneficial to viewers, it is important to note these activities are taking place surrounded by advertisements on a corporate Web site designed to be profitable. Media industry studies' theories and methods are useful in the recognition and examination of such tensions.

ABOUT *ARMY WIVES*

Lifetime made strategic changes in 2007 and 2008 to redefine the brand and attract a younger audience. Integral to this effort was the launch of a redesigned Web site that was renamed MyLifetime.com and the aim of which was to epitomize the new slogan, "My Story is on Lifetime." The Web site was intended to serve as a forum where viewers can interact with other viewers, talk about their favorite shows, interact with producers and actors, view videos about the shows and characters and provide feedback to the network. The site was renovated to better serve its viewers and broadened its content to better compete with Internet rivals, such as iVillage and About.com. In a 2007 interview, Andrea Wong stated that the goal was "to be everywhere a woman could want [them] to be, providing the information and the entertainment they want."[18] Rather than focusing solely on content that promotes Lifetime's programming and serves as a companion to its shows, the new Web site also includes information on topics women are interested in and utilized content from parent company Hearst's magazines. Wong said, "We have health and wellness information, we have recipes, we have shopping opportunities, we have community."[19] Lifetime also formed partnerships with: Glam Media, the top aggregator of women-centered Web sites; RealArcade, an online game service; and RevolutionHealth.com, a Web provider of health information. Lifetime Digital expanded its online properties with the purchases of ParentsClickNetwork, a social networking and blog service; DressUpChallenge.com; Roiworld and, in 2008, posted its highest traffic in history. The Web site offers users different ways to interact and participate with horoscopes, games, original Web series and even a way "to purchase items they've created through a dress up game, viewed in a celebrity photo gallery or . . . a sponsor's . . . customized shop."[20] Dan Suratt, vice president of digital media and business development at Lifetime, said the chief aim was to make the Web site "the ultimate digital destination for women's entertainment and escape" (Lafayette, November 12, 2007).[21] He is not the only Lifetime executive who shared the belief that female consumers look to media for escape. Wong echoed him when she said: "Our audience looks to Lifetime to provide an escape—in the digital space and on all platforms."[22]

In addition to the network's new online presence, *Army Wives* was to serve as the foundation for Lifetime's rebranding initiative, and its success encouraged execs that they were on the right track. Of the show's ratings success, Wong said, "It gave the whole organization a great lift and a great optimism about where we were going. It was just the wind against our backs that we needed to propel us forward."[23] Later on in the interview, she added, "*Army Wives* is a classic example of where we want Lifetime to go."[24] The network's promotion of the show reinforced the traditional buzz words and themes that Lifetime historically used to promote—such as stressing the relevance of the characters and plotlines to "real women," the focus on friendships and a message of female empowerment. Wong reiterated the show's relevance to women and Lifetime's mission to do the same: "We have tried very hard to make sure we deal with real life issues in ways that real life people would deal with them."[25] Finding ways to connect programs to their own lived experiences is an important factor to female audiences' derived pleasure. The show's high ratings offer support to this supposed resonance, and while it is not the first to address issues related to military life, the program's success is certainly in part due to the network's promotional efforts in addition to the particularities of the genre and themes.

The official Lifetime press release that announced the show's production described its plot succinctly: "The new ensemble drama series about the struggles, dreams and friendships of a diverse group of women—and one man—living with their spouses and families on an active Army post and the pressures and traditions of the military on those who are left behind while their partners serve their country."[26] The show was filmed in Charleston, South Carolina, on an unused naval base to add to the program's authenticity. It starred: Catherine Bell as "Denise," Bridget Brannagh as "Pamela," Wendy Davis as "Joan," Kim Delaney as "Claudia Joy" and Sally Pressman as "Roxy." The executive producer, Mark Gordon, is also an executive producer for ABC's *Grey's Anatomy* and CBS's *Criminal Minds*. The program was loosely based on a book by journalist, Army wife, and series consultant Tanya Biank about violent crimes that occurred at Fort Bragg, North Carolina, in 2002. Within six weeks, "four Fort Bragg soldiers killed their wives; then two of the men committed suicide; and a third would hang himself in jail . . . [and] an officer's wife was charged with killing her husband."[27] While this is more typical of content covered in Lifetime's classic "woman-in-peril" movies (exactly the reputation that executives hoped the show would move the network away from), the series took a different approach and did not focus on the violence in Biank's book. Maria Grasso, senior vice president of series development for Lifetime, acknowledged this fact when she said: "The book has a darker tone than the series and deals with more serious issues . . . we wanted to balance [serious issues] with optimism and qualities of inspiration."[28] Grasso's statement is a signal that network executives purposely steered the series away from such narratives in a strategic effort to distance Lifetime from the type of content it had become notorious for featuring.

Media coverage for the show drew attention to fans who were actual "Army wives" and asked their assessment of its accuracy and resemblance to their lives. An article in the *Washington Post* called it a "military soap opera" and attributed its success "to a loyal fan base among military wives and girlfriends."[29] The article also quoted a fan who said, "It's nice someone wants to share our side of the story." The fans interviewed in such pieces were pleased that their experience of the war was being shared, and one commended its truthfulness: "They've really done a good job at portraying the Army lifestyle."[30] The Web site Yahoo! Voices, which publishes stories submitted by users on almost any topic of their choosing, contains several reviews of the show written by real-life Army wives. One such article is titled: "Lifetime's *Army Wives* Gets It Right: Show Finally Portrays the Lives of Army Wives Correctly."[31] Jones is a self-proclaimed "Army wife" who particularly mentioned that the show was "realistic" in its representation of the military rank system and how that affected Army wives' relations, how being deployed overseas impacts families and how frequently rumors are spread. She said, "The show gives viewers an insight as to what the true-life Army wife goes through."[32]

While all reviews from actual military spouses were positive, some fans did admit that the show sometimes privileged drama and plots to improve ratings over a commitment to authenticity. One criticized, "Although many people feel that *Army Wives* is one of the most accurate shows out there, most would agree that it is more focused on emotions than it is on reality."[33] Other fans, such as Jenny Trolley,[34] said that despite the melodrama, the show was entertaining, fun, and brought attention to their real experiences. Trolley praised the program for not being "very political," although early critics pondered and feared that if the series focused too much on the controversial wars, it would alienate viewers. Instead, Lifetime took care to avoid politics and kept the focus on the friendships and families of the women. Executive producer Mark Gordon said that the producers did not want the show to resemble the book closely or draw attention to current wars. Gordon said, "We want to try and make it real, and at the same time make it entertaining."[35] The creator and writer of the show, Katherine Fugate, said her intent was to highlight the experiences of real Army wives: "It's extremely important that I portray them accurately. I have great admiration for the wives. It's the last untold story, about how they maintain relationships and how they are single mothers much of the time. That story is why I created the series."[36] Fugate maintained her commitment to accuracy by using two Army consultants for the writing of every episode. Another part of this strategy included shooting the show on location, using locals from Charleston and routinely consulting the book's author, Tanya Biank, for authenticity.

ARMY WIVES' PROMOTIONS AND RATINGS

Lifetime strategically planned the premiere episode of *Army Wives* after Memorial Day weekend 2007 and a week-long celebration of men and women serving in the military. Emphasizing the relationship between the show and its real world counterparts, the premiere followed a brief film titled *Wives on the Homefront* about the cast members meeting military families involved with Operation Homefront, a nonprofit organization serving troops and their families. The second season premiere took place at a red carpet screening on July 4, 2008 at Walter Reed Army Medical Center. Also at the premiere, Lifetime launched another outreach event with Operation Homefront, which was a postcard drive that encouraged fans to send messages of support to families of troops deployed overseas. Throughout its history, Lifetime has incorporated social activism and advocacy into its network practices, including playing an active role in lobbying for legislation. This has proven a successful avenue for Lifetime to promote its shows and the network's image as being the place *for* women, to address issues important to women, and the use of outreach to military families is another example of the network's usual promotional approach. It also used the strategy as a way for its sponsor, Ford Motors, to take place in these efforts and publicize their corporate philanthropy. Ford launched "Ford Cares," which involves advocacy for breast cancer awareness in a project called Warriors in Pink, and *Army Wives* featured a story line in season 2 when a character diagnosed with breast cancer purchased a Warriors in Pink Signature Ford. The automobile company also partnered with the show in a contest to present real Army wives with Ford cars. This was all part of a campaign for Lifetime's special event tie-in show titled "*Army Wives* Gives Back," which collected products and services from corporations such as JC Penny, Big Lots, and T-Mobile for military families and used the show's stars in promotional events to present the gifts to the families. In 2008, "Lifetime surprised real women with a home makeover, a dream honeymoon, Vespas, a new car and a chance to meet the show's stars."[37] Again, this was part of a strategic promotional campaign to garner favorable media coverage and a positive image for the program while also reinforcing the relationship between the show's plots and characters to real military families. It further highlights the unique functionality and goals of the network Web site, which sought to be everything for every woman.

The Web section for *Army Wives* on MyLifetime.com provided unique ways to promote the show and interact with viewers. It offered users access to information about the show, its cast, lists of songs used, and an episode guide along with a trivia game for fans to test their knowledge about the show. Video extras included interviews with cast members, clips from previous episodes, and previews of future shows. An interesting video series called "*Army Wives* Confidential" featured interviews with cast members about decisions made by the characters they played or important scenes that affected the characters' lives. This feature helped viewers better understand the characters and story lines, providing information that helped them negotiate meaning from the media text. Sharon Marie Ross argued this is an important way that TV and the Internet converge and mediate between the storytelling of television series and the audience's meaning making process. She wrote, "In short, the Internet's placement 'between' sites of production and sites of reception creates a sense of proximity among those at work in these sites that in turn encourages a sense of reciprocity and closeness between industry professionals and viewers."[38] The *Army Wives* blog did this with articles that helped viewers gain a deeper understanding of the show and its characters while also providing relevant news about the show and articles about the "*Army Wives* Gives Back" campaign. Such features promoted the show and Lifetime's activism while also providing content to fans who wanted more information about *Army Wives* and offering them another way to participate with the television text. In accordance with the desire to reiterate the connection between the show's fictional Army wives and real military families, the *Army Wives* Web site had a section encouraging actual Army wives to interact with each other:

> The network has also created online communities so military wives around the country can connect with each other. Posts deal with issues ranging from spouses and family members who are on or who have left for tours of duty in other parts of the world to issues that crop up when they come home.[39]

The Web features proved successful as web traffic for MyLifetime.com steadily rose, and the viewing of *Army Wives* episodes online increased twelve times during a ten-day period after the 2008 season 2 premiere than it was at the same time in 2007. In a 2009 interview, Wong revealed that Lifetime's Web traffic had tripled in just that year.[40]

The coordinated promotional efforts were also apparent in the show's ratings, which smashed records and brought in more young viewers to the network. The series premiere in June 2007 received 3.5 million viewers, a record high for Lifetime at the time, was cable's top show for women eighteen to forty-nine, and achieved Lifetime's highest rating for women eighteen to thirty-four years old in five years.[41] In season 1, it became basic cable's top drama for eighteen- to thirty-four-year-old women, a key indication that Lifetime's efforts to use the show to bring in a younger audience was a success. The season 1 finale was Lifetime's most watched broadcast in its history at 4.1 million viewers while also being the top cable show in primetime for that night.[42] The show's success increased web traffic to Lifetime's redesigned Web site with gains of 135 percent during its first season, which suggested that the Web site's new features correlated with viewers' enjoyment of the show. Season 2 of *Army Wives* was just as successful with a premiere night attracting 4.5 million viewers and a season average of 3.9 million viewers.[43] Season 3's finale also brought in 3.9 million viewers, and though the ratings remained steady, the show was a steady performer for Lifetime until it was canceled in 2013 after seven seasons. [44]

Online, fans of the show utilized Lifetime's newly refurbished Web site to discuss the show and connect with one another. At the time of analysis, there were almost three hundred different discussions with 6,187 members. While anyone could view the postings, only members with registered accounts on the site could post comments. Discussions on the site had anywhere from one to over three thousand individual comments. Topics ranged from "Season 3" to "Surviving Deployment" and "Here Comes Baby . . . Bye Bye Daddy." While most of the discussions centered on events on the show, its characters, and how to access episodes and seasons online, a large portion focused on discussions with self-identified "real military wives." Within these discussions, two major themes were identified in the analysis: (1) advice/support and (2) criticism of the show's depiction of military life. Such patterns are significant because they demonstrate the ways that commenters made connections between the show's content and their lived experiences. There are important distinctions to consider between those commenters who used the Web forum to talk about the show (not included in this analysis), others who used it to offer or seek out encouragement, and the viewers who criticized the program for unrealistic depictions. Although a majority of the women were themselves military wives who approached the text from personal perspectives, it did not necessarily follow that their participation on the web forum would engage directly or indirectly with the text or involve a blind acceptance of the program's content or portrayals. Bronwyn Williams stressed the presence of audience agency in viewers' interpretations of media content: "While acknowledging that popular culture reflects and reproduces dominant cultural ideologies, individuals in the audience who interpret such texts in the contexts of their own experiences do not accept them without question." [45] Though this certainly proved true in the current analysis, *Army Wives* commenters did not address their role as consumers using a corporate web forum or how these activities were exploited by Lifetime and its advertisers. Despite a lack of such acknowledgments, this analysis revealed that viewers who contributed to the message boards engaged in active meaning-making processes, used the online space to empower one another and drew attention to the unique experience and perspectives of military spouses.

ADVICE/SUPPORT

Many of the women mentioned how difficult military life was and how challenging it was to worry about their partner while taking care of work and children. While almost all of the women described feeling admiration for her husband's service, many also talked about dealing with repercussions from a life they did not choose. XSoldiersGirlX wrote: "I mean, I fell in love before the army. The army is something that I'm choosing to 'deal' with or 'work' with."[46] The stress of worrying about whether their spouses would be injured, being a single mom while their partners are gone, and being unable to control when they could even talk to their husbands understandably takes an emotional toll on a relationship and an individual. A woman who called herself Mary1986 shared: "My father was killed. And my husband is deployed, this is not something I really want to do alone. But what can we do? We love are [sic] husbands and it's how they take care of us."[47] Many of the women mentioned feeling lonely or alone in their situation, which is perhaps why they went online looking for solace and friendship. It was a testament to Lifetime's marketing that the fans used the Web site to forge connections and talk about their lives—or, to reiterate Lifetime's slogan, their "stories." The usage of the message boards to forge friendships supports the theories identified by Baym[48] and Ross[49] about how female fans use a shared interest to bond with one another online.

Demonstrating the strength of these bonds, many women sought advice from one another on a variety of topics ranging from emotional support to specific information regarding military policies. The most frequent sentiment expressed involved coping with their husbands' deployment—from taking care of the kids while addressing their concerns for their fathers to finding inventive ways to deal with their own apprehension and even asking for different ideas to raise their soldiers' spirits. One woman asked for help in understanding the disbursement of her husband's deployment pay. Another, Bridget22, asked for parenting tips:

> I have noticed since my husband has been gone my 3 1/2 year old and 2 year old have been acting out in bad ways. They are usually really well behaved, it's just lately I don't know what to do with them anymore. My three-year-old is constantly telling me she misses her daddy, and I just keep telling her that he will be back but it won't be for a really long time. Does anybody have any advice on how to deal with this type of behavior?[50]

The responses she received included suggestions to involve her children in dance and gymnastics, which will "give them something to focus on and be proud of while expending energy."[51] One response suggested making a calendar counting down the days until their father returns from war. Another member said that the military offers classes that help children deal with the deployment of a parent and that Sesame Street even has a DVD on the subject. Of course, it's not only the children that experience difficulty. MissPatriot asked for advice from other women on how to keep herself busy while she's waiting for the phone to ring or an email to arrive. MedicWife suggested: "Clean, clean, clean, and if there is nothing to clean . . . find it!"[52] Another comment asked for a "Chicken Murphy" recipe and received a detailed recipe in response. The friendships forged by *Army Wives* fans on message boards went beyond simple fandom and influenced their lives offline. The women used a common interest in the same TV program to discuss serious issues and real-world problems. Several women tell what military base they are stationed at and ask others for ideas about restaurants and activities in the area. From these examples, it is evident that military wives were brought together by their shared fandom and utilized Lifetime's Web forum in ways that supported the network's design and goals just as Ross[53] described. An article that previewed the newly designed Web site's launch detailed executives' vision:

> Lifetime has big plans for its Web site . . . intended to offer women a community for all things female, not just Lifetime programming. With Lifetime's "My Story" branding, the site is intended to be a place for women to share stories, with tagged topics, articles and discussion forums on such subjects as health and lifestyle.[54]

Dan Suratt, executive vice president of digital media and business development for Lifetime, said viewers came to the site hoping to "get that decompression they're looking for."[55] Many fans, however, wanted to continue online discussions off the Web site forum.

Some women posted their personal email addresses in comments, others asked commenters to contact them through the messaging system on the Lifetime Web site, and many made plans to meet up in person to further conversations and friendships made in the online community. While the friendships were developed online, it seemed as though numerous participants took active efforts to form real, offline relationships with one another. Baym[56] and Ross[57] did not report such instances in their studies, but it is reasonable to assume some of the *Army Wives* fans converging on the message boards moved their online engagement into the "real-world." Haydee113 posted that she and her family recently moved to Fayetteville, North Carolina, and wrote that she was homesick and bored. This post received four responses—three out of four of them said they also lived in that area and asked Haydee to message them online. One of those suggested a museum Haydee's children might enjoy. Another woman, who identifies herself as "HarveysWife," asked if anyone is posted in Fort Carson. One response she received from mikes_booboo says she lived there as well and shared her email address and information about her profile on the social networking site, MySpace. The questions and advice sought by posters seemed heartfelt and as though they felt they were seeking advice from friends, not strangers.

The discussion participants shared personal feelings and details about their lives, such as one woman who confessed she thought about leaving her husband. "My husband went to Korea 2 weeks after we got married and something happen[ed]. He's started being really mean to me about 3 months after he got there, now he says he wants a divorce. I don't know what to do."[58] Other women comforted her by sharing stories about how they made it through similar situations, and another woman said that her first husband "did the same thing."[59] There are frequent discussions where women revealed their marital problems and asked for advice from other army wives. In one such discussion, a woman offering counsel suggested that the husband in question might suffer from post-traumatic stress disorder, a common syndrome afflicting many returning soldiers. In fact, there were several mentions of PTSD in different discussions with commenters listing symptoms to help women identify the disorder in their husbands. One woman responded to a discussion about PTSD by pleading with the women to do something if they believe a soldier has symptoms. She wrote, with her own emphasis: "If it seems like this is what he has, get him help RIGHT away! DON'T WAIT! Do whatever you have to do to get him seen, it DOES NOT go away on it's own. He can recover, but he'll need more than just your help if PTSD is the case."[60] Such discussions revealed the severity of the stress on members of the armed services and their families as well as the extent to which women were using the Lifetime message boards to go beyond the text and discuss personal and intimate experiences.

The women frequently talked about how they coped with their husbands away at war and how their children responded to the stress. Pfclogan counseled a fellow message board commenter about how hard it is to be a military wife. She wrote: "Things will happen that you will have to go through alone."[61] Others also described the loneliness they felt. MissIssiee described the pressures and complexities that come with being a military wife:

> As military wives we are expected to be flawless with our understanding, love, humility, and waiting. Flawless an [sic] enduring lonely nights and loneliness when your husband is deployed or in training. You are expected to be strong in the face of uncertainty and fear and be the ambassador of good will.[62]

"Loneliness" is persistently mentioned in the *Army Wives* discussions. Perhaps it is the loneliness that sends such women online seeking friendship and compassion. Many also discussed how hard it was to make friends with other military wives, a reason often given when they asked if the other message board participants lived nearby and wanted to meet. Iemomi wrote that she is "a nice person" who said she tried to make friends on base, but had a hard time finding any: "We have been here for going on 8 months, and I have no friends here and it's hard."[63] While Usmc_cowgirl_00 mentioned it was hard to find a trustworthy friend in other Army wives on her base. She said: "You have to be careful whom you talk to because there are some wives out there that are vindictive. And they like to make sure your significant other gets in trouble."[64] Although many women expressed difficulty forming friendships offline, the frequent requests to meet commenters in person signals that, as Baym[65] and Ross[66] posited, message board participants felt that they were members of a community online that fulfilled some of the emotional support lacking in their everyday lives.

It is evident from several posts that due to having spouses in the military, the women were regularly moving around the country with their families, which made it difficult to meet new friends. With husbands overseas, they spent much of their time caring for their children as single parents in many ways, but used the online message boards as a channel for their frustrations and worries. Some women shared that their husbands were serving abroad when their children were born or experience important milestones—what Laciraye referred to as "firsts," such as her daughter's first day of kindergarten and her son's first day of preschool. Such difficult situations also contributed to their feelings of loneliness. Frequently, the women thanked each other for friendship and expressed how much the other posters mean to them—this also suggests the strong feelings of community and social acceptance the message boards created for fans. Ashprivatecp writes: "Somedays i feel like i have nothing left of me, cry hysterically, and mope around like my world has come to a end. But reading things like this really keeps me positive."[67]

The women often conveyed frustration at people who are not familiar with the military or who are not married to someone in the armed services. That seemed to be another reason they found such comfort in talking with one another—because they were able to talk to other women who have gone through what they were experiencing or were going through similar situations at the same time. Unlike fans studied by Baym,[68] Ross[69] and others, the fans who interacted in many of the online discussions on the *Army Wives* message boards were already part of a community by simply being military spouses. This may explain why so many of the commenters formed such strong bonds and feelings of attachment with one another and why these connections seem to be stronger for this online group of fans compared to others.

ArmyWife2009 responded to a comment by a commenter who was upset at friends who don't understand her situation: "They don't kno [sic] how it is being with someone in the army they don't feel our pain because their [sic] not the ones going through it. Just stay strong, and believe me, I know it is hard."[70] This comment offered advice while also validating the original commenter's feelings. Usaf-gf0801 felt the same way: "Like yourself, I have no one else in my group of friends who have gone through this, and it does indeed make it hard to find support."[71] Again, this is another instance where commenters found a community on the Lifetime message boards that they were unable to find elsewhere. Although the women used the message boards to talk about the difficulties of their circumstances, they often reaffirmed their life choices by saying all the heartache and sacrifice was worthwhile. Lilmspl responded to a discussion about encountering people who don't understand what military wives have to go through:

> I honestly do believe that God picked the strongest women to be military wives. . . . [Other people] don't understand why we chose this life and the [sic] never will. Yes, we might be young, but we are probably stronger than the average wife. All we have to do is keep our heads up and ignore what every-one else thinks because they will never know wat [sic] it's like to be a military spouse . . . we do it because we love our men![72]

She said that even though it was difficult, she chose the Army lifestyle because she loves her husband. Others shared similar sentiments. One com-menter said that her boyfriend, who was being deployed to Iraq, wanted to get married, but she worried that they were too young. But just reading the message board changed her mind: "Reading your post really made me realize alot [sic] with my own realtionship! [sic]"[73] This woman said she made a life-altering decision based upon comments written on the Lifetime network Web site—more than any examples thus far, this revelation underscores the importance of the *Army Wives* message board community to fans of the show. The series itself has great value to commenters as well. Tatterbugg shared: "What they go thru in this show is a lot [sic] like what we 'Army Wives' really do deal with, and I am thankful I have this to kinda lean on."[74] Spc_Deatons said the show "gives [her] comfort" as "a Real Army Wife," and Arynamber wrote that *Army Wives* "shows what its all about. . . . I believe this show is helping my family kind of understand my struggles as an [army wife]."[75] Despite the difficulties and the stress, the women report to have found support in each other as well as how the show depicted situations they experienced or were familiar with.

Though there were countless times participants reported feeling alone, they attributed the show and message boards with increasing their happiness and overall well-being. Radway's[76] famous study suggested that female au-diences seek media texts that allow them to escape their everyday lives, and Lifetime's executives repeatedly claim in the network's marketing that its shows allow women to escape. This is not the case with the *Army Wives* fans who used the message boards for advice and counsel—they are not escaping their life. Instead, they are watching their life depicted onscreen and escape it when they interact online. Feeling supported and cared for is what helps them escape their troubles, not by simply watching the show but by interacting with the community they found through their shared fandom. Although Life-time certainly profited by such behavior on its Web site, the escape it pro-vided its viewers took place off the TV screen.

CRITICISM OF *ARMY WIVES*

Not every comment on the message boards was positive, however, about how the series portrayed the life of a military spouse. This blurring between their own lives and the lives of the characters in the program supports Wood's[77] theory about referential viewing. In this particular case, the referential viewing that the message board participants engaged in was especially powerful because the characters actually were meant to represent the fans and depict their unique experiences as military wives. This is not typical of other television programs or fan engagement. As a result of this heightened referential viewing, most of those who criticized *Army Wives* described themselves as fans of the show while still expressing frustration at onscreen instances that were inconsistent with their own lives.

The majority of the issues noted on the message boards involved perceived inaccuracies and errors. The most commonly cited problem was the show's depiction of friendships between women whose husbands have positions with different ranks within the Army. Many comments referred to this as "fraternization" and several suggested it could end an officer's career. Usmc_cowgirl_00 described the severity of this offense and advised wives to avoid socializing with wives of members outside their husband's ranking:

> The reason is that if you intermingle with the other ranks, enlisted with officers and vice versa, is that more than likely your spouse might intermingle with them also. In the military, it will be misconstrued as unprofessional and fraternization. The military member can be charged with Articles 92, 133, and 134. . . . What you do reflects your significant other.[78]

The characters on the show were each married to men of varying ranks, and this is likely due to providing writers with more opportunities for story lines rather than focusing solely on the experiences of the wives of enlisted soldiers in only one rank. Some fans married to men in the military found these portrayals offensive; however, some simply found such unrealistic depictions amusing. Tonnegirl joked: "Most of the Officer wives I know would rather eat dirt that associate with lowly Jr. NCO's wives."[79] Afwife13 agreed: "in very many years, I've never seen enlisted wives as confidants of officer's wives or vice versa. It just isn't done."[80] I interpret some of the dissatisfaction with the show's portrayal of female friendships as stemming from the difficulty forming friendships that many commenters acknowledged in other posts. This is an excellent example of referential viewing.

Other critiques of the show did not mention problems with the plot or characters, but rather focused on mistakes with costumes or script glitches. Militarywife1995 pointed out that an oath taken by the character General Holden was incorrect. She wrote, with emphasis: "No 'In accordance with the Uniform Code Of Military Justice so help me God'—see online for info!"[81] It would take a keen eye to notice such an error, but perhaps again, the writers took creative freedom to express a sentiment for a specific narrative purpose. Becca_Baca22 noticed another mistake, but seemed to enjoy that her experience as a military wife allowed her to pick up on such instances. Possessing inside knowledge about the military may have made her feel superior to other viewers. She said:

> I noticed in a few episodes the men would approch [sic] the female soldiers and greet them by 'sir.' My husband had a female TI and one Air Men called her Sir instead of Mam [sic], and she about tore that poor Air Men up. That was a little offinsive [sic] in the female perspective; however, it was still a good show. As actual Military wives we can catch that stuff pretty fast lol.[82]

Note that she also admitted that while such inaccuracies exist, she still enjoyed watching *Army Wives*. Some fans enjoyed finding errors and utilize their experience with the military as a kind of social capital. Jessa1977 noticed a discrepancy with character Frank Sherwood's uniform and that the major insignia worn on his blouse did not correspond with the insignia on his cap. She wrote that his shirt's insignia represented his ranking as a major, but the insignia on his hat was for a lieutenant colonel. Though a very small number of posters mentioned things they disliked about the military, one comment criticized both the military and *Army Wives*. Lambsown15: "I have to say, seeing all of these idealistic portrayals of the lives of military wives disturbs me on a very personal level. The army knew my partner was abusing me, and supported him."[83] She was dissatisfied with the positive portrayals of Army life on the program because a solider victimized her. This comment is interesting in the context of the decision by network executives not to delve into the instances of domestic partner violence involving four soldiers murdering their wives detailed in the book the series was based upon. This was the only comment on the subject of abuse found on the *Army Wives* message boards.

   Despite drawing attention to inaccuracies, fans conceded that such mistakes "make for good TV," as Devotedarmywife922 explained.[84] After she shared a story from personal experience regarding improper fraternization, Marine_wife wrote: "I will continue to watch the show because I enjoy it, however it is after all a TV show."[85] Commenters expressed they understood such things occur when creating a fictional television program, despite its dissonance with their own lived realities. Jlybn210 admitted, with emphasis: "The themes are SO real. In fact, I laugh at how close they get to real sometimes, but life never resolves itself in 60 minutes, so in that way, no. Also, they have to exaggerate certain aspects of life in order to make it TV worthy."[86] Though it was never expressed on the message boards, it is reasonable to assume that military wives who were fans of the show may have enjoyed feelings of escapism in watching complex problems they struggled with daily solved so quickly and easily for the characters on the show, despite how unrealistic such resolutions seemed. Like many fans of fictional television programs, errors that require a suspension of disbelief are not reason enough to stop watching a program. Awife13 criticized mistakes she spotted, and then admitted: "After having said all this, I really enjoy the show. There's alot [sic] of unrealistic parts . . . but, there's also much that rings true."[87] Just as the women who sought support and advice appreciated how *Army Wives* gave a voice to difficulties that sometimes come with being a military wife, fans who criticized the show's errors also appreciated the attention it brought to their unique lived experience. Chastising and dissecting the mistakes in *Army Wives* allowed fans to bond with one another online over their private knowledge of military life. In addition to seeking advice and support, engaging in critical discussions about the show also helped strengthen the feelings of being part of fan community as well as a larger community of military spouses.

CONCLUSION

Marketing for *Army Wives* reinforced the new network identity "My Story is on Lifetime" by emphasizing the characters' relevance to real women in promotional campaigns. The show's creator and writer, Katherine Fugate, wrote in a 2008 *Television Week* guest commentary article that she had been told by military wives of the program's ability to comfort them and their gratitude for the way it calls attention to the experience of military families. She said, "But we believe our show is more than just entertainment. We believe that on a higher level, we are embracing the stories of real people, real sacrifice."[88] Interestingly, while many commenters used the *Army Wives* message boards to comment on the show's plotlines and characters, about half of the total posts came from women who self-identify as women with husbands in the military, proving the network's marketing tactics were effective. In these strategic marketing and promotional plans, Lifetime played up the show's ability to give voice to real women and families. Indeed, the collective meaning-making taking place by military spouses on Lifetime's Web site was not with regard to the television text, but rather, a large number of commenters discussed their shared experiences and everyday lives.

Fans of *Army Wives* who were Army wives themselves enjoyed the show's portrayal of their unique lived experiences, but appropriated the provided space to seek advice, support, and friendship. Members of the U.S. military and military families do endure great stress and hardship because of the obligations and sacrifices required of them. The military wives who participated on *Army Wives* message boards expressed an appreciation of the attention the show gave to their experiences, even when the show's producers made errors. Many of the women mentioned feeling isolated and alone, yet finding common ground with others and being part of a supportive online message board community helped ease their anxieties. Researchers, such as Summers, have reported on the positive ways that women use the Internet to collaborate, support one another and "find validation for their opinions and feelings."[89] Based upon this analysis, it is clear that women on the message boards did feel empowered and encouraged by their participation and likely used the site for these purposes.

While commenters expressed that they were helped by such conversations, such behavior contributed to accomplishing a key economic goal for the network: favorable media coverage, viewer participation on the network Web site, and high ratings translated in an increase in television and online advertising revenue. Lifetime is, above all, a corporation, and the function of every industrial strategy network executives employed was to maximize profit. *Army Wives* fans who used the message boards to offer support and advice to one another were doing so on the network Web site, which increased traffic, and in turn, advertising revenue. As Turow[90] contended, media firms shape web and television content to attract audiences desired by advertisers and construct online spaces, or "walled gardens," that encourage consumer participation and cultivating a relationship between the media company and viewers in ways that create and manage consumer trust. Turow described a walled garden as "an online environment where consumers go for information, communications, and commerce services and that discourages them from leaving for the larger digital world."[91] The Lifetime Web site is an example of a walled garden, and the purpose for its design and efforts to connect with viewers is to keep consumers on the site where they will be exposed to ads while their data is captured and sold to advertisers. The goal of this research project is to remain mindful of the economic imperatives at work while drawing attention to ways television viewers usurped message boards on a corporate Web site for their own personal purposes. Such efforts reflect a limited agency and raise questions regarding participatory culture and free labor that require further examination.

NOTES

1. Eileen R. Meehan; Jackie Byars. "Telefeminism: How Lifetime Got Its Groove, 1984–1997," *Television and New Media* 1, no. 1 (2000): 33–51.

2. Lacey Rose, "Talking Television with Lifetime's Andrea Wong," *Forbes*, August 17, 2009, http://www.forbes.com/2009/08/14/lifetime-project-runway-business-entertainment-television.html (accessed June 1, 2011).

3. Catherine Johnson, *Branding Television* (New York: Routledge, 2011), 168.

4. Message board comments included in this analysis come from Lifetime's message boards on the *Army Wives* portion of the network Web site (http://www.mylifetime.com/community/army-wives/discussions).

5. Janice Radway, *Reading the Romance: Women, Patriarchy and Popular Literature* (Chapel Hill: University of North Carolina Press, 1984).

6. Ibid., 61.

7. Allison J. Waldman, "Lifetime Grabs Headlines, Ratings," *Television Week*, April 20, 2008, http://www.tvweek.com/news/2008/04/lifetime_grabs_headlines_ratin.php (accessed January 11, 2011).

8. Jon Lafayette, "Wong's New Lifetime Image to Take Flight," *Television Week*, April 14, 2008, http://www.tvweek.com/news/2008/04/wongs_new_lifetime_image_to_take_flight.php (accessed January 11, 2011).

9. Ien Ang, *Watching Dallas: Soap Opera and the Melodramatic Imagination* (York: Methuen & Co., 1985).

10. Ibid., 83.

11. Helen Wood, "Texting the Subject: Women, Television and Modern Self-Reflexivity," *Communication Review* 8, no. 2 (2005): 115–35.

12. Nancy K. Baym, *Tune In, Log On: Soaps, Fandom and Online Community* (Thousand Oaks, CA: Sage, 2000).

13. Sharon Marie Ross, *Beyond the Box: Television and the Internet* (Oxford: Blackwell Publishing, 2008).

14. Ibid., 44.

15. Jennifer Holt and Alisa Perren, *Media Industries: History, Theory, and Method* (Malden: John Wiley & Sons, 2011).

16. Paul du Gay, Stuart Hall, Linda James, Hugh Mackay, and Keith Negus, *Doing Cultural Studies: The Story of the Sony Walkman* (London: Sage, 1997).

17. Timothy Havens, Amanda Lotz, and Serra Tinic, "Critical Media Industry Studies: A Research Approach," *Communication, Culture & Critique* 2, (2009): 234–53.

18. Jon Lafayette, "Q&A: Lifetime's Andrea Wong," *Television Week*, September 2, 2007, http://www.tvweek.com/in-depth/2007/09/qa_lifetimes_andrea_wong (accessed January 11, 2011).

19. Ibid.

20. Mariel Bird, "Mylifetime.com Partners with Shop.com," *Broadcasting and Cable*, November 11, 2008, http://www.broadcastingcable.com/news/programming/mylifetimecom-partners-shopcom/33625 (accessed January 11, 2011).

21. Jon Lafayette, "Lifetime Relaunches New Web Site, Offers New Features," *Television Week*, November 12, 2007, http://www.tvweek.com/in-depth/2007/09/lifetime_relaunches_web_site (accessed January 11, 2011).

22. Sarah Armaghan, "Lifetime Acquires Women's Gamer RoiWorld," *Broadcasting and Cable*, November 10, 2008, http://www.broadcastingcable.com/news/technology/lifetime-acquires-womens-gamer-roiworld/46398 (accessed January 11, 2011).

23. Quoted in Jon Lafayette, "Q&A: Lifetime's Andrea Wong," *Television Week*, September 2, 2007, http://www.tvweek.com/in-depth/2007/09/qa_lifetimes_andrea_wong (accessed January 11, 2011).

24. Ibid.

25. K.C. Neel, "*Army Wives*, Lifetime," *Multichannel News*, November 17, 2008, 29, no. 45.

26. "Lifetime to Premiere *Army Wives*, from the Mark Gordon Company and Touchstone Television," TheFutonCritic.com via press release from Lifetime, January 11, 2007, http://www.thefutoncritic.com/news/2007/01/11/lifetime-to-premiere-army-wives-from-the-mark-gordon-company-and-touchstone-television--23299/20070111lifetime05/.

27. Tanya Biank, *Army Wives: The Unwritten Code of Military Marriage* (New York: St. Martin's Press, 2006), xi.

28. Jarre Fees, "A Lifetime Wives' Tale," *Television Week*, June 1, 2008, http://www.tvweek.com/in-depth/2008/06/a_lifetime_wives_tale (accessed March 29, 2011).

29. Amy Argetsinger; Roxanne Roberts. "A Real-Wife Adventure." *Washington Post*, July 3, 2008. http://www.highbeam.com/doc/1P2-16815000.html (accessed March 29, 2011).

30. Ibid.

31. Kristina Jones, "Lifetime's *Army Wives* Gets It Right," Yahoo.com, June 20, 2007, http://voices.yahoo.com/lifetimes-army-wives-gets-right-392844.html?cat=41 (accessed March 29, 2011).

32. Ibid.

33. Jodi Morse, "Army Wives: Should New Military Spouses Watch this Lifetime Series?" Yahoo.com, July 11, 2008, http://voices.yahoo.com/army-wives-military-spouses-watch-this-1625193.html (accessed March 29, 2011).

34. Jenny Tolley, "Lifetime's Army Wives: A Delightful Mix of Truth and Fiction," Yahoo.com, June 23, 2008, http://voices.yahoo.com/lifetimes-army-wives-delightful-mixture-truth-1574297.html (accessed March 29, 2011).

35. Fees, 2008.

36. Felicia R. Lee, "Watching Army Wives Watching *Army Wives*," *New York Times*, June 28, 2007. http://www.nytimes.com/2007/06/28/arts/television/28wives.html (accessed January 11, 2011).

37. Neel, *Army Wives*.

38. Ross, *Beyond the Box*, 10.

39. Neel, *Army Wives*.

40. Daniel Frankel, "Lifetime's Andrea Wong: Grilled," *TheWrap.com*, August 17, 2009, http://www.thewrap.com/tv/article/lifetimes-andrea-wong-grilled-5263/ (accessed June 1, 2011).

41. Jon Lafayette, "Lifetime Deploys *Army Wives* to Good Effect," *Television Week*, June 5, 2007, http://www.tvweek.com/in-depth/2007/06/lifetime_deploys_army_wives_to (accessed June 1, 2011).

42. Marisa Guthrie, "*Army Wives*' Finale Lifetime's Most Watched Telecast," *Broadcasting & Cable*, September 2007, 137, no. 35.

43. John Eggerton, "Lifetime Hits New High with Army Wives Return," *Broadcasting & Cable*, June 9, 2008, http://www.broadcastingcable.com/news/programming/lifetime-hits-new-high-army-wives-return/32524 (accessed March 29, 2011).

44. Robert Seldman, "Cable Ratings," *Tvbythenumbers.com*, October 13, 2009, http://tvbythenumbers.zap2it.com/2009/10/13/cable-ratings-record-breaking-monday-night-football-mlb-playoffs-on-tbs-ncis-wwe-raw-and-wizards-of-waverly-place-top-weekly-cable-charts/30298/ (accessed March 29, 2011). Lynette Rice, "*Army Wives* Canceled by Lifetime," *EW.com*, September 24, 2013, http://insidetv.ew.com/2013/09/24/army-wives-to-conclude-seven-year-run/ (accessed May 29, 2014).

45. Bronwyn Williams, "What *South Park* Character Are You?: Popular Culture, Literacy, andOnline Performances of Identity," *Computers and Composition* 25 (2008): 25.

46. xSoldiersGirlx, December 7, 2009, comment on Real Military Wives, myLifetime.com Community, http://www.mylifetime.com/community/army-wives/discussions/real-military-wives?

47. Mary1986, August 12, 2009, comment on Real Military Wives, myLifetime.com Community, http://www.mylifetime.com/community/army-wives/discussions/real-military wives? page=2.

48. Nancy K. Baym, *Tune In, Log On: Soaps, Fandom and Online Community* (Thousand Oaks, CA: Sage, 2000).

49. Ross, *Beyond the Box*.

50. Bridget22, June 27, 2009, comment on Real Military Wives, myLifetime.com Community, http://www.mylifetime.com/community/army-wives/discussions/real-military-wives? page=2.

51. Ibid.

52. MedicWife, July 12, 2009, comment on Real Military Wives, myLifetime.com Community, http://www.mylifetime.com/community/army-wives/discussions/real-military-wives? page=2.

53. Ross, 2008.

54. Anne Becker, "Betting on Summer and Beyond," *Broadcasting & Cable*, April 2, 2007, 137, 14.

55. Anne Becker, "Cable Networks Redesign Sites," *Broadcasting & Cable*, March 26, 2007, 137, No. 13.

56. Baym, *Tune In, Log On*.

57. Ross, *Beyond the Box*.

58. GaGirl2, April 28, 2009, comment on Real Military Wives, myLifetime.com Community, http://www.mylifetime.com/community/army-wives/discussions/real-military-wives?page=3.

59. emmy2blonde, April 29, 2009, comment on Real Military Wives, myLifetime.com Community, http://www.mylifetime.com/community/army-wives/discussions/real-military-wives?page=3.

60. privatewife, May 4, 2009, comment on Real Military Wives, myLifetime.com Community, http://www.mylifetime.com/community/army-wives/discussions/real-military-wives?page=3.

61. Pfclogan, June 30, 2009, comment on Real Military Wives, myLifetime.com Community, http://www.mylifetime.com/community/army-wives/discussions/real-military-wives?page=2.

62. MissIssiee, April 18, 2009, comment on Real Military Wives, myLifetime.com Community, http://www.mylifetime.com/community/army-wives/discussions/real-military-wives?page=3.

63. lemomi, March 2, 2009, comment on Real Military Wives, myLifetime.com Community, http://www.mylifetime.com/community/army-wives/discussions/real-military-wives?page=4.

64. Usmc_cowgirl_00, August 1, comment on Real Military Wives, myLifetime.com Community, http://www.mylifetime.com/community/army-wives/discussions/real-military-wives?page=2.

65. Baym, *Tune In, Log On*.

66. Ross, *Beyond the Box*.

67. Ashprivatecp, June 18, 2009, comment on Real Military Wives, myLifetime.com Community, http://www.mylifetime.com/community/army-wives/discussions/real-military-wives?page=2.

68. Baym, *Tune In, Log On*.

69. Ross, *Beyond the Box*.

70. ArmyWife2009, August 1, 2009, comment on Real Military Wives, myLifetime.com Community, http://www.mylifetime.com/community/army-wives/discussions/real-military-wives?page=2.

71. Usaf-gf0801, May 13, 2009, comment on Real Military Wives, myLifetime.com Community, http://www.mylifetime.com/community/army-wives/discussions/real-military-wives?page=3.

72. Lilmspl, June 22, 2009, comment on Real Military Wives, myLifetime.com Community, http://www.mylifetime.com/community/army-wives/discussions/real-military-wives?page=2.

73. dancingqueen, March 30, 2009, comment on Real Military Wives, myLifetime.com Community, http://www.mylifetime.com/community/army-wives/discussions/real-military-wives?page=3.

74. Tatterbugg, March 8, 2009, comment on Real Military Wives, myLifetime.com Community, http://www.mylifetime.com/community/army-wives/discussions/real-military-wives?page=4.

75. Spc_Deatons, June 27, 2009, comment on Real Military Wives, myLifetime.com Community, http://www.mylifetime.com/community/army-wives/discussions/real-military-wives?page=2.Arynamber, February 20, 2009, comment on Real Military Wives, myLifetime.com Community, http://www.mylifetime.com/community/army-wives/discussions/real-military-wives?page=4.

76. Radway, *Reading the Romance*.

77. Wood, "Texting the Subject."

78. Usmc_cowgirl_00, July 25, 2009, comment on Real Military Wives, myLifetime.com Community, http://www.mylifetime.com/community/army-wives/discussions/real-military-wives?page=2.

79. Tonnegirl, July 26, 2009, comment on Real Military Wives, myLifetime.com Community, http://www.mylifetime.com/community/army-wives/discussions/real-military-wives?page=2.

80. Afwife13, July 25, 2009, comment on Real Military Wives, myLifetime.com Community, http://www.mylifetime.com/community/army-wives/discussions/real-military-wives?page=2.

81. Militarywife1995, May 6, 2009, comment on Real Military Wives, myLifetime.com Community, http://www.mylifetime.com/community/army-wives/discussions/real-military-wives?page=3.

82. Becca_Baca22, March 15, 2009, comment on Real Military Wives, myLifetime.com Community, http://www.mylifetime.com/community/army-wives/discussions/real-military-wives?page=4.

83. Lambsown15, August 3, 2009, comment on Real Military Wives, myLifetime.com Community, http://www.mylifetime.com/community/army-wives/discussions/real-military-wives?page=2.

84. Devotedarmywife922, April 19, 2009, comment on Real Military Wives, myLifetime.com Community, http://www.mylifetime.com/community/army-wives/discussions/real-military-wives?page=3.

85. Marine_wife, June 3, 2009, comment on Real Military Wives, myLifetime.com Community, http://www.mylifetime.com/community/army-wives/discussions/real-military-wives?page=3.

86. Jlybn210, August 12, 2009, comment on Real Military Wives, myLifetime.com Community, http://www.mylifetime.com/community/army-wives/discussions/real-military-wives?page=2.

87. Awife13, March 21, 2009, comment on Real Military Wives, myLifetime.com Community, http://www.mylifetime.com/community/army-wives/discussions/real-military-wives?page=4.

88. Katherine Fugate, "*Wives*' Puts a Face on Military's Unsung Heroes," *Television Week*, 27, No. 21.

89. Sarah Summers, "'*Twilight* Is so Anti-Feminist That I Want to Cry': *Twilight* Fans Finding and Defining Feminism on the World Wide Web," *Computers and Composition* 27, (2010): 318.

90. Jospeh Turow, "Audience Construction and Culture Production: Marketing Surveillance in the Digital Age," *Annals of the American Academy of Political and Social Science* 597, (2005): 103–21.

91. Ibid., 116.

REFERENCES

Ang, Ien. *Watching Dallas: Soap Opera and the Melodramatic Imagination*. York: Methuen & Co., 1985.

Argetsinger, Amy and Roxanne Roberts. "A Real-Wife Adventure." *Washington Post*, July 3, 2008. http://www.highbeam.com/doc/1P2-16815000.html (accessed March 29, 2011).

Armaghan, Sarah. "Lifetime Acquires Women's Gamer RoiWorld," *Broadcasting and Cable*, November 10, 2008, http://www.broadcastingcable.com/news/technology/lifetime-acquires-womens-gamer-roiworld/46398 (accessed January 11, 2011).

Baym, Nancy K. *Tune In, Log On: Soaps, Fandom and Online Community*, Thousand Oaks, CA: Sage, 2000.

Becker, Anne. "Betting on Summer and Beyond," *Broadcasting & Cable*, April 2, 2007, 137, no. 14.

———. "Cable Networks Redesign Sites," *Broadcasting & Cable*, March 26, 2007, 137, no. 13.

Biank, Tanya. *Army Wives: The Unwritten Code of Military Marriage*. New York: St. Martin's Press, 2006.

Bird, Mariel. "Mylifetime.com Partners with Shop.com," *Broadcasting and Cable*, November 11, 2008, http://www.broadcastingcable.com/news/programming/mylifetimecom-partners-shopcom/33625 (accessed January 11, 2011).

Eggerton, John. "Lifetime Hits New High with Army Wives Return," *Broadcasting & Cable*, June 9, 2008, http://www.broadcastingcable.com/news/programming/lifetime-hits-new-high-army-wives-return/32524 (accessed March 29, 2011).

Fees, Jarre. "A Lifetime Wives' Tale," *Television Week*, June 1, 2008, http://www.tvweek.com/in-depth/2008/06/a_lifetime_wives_tale (accessed March 29, 2011).

Frankel, Daniel. "Lifetime's Andrea Wong: Grilled," *TheWrap.com*, August 17, 2009, http://www.thewrap.com/tv/article/lifetimes-andrea-wong-grilled-5263/ (accessed June 1, 2011).

Fugate, Katherine. "*Wives*' Puts a Face on Military's Unsung Heroes," *Television Week*, 27, no. 21.

Guthrie, Marisa. "*Army Wives*' Finale Lifetime's Most Watched Telecast," *Broadcasting & Cable*, September 2007, 137, no. 35.

Johnson, Catherine. *Branding Teleivision*, New York: Routledge, 2011.

Jones, Kristina. "Lifetime's *Army Wives* Gets It Right," Yahoo.com, June 20, 2007, http://voices.yahoo.com/lifetimes-army-wives-gets-right-392844.html?cat=41 (accessed March 29, 2011).

Lafayette, Jon. "Wong's New Lifetime Image to Take Flight," *Television Week*, April 14, 2008, http://www.tvweek.com/news/2008/04/wongs_new_lifetime_image_to_take_flight.php (accessed January 11, 2011).

———. "Lifetime Relaunches New Web Site, Offers New Features," *Television Week*, November 12, 2007, http://www.tvweek.com/in-depth/2007/09/lifetime_relaunches_web_site (accessed January 11, 2011).

———. "Q&A: Lifetime's Andrea Wong," *Television Week*, September 2, 2007, http://www.tvweek.com/in-depth/2007/09/qa_lifetimes_andrea_wong (accessed January 11, 2011).

———. "Lifetime Deploys *Army Wives* to Good Effect," *Television Week*, June 5, 2007, http://www.tvweek.com/in-depth/2007/06/lifetime_deploys_army_wives_to (accessed June 1, 2011).

Lee, Felicia R. "Watching Army Wives Watching *Army Wives*," *New York Times*, June 28, 2007. http://www.nytimes.com/2007/06/28/arts/television/28wives.html (accessed January 11, 2011).

"Lifetime to Premiere *Army Wives*, from the Mark Gordon Company and Touchstone Television," TheFutonCritic.com via press release from Lifetime, January 11, 2007, http://www.thefutoncritic.com/news/2007/01/11/lifetime-to-premiere-army-wives-from-the-mark-gordon-company-and-touchstone-television--23299/20070111lifetime05/.

Meehan, Eileen R. and Jackie Byars. "Telefeminism: How Lifetime Got Its Groove, 1984–1997," *Television and New Media* 1, no. 1 (2000): 33–51.

Morse, Jodi. "Army Wives: Should New Military Spouses Watch This Lifetime Series?" Yahoo.com, July 11, 2008, http://voices.yahoo.com/army-wives-military-spouses-watch-this-1625193.html (accessed March 29, 2011).

Neel, K. C. "*Army Wives*, Lifetime," *Multichannel News*, November 17, 2008, 29, no. 45.

Radway, Janice. *Reading the Romance: Women, Patriarchy and Popular Literature*, Chapel Hill: University of North Carolina Press, 1984.

Rice, Lynette. "*Army Wives* Canceled by Lifetime," *EW.com*, September 24, 2013, http://insidetv.ew.com/2013/09/24/army-wives-to-conclude-seven-year-run/ (accessed May 29, 2014).

Rose, Lacey. "Talking Television with Lifetime's Andrea Wong," *Forbes*, August 17, 2009 http://www.forbes.com/2009/08/14/lifetime-project-runway-business-entertainment-television.html (accessed June 1, 2011).

Ross, Sharon Marie. *Beyond the Box: Television and the Internet*. Oxford: Blackwell Publishing, 2008.

Seldman, Robert. "Cable Ratings," *Tvbythenumbers.com*, October 13, 2009, http://tvbythenumbers.zap2it.com/2009/10/13/cable-ratings-record-breaking-monday-night-football-mlb-playoffs-on-tbs-ncis-wwe-raw-and-wizards-of-waverly-place-top-weekly-cable-charts/30298/ (accessed March 29, 2011).

Tolley, Jenny. "Lifetime's Army Wives: A Delightful Mix of Truth and Fiction." Yahoo.com, June 23, 2008, http://voices.yahoo.com/lifetimes-army-wives-delightful-mixture-truth-1574297.html (accessed March 29, 2011).

Waldman, Allison J. "Lifetime Grabs Headlines, Ratings." *Television Week*, April 20, 2008, http://www.tvweek.com/news/2008/04/lifetime_grabs_headlines_ratin.php (accessed January 11, 2011).

Williams, Bronwyn. "What *South Park* Character Are You?: Popular Culture, Literacy, and Online Performances of Identity." *Computers and Composition* 25 (2008): 24–39.

Wood, Helen. "Texting the Subject: Women, Television and Modern Self-Reflexivity." *Communication Review* 8, no. 2 (2005): 115–35.

Chapter Fourteen

"Butter," Facebook, and Paula Deen

Examining Fans' Use of Social Media in Crisis

Michel M. Haigh and Shelley Wigley [1]

How many people save their butter wrappers? Not many. However, in July of 2013 Paula Deen fans not only saved their butter wrappers, they mailed them to the *Food Network* in support of Deen. The *Food Network* announced it would not renew Deen's contract in June of 2013 following a leaked deposition in which Deen admitted to having used the N-word when she was held at gunpoint by an African American man in 1986.[2] Supporters also called the *Food Network* "hypocrites" for firing Deen for something that happened more than thirty years earlier.

The information about Deen came to light when a former employee sued Deen and her brother. The *National Enquirer* printed part of Deen's deposition in the lawsuit in which she admitted to using the N-word. Two days later, the *Food Network* announced it would not renew Deen's contract. In a matter of days, Deen lost her *Food Network* contract and most of her corporate sponsors.

But Deen's fans and supporters rallied. They not only mailed in butter wrappers, they also took to social media, primarily Facebook, to express their love and support for Deen and their discontent with the *Food Network*'s decision. Fans posted on the *Food Network*'s Facebook page and Deen's own Facebook page, which is independent of the *Food Network*. Additionally, fans started a "We Support Paula Deen" page on Facebook.[3] By July 4, roughly two weeks after Deen was fired, more than 500,000 supporters had "liked" the "We Support Paula Deen" page. Two months later, the lawsuit was dismissed, and by that time more than 612,000 supporters had "liked" the "We Support Paula Deen" Facebook page. Deen's own page gained more than 400,000 new "likes" during the same time period.

The current study examines what fans were posting during the crisis. Fans' Facebook posts were examined to determine what they were saying about Deen and the *Food Network*.

LITERATURE REVIEW

Reality TV and Social Media

It is not a surprise Deen's fans would rally around her during the controversy. Viewers of reality TV, which includes everything from *Survivor* to *Sister Wives*, say they enjoy the genre because it provides authenticity, intimacy, and focuses on "real" people.[4] The reality TV genre has increased exponentially over the past decade. In 2010, reality TV programs numbered approximately six hundred and accounted for around 40 percent of primetime programming.[5] The shows are popular because they focus on everyday people and their everyday lives; they are people viewers can identify with and who they believe are "just like them."

Reality TV celebrates real people and their flaws, obstacles, and issues. Producers of these programs acknowledge the connection viewers often form with cast members of reality TV programs and do their best to exploit features of the genre that strengthen the bond between viewers and personalities. For example, producers of *TLC*'s *Extreme Couponing* exploit the perceived relationships that exist between viewers and TV personalities by featuring personalities co-existing with their families and friends sharing their most intimate thoughts and feelings with viewers. Additionally, each personality shares his or her personal story on what led to couponing; many of these stories are wrought with tales of financial hardships designed to draw in viewers.[6]

With *A&E*'s *Duck Dynasty*, members of the Robertson family exemplify togetherness. The Robertson men and their sons are often seen hunting together and each episode ends with the entire Robertson clan sitting down together to enjoy a home cooked meal.[7] Family and togetherness is a common theme throughout *Duck Dynasty*. The same is true for *Here Comes Honey Boo Boo*. The entire *HCHBB* family is often seen doing everyday activities together, and a first person confessional gives family members the opportunity to address the audience, thereby creating intimacy with viewers.[8]

All the disclosure, togetherness, and behind the scenes access that occurs in reality TV programming naturally leads to viewers' immersion into TV personalities' lives. However, weekly viewing may not be enough for some viewers who wish to extend the bond beyond TV watching. Because of social media, viewers of reality TV have even greater access to reality TV personalities. Reality TV programs often offer viewers the opportunity to engage with the program's personalities, story lines and each other through Facebook pages, message boards, and Twitter feeds dedicated to the show. Viewers take advantage of the additional access by posting comments and opinions, some addressed directly to the personalities themselves.

A review of comments and posts by researchers suggest how banter on social media sometimes mimics a perceived relationship in the eyes of the viewer. One viewer of the reality show *Here Comes Honey Boo Boo* posted "I love this family!" on the show's official Facebook page. Other posts often reference the family's realness and authenticity. As one fan pointed out, the family members are not actors. "They are their true selves and that is the best kind of reality."[9] Still another viewer posted: "I love the whole family— working class real people who ain't afraid to just be themselves!"[10] Other posts on the *HCHBB* page included: "They are ordinary people"; "It's real sh*t. That's why I love them"; and "Luv 'em."[11]

In a study of the official message board affiliated with *The Real House-wives* series, viewers sometimes addressed the women on the show directly. For example, one person posted a comment addressed directly to one of the personalities who had made a game of finding homeless people (or bums). The post stated: "Hey Jeana, they aren't bummers, they're people."[12] Still another addressed Lisa directly, who had been complaining about the Depart-ment of Motor Vehicles. "Lisa at the DMV office, welcome to our world honey."[13] These posts reveal some viewers feel as if they know the people featured in reality TV programming intimately. While these reality personal-ities were not in crisis, it is no surprise fans of Deen would use social media to show their support for her and disdain for the entity threatening to end the relationship between them and one of their favorite TV personalities.

Parasocial Interaction (PSI)

Horton and Wohl developed the concept of Parasocial Interaction (PSI) as an explanation for the way the media and mediated performers present the illu-sion of an interpersonal relationship.[14] Early research focused on personae, or performers, who appeared on radio and TV, the primary modes of pro-gramming at the time. The persona, or performer, offers a continuing rela-tionship whose appearance is a regular and dependable event that is planned for and integrated into the routine of daily life.[15] This continued association results in a history of shared past experiences and a bond between viewer and persona. Over time, the viewer, or "fan," comes to believe he or she knows the persona intimately and better than others do. The bond of intimacy is a primary component of PSI. This "illusionary" bond allows viewers and per-sonae to seemingly grow closer to one another, at least in the minds of the viewers. The intimacy is only an illusion because: it is one-sided; the media-user's reaction cannot reach the persona; and the bond is not reciprocated from the persona.[16]

As evidenced by the posts appearing on reality TV fan sites and message boards, PSI appears to be taking place between viewers of reality TV programs and those programs' personae. The intimate, behind-the-scenes access and full disclosure viewers get by watching reality TV programs easily translates into a strong bond between viewers and personae. The authenticity and realness offered up by reality TV programming lends itself naturally to PSI.

Although Horton and Wohl[17] developed the concept of PSI to explain both radio and television consumption, the majority of PSI research has focused on television and explored soap operas,[18] news anchors,[19] athletes,[20] politicians,[21] primetime TV characters,[22] and celebrity pitchmen.[23]

The perceived relationship that develops between audience members and personae because of PSI is sometimes referred to as a Parasocial Relationship (PSR). Over the years, there has been some blurring of the lines between the two concepts, with some scholars using the terms interchangeably.[24] Schramm and Hartmann offer the following distinction between the two concepts: PSI is bound by interpersonal type processes that take place during the media exposure; whereas PSR refers to the cross-situational relationship between the viewer and persona that may include cognitive, affective, and behavioral components.[25] The researchers explain PSI can only occur during the actual media exposure, whereas PSR may endure beyond a single media exposure, "like a friendship that exists between two persons beyond their face-to-face communication sequences."[26]

Nordlund emphasized PSI can extend beyond the actual media encounter and occurs when viewers show interest in characters, "participate" in what happens to the characters, and "know" the characters.[27] Clearly, social media offers many opportunities for PSI to develop, strengthen and endure beyond viewers' exposure to reality TV programming. Social media presents viewers additional ways to interact with reality TV personae even after the TV program has ended. This increased PSI that occurs through social media most likely leads to stronger PSR between viewers and personae.

Schramm and Hartmann describe parasocial processing as capturing viewers' responses toward personae, even if viewers are not motivated to be a part of a reciprocal encounter. "Parasocial processing may simply be seen as processes of person perception that set in as soon as a user encounters a persona."[28] This processing manifests in several forms including: increased interest in a persona, intense thoughts and deliberations, gestures and facial expressions, and words spoken to the persona displayed on TV. In summary, PSI as parasocial processing is about viewers' cognitive, affective, and behavioral responses to media characters. It is about viewers interacting psychologically with media characters.

Cognitive responses include persona perception and evaluation (e.g., judging Deen as a warm, sweet, southern lady), activation of memories or life experiences (e.g., Deen's dishes remind me of my mother's cooking) or social comparisons (e.g., Deen likes home cooking just like me). Affective responses include positive and negative feelings toward the persona or emotions evoked by the persona (e.g., praying for Deen, forgiveness, regret, "we love you"). Behavioral responses include verbal and nonverbal behaviors as well as behavioral intentions (e.g., threatening to boycott the *Food Network* or to mail in butter wrappers to the network).[29]

In their study, Schramm and Hartmann devised a set of PSI-Process scales that break the processes up into cognitive, affective, and behavioral dimensions.[30] For the current study, the authors looked at these same types of characteristics in a content analysis instead of a scaled survey by examining the type of information presented in the posts on social media sites. In this instance, the posts on Deen's Facebook page and those on the *Food Network*'s Facebook page were examined for insight into the PSI that took place between Deen and her fans during the crisis she faced in 2013.

Using the PSI-Process Model, the following research questions were posed for this study:

RQ1: Did fans discuss Deen's actions, evaluate her actions, or build a relationship between Deen and themselves based on growing up and living in the south (cognitive components of PSI)?

RQ2: Did fans discuss sympathy, forgiveness, or love (affective components of PSI)?

RQ3: What were the fans behavioral intentions?

RQ4: When examining posts, did the type of fan post (cognitive, affective, or behavioral) change from June, when Deen was fired, to August, when the lawsuit was dismissed?

RQ5: When examining posts, was there a difference in the type of fan post (cognitive, affective, and behavioral) on Paula Deen's Facebook page compared to the type of fan post on the *Food Network*'s Facebook page?

Method

The dates selected for the sample spanned forty-eight hours after the *Food Network* fired Deen, and another forty-eight hours after the judge dismissed the lawsuit. The unit of analysis was each Facebook post (N = 14,954) made to the Deen Facebook page (n = 8,591) or the *Food Network* Facebook page (n = 6,363). The unit of analysis was the original post, not the comments made to the post.

Demographics

The number of "likes" (M = 2.03, SD = 3.64), shares (M = 0.02, SD = 0.96), and comments (M = 0.83, SD = 2.43) varied for each post. Very few posts included an external link (3.4 percent), or a photo of Deen (0.7 percent). Usually the photo of Deen was a "meme" (0.5 percent). Females (75.5 percent) tended to post more frequently than males (23.2 percent).

Coder Training

Five undergraduate students enrolled at a Mid-Atlantic university conducted the content analysis. A written coding instrument was developed to code the sample. Coding norms were established during a supervised training session. Five percent of the sample was coded during the training phase. Coders established a high degree of standardization during the training phase, resulting in effective inter-coder reliabilities between 0.85 and 1.0 when employing Cohen's Kappa.

Categories Coded

The posts were either marked "present" or "absent" if the post included information about the topic. See Table 14.1 for a complete list of categories and the frequency each occurred.

Schramm and Hartmann discuss the properties of Parasocial Interaction. The categories developed included cognitive, affective, and behavioral processes.[31] *Cognitive categories* coded included: fan comments about Deen growing up in the south and that "was the way it was," the fan commented he/she relates because of living or growing up in the south, the fan commented about Deen not lying under oath, the fan commented Deen "made a mistake," or the fan commented he/she and Deen are "friends" or "family." Another cognitive category was the cognitive construction of the fan "connecting to Deen through her show."

Affective categories coded included comments about: praying for Paula, references to Deen "hanging in there/keep her head up/chin up," references to forgiveness, or posts stating, "we love ya/love ya." Behavioral categories coded included: boycotting the *Food Network*, canceling a *Food Network* magazine subscription, boycotting the sponsors that dropped Deen, and telling Deen to "call them" and posting their phone number.

RESULTS

Descriptive statistics and Chi-Square analyses were used to answer the research questions. Frequencies were used to answer Research Questions 1–3. Research Question 1 examined the cognitive element of PSI by examining if fans discussed Deen's actions, evaluated her actions, or had a relationship with Deen. Cognitive categories coded included: fan commented about Deen growing up in the south and that "was the way it was" (2.9 percent); the fan

Table 14.1.

	Cohen's Kappa	Percent Present
Canceling the *Food Network*	0.99	19.8%
Canceling the *Food Network* magazine	1.0	1.2%
Paula *growing up* in the south and that is the way it was	0.96	2.9%
The person posting is relating because they grew up in the south	0.95	2.7%
Praying for Paula	0.98	4.2%
"Hanging in there/keep head up/chin up"	0.98	15.6%
Reference Paula's sponsors	0.96	1.1%
Paula honest under oath	0.98	6.1%
Forgiveness	1.0	4.8%
Reference to making mistakes	0.98	10.6%
"We love ya/love ya"	0.85	18.9%
Say "call me"	0.88	0.1%
Mention "we're friends" or "like friends or family"	0.85	0.6%
Connecting to her via the show	0.90	1.1%

commented he or she related because of living or growing up in the south (2.7 percent), the fan commented about Deen not lying under oath (6.1 percent); the fan evaluated Deen's action as "making a mistake" (10.6 percent); the fan commented he/she and Deen are like "friends" or "family," (0.6 percent); and the fan commented about "connecting to Deen through her show" (1.1 percent).

Research Question 2 asked what the affective PSI posts might discuss. Affective posts include fan comments about: praying for Paula (4.2 percent); referenced Deen "hanging in there/keep her head up/chin up/ because she has a lot of support," (15.6 percent); referenced forgiveness (4.8 percent), or stated, "we love ya/love ya" (18.9 percent).

Research Question 3 examined fans' behavioral intentions. Almost a fifth of the posts (19.8 percent) discussed not watching the *Food Network* because Deen had been fired, and 1 percent of the fans said they were going to cancel their subscription to the magazine, as well as boycott the sponsors that dropped Deen (1.1 percent). Others told Deen to "call them" (0.1 percent) and posted their phone numbers.

Chi-Square analyses were used to examine Research Questions 4 and 5. Most of the fans posted or commented at the beginning of the crisis (89.6 percent, n = 13,401 posts) when the *Food Network* fired Deen. The Chi-

Square analyses for Research Question 4 found significant differences for the *cognitive categories* of: Deen growing up in the south and that "was the way it was" $\chi^2(1, N = 14,954) = 41.79$, $p < 0.001$; the fans commented about Deen not lying under oath χ^2 $(1, N = 14,954) = 86.73$, $p < 0.001$; and the fans evaluated Deen's action as "making a mistake" χ^2 $(1, N = 14,954) = 148.20$, $p < 0.001$.

Significant differences were found for the *affective categories*: praying for Paula χ^2 $(1, N = 14,954) = 4.00$, $p < 0.05$; referenced Deen "hanging in there/keep her head up/chin up," χ^2 $(1, N = 14,954) = 68.20$, $p < 0.001$; referenced forgiveness χ^2 $(1, N = 14,954) = 64.66$, $p < 0.001$; or stated, "we love ya/love ya" χ^2 $(1, N = 14,954) = 12.08$, $p < 0.001$.

Significant differences for the *behavioral categories* included: not watching the *Food Network* χ^2 $(1, N = 14,954) = 306.10$, $p < 0.001$, canceling the subscription to the *Food Network* magazine χ^2 $(1, N = 14,954) = 20.29$, $p < 0.001$; and boycotting the sponsors that dropped Deen χ^2 $(1, N = 14,954) = 147.70$, $p < 0.001$. So there were differences in the type of post a fan made when comparing June to August.

Research Question 5 asked if there was a difference in the type of fan post (cognitive, affective, or behavioral) when comparing Deen's Facebook page to the *Food Network*'s Facebook page.

Chi-Square analyses found significant differences for the *cognitive categories* of: the fans grew up in the south and that "was the way it was" $\chi^2(1, N = 14,954) = 97.90$, $p < 0.001$; the fans commented about Deen not lying under oath χ^2 $(1, N = 14,954) = 13.81$, $p < 0.001$; the fans evaluated Deen's action as "making a mistake" χ^2 $(1, N = 14,954) = 11.92$, $p < 0.001$; they were like "friends or family" χ^2 $(1, N = 14,954) = 68.85$, $p < 0.001$; and they were able to connect to Deen via the show χ^2 $(1, N = 14,954) = 103.35$, $p < 0.001$.

Significant differences were found for the *affective categories*: praying for Paula χ^2 $(1, N = 14,954) = 445.00$, $p < 0.001$; referenced Deen "hanging in there/keep her head up/chin up" χ^2 $(1, N = 14,954) = 1337.69$, $p < 0.001$; referenced forgiveness χ^2 $(1, N = 14,954) = 64.66$, $p < 0.001$; or stated "we love ya/love ya" χ^2 $(1, N = 14,954) = 1850.93$, $p < 0.001$.

Significant differences for the *behavioral categories* included: not watching the *Food Network* χ^2 $(1, N = 14,954) - 752.98$, $p < 0.001$; canceling the subscription to the *Food Network* magazine χ^2 $(1, N = 14,954) = 53.71$, $p < 0.001$; and boycotting the sponsors that dropped Deen χ^2 $(1, N = 14,954) = 31.72$, $p < 0.001$. There were also significant differences found for the category "call me" χ^2 $(1, N = 14,954) = 10.38$, $p < 0.001$.

DISCUSSION

This study examined fans' Facebook posts during a time of crisis. The crisis was unique in the sense that it was a form of a "breakup." The "breakup" meaning the fans' parasocial relationship with Deen was coming to an end since her show was canceled. A quantitative content analysis was conducted of two different Facebook pages in order to gain a complete picture of the content being posted. Cohen[32] proposed the parasocial breakup concept. The parasocial breakup is similar to real-life breakups. The stronger the perceived relationship, the more likely the person will feel loss when the character is taken away. Granted, Deen is not a character, she's a real person, but the same effect may occur.

Research Question 1 examined the cognitive element of PSI by examining if fans discussed Deen's actions, evaluated her actions, or had a relationship with Deen. Deen's fans rarely commented about Deen or themselves growing up in the south. When making a cognitive connection with Deen, the fans were more likely to discuss her telling the truth and not lying while under oath. This is an interesting finding because it indicates that when fans were discussing the leaked testimony, most would make a comment (more than 10 percent) that more than thirty years had passed since the incident and the use of the N-word. The time variable could have also influenced the fans commenting about Deen "making a mistake," another common cognitive evaluation fans made. Rarely did Deen's fans comment they were "friends" or "family."

Research Question 2 asked how often fans may discuss the affective part of PSI. The most frequent way fans related affectively was by telling Deen "we love ya/love ya," followed by "keep her head up or hang in there." Another way fans posted and showed emotional support was to state "praying for you Paula" and offering forgiveness. These findings support Schramm and Hartmann's[33] idea that parasocial processing includes intense thoughts and deliberations. It also supports the idea PSRs develop as a result of increased involvement.[34]

Research Question 3 examined what fans' behavioral intentions were after Deen's show was canceled. Taking to social media to express disappointment in an organization is common. For example, Netflix, the movie rental service decided to divide its one service into two—one offering streaming movies over the Internet, the other offering old-fashioned DVDs in the mail.[35] A review of the company's Facebook page immediately following Netflix's announced changes to the service and pricing structure in July 2011 revealed more than 56,520 people responded to the changes by posting comments on the Netflix Facebook wall. The company reported it lost 800,000 US subscribers in the third quarter, and the company's stock plummeted more than 25 percent.[36]

In Deen's case, almost a fifth of the posts discussed not watching the *Food Network* because Deen had been fired, canceling their subscription to the magazine, as well as boycotting the sponsors that dropped Deen. The current study supports the idea of the "illusionary" relationship between viewers and the persona since some fans actually posted "call me" with their actual phone numbers.[37]

Most of the fans posted or commented at the beginning of the crisis (89.6 percent, $n = 13,401$ posts) when the *Food Network* fired Deen. The type of post a fan made changed over time. For the *cognitive categories*, fans were more likely to post about Deen growing up in the south and that "was the way it was" in June when the crisis first occurred. The fans also commented more about Deen not lying under oath and "making a mistake" in June when compared to August posts. For the affective categories, fans were more likely to state they were praying for Paula, telling her to "hang in there," referencing forgiveness, or "we love ya/love ya" in June when compared to August. When examining the behavioral categories fans stated more often they were not going to watch the *Food Network*, cancel their subscriptions to the *Food Network* magazine, and boycott the sponsors that dropped Deen in June when compared to August. So there were differences in the type of post a fan would make when comparing June to August.

Research Question 5 asked if there was a difference in the type of fan post (cognitive, affective, or behavioral) when comparing Deen's Facebook page to the *Food Network*'s Facebook page. There were differences in the type of post a fan made to each page.

Deen's fans were more likely to post to Deen's page about the fan growing up in the south, the fact that Deen did not lie under oath, and the fact Deen made a mistake. They also were more likely to talk about being "friends or family" and connecting to Deen via the show.

They posted more affective posts such as praying for Paula, telling her to "hang in there," offering forgiveness, or stating, "we love ya" on Deen's page than the *Food Network* page. When looking at the three categories (cognitive, affective, or behavioral), the only category that appeared more frequently on the *Food Network* page were the behavioral intentions. Fans were more likely to post on the *Food Network* page that they were not going to watch the network, cancel their subscriptions to the magazine, as well as boycott the sponsors that dropped Deen. This indicates fans can differentiate between pages maintained by the network and pages maintained by the celebrity (persona). The cognitive and affective components of PSI were more apparent when the fans were "directly" communicating with Deen. The behavioral components were present when the fans were expressing their displeasure about the breakup to the management, or the group who made the decision about canceling the show.

LIMITATIONS AND FUTURE DIRECTIONS

All studies have limitations. One of the biggest limitations for this study was the amount of data to code and collect. When the author tried to gather the data immediately after Deen was fired, Facebook was overwhelmed by the amount of people posting. In order to gather the data, the author had to click on "posts by others" which opened a window displaying the posts by fans. The same procedure was used to gather data from both the Deen page and the *Food Network* page. When the author was saving the data in June 2013, scrolling through the posts indicated posts from certain times were missing. The posts on the Deen page went from stating "24 hours ago" or "Friday, June 22" to posts with April or May dates. The author ended up gathering the data over a period of several days to make sure as much data as possible was collected. This only occurred on the Deen page and not on the *Food Network* page.

This chapter represents only a portion of the data coded. Besides coding for parasocial interaction categories, the coders also coded for posts discussing the First Amendment (freedom of speech), civil rights, and the use of the N-word in entertainment. During the training period, mutually exclusive and exhaustive categories were developed to make sure the coders agreed on the content of the post as well as the category each post fit into. Due to all the training sessions, the "other" category was not employed, but this also led to small percents of "present" in the other categories. The additional categories will be examined in future writings.

Several things became apparent while the research was being conducted. First, fans post different types of things to the organization's page compared to the celebrity's page. Also, the organization is not likely to respond to or delete the negative fan posts on its page. Something else to consider, when Facebook updates how the page looks, none of the posts from the fans appear on the main page. The only way an observer would know the organization or celebrity (such as the *Food Network* or Deen) is going through a crisis is by clicking on the "posts by others" section of the page. None of the negative comments or discussion taking place is present on the Facebook page viewable to the novice visitor. This is obviously a benefit to any celebrity or organization experiencing negative media attention or going through a crisis.

This study only examined the posts fans made to Facebook, but it can be anticipated that fans also took to Twitter to lament about the crisis. Additional research could examine Twitter data to see if the same themes and parasocial interaction categories appeared in that medium as well.

In the end, fans' support did not save Deen. Deen was not reinstated. Even re-runs of her show stopped airing after her contract expired. However, one can still access her recipes on the *Food Network* Web site more than a year later. Fans made their voices heard thanks to social media. When a celebrity or persona has a crisis, particularly one that threatens an established parasocial relationship, those invested in the relationship will take to social media to voice their support and vow to take action against those who threaten the relationship. In this case, Deen's supporters took to Facebook to threaten the *Food Network* with boycotting the network and canceling subscriptions to the magazine because of Deen's firing. Entities that interfere with a parasocial relationship between a celebrity/persona and fans can expect backlash in the form of negative behavioral intentions, and in today's digital world, rest assured those behavioral intentions will be posted for all to see. In this particular case, Deen really did acknowledge the support she had online—on her own page as well as the "We Support Paula Deen" page.

Her responsiveness to fans did not end there. In September of 2014, Deen launched her online *Paula Deen Network*. She said the reason she started the network was because of fan support. The site is ad-free and funded by subscriptions ($9.99 a month or $7.99 a month if one purchases a year subscription). The Web site includes a vintage section that features Deen's *Food Network* shows, which the financial backer for the new network acquired. Deen's continued fan support on social media led her to start the network. She told the *Huffington Post* "One of my salvations in that year and three or four months when I was out of the public eye was a Web site that one of my family members showed me—it was 'We support Paula Deen.' [referring to the Facebook page] And I saw that my Web site had grown to over four and a half million people," she said, referring to her popular Facebook page. "That was staggering to me, that my Web site had actually grown rather than decreasing."[38]

NOTES

1. This project was funded in part by a Penn State Research in Undergraduate Education Grant. Thanks to: Kaitlin Eckrote, Kelsey Thompson, Paige Whiteley, Maya Kouassi, and Amelia Friedrichs undergraduates in the college for their help in coding the Facebook posts.

2. Sadie Gennis. "A Timeline of Paula Deen's Downfall," *TV Guide*, June 27, 2013, accessed, June 30, 2014, http://www.tvguide.com/news/paula-deen-scandal-timeline-1067274. aspx.

3. Gennis, "A Timeline of Paula Deen's Downfall."

4. Andre Cavalcante. "You Better 'Reneckognize'!: Deploying the Discourses of Realness, Social Defiance, and Happiness to Defend *Here Comes Honey Boo Boo* on Facebook," in *Reality Television: Oddities of Culture*, ed. Alison F. Slade et al. (Lanham, MD: Lexington Books, 2014), 39–58. Leandra H. Hernandez, "'I Was Born This Way': The Performance and Production of Southern Masculinity in A&E's *Duck Dynasty*," in *Reality Television: Oddities of Culture*, ed. Alison F. Slade et al. (Lanham, MD: Lexington Books, 2014), 21–37.

5. Pamela L. Morris and Charissa, K. Niedzwiecki. "Odd or Ordinary: Social Comparisons Between Real and Reality TV Families," in *Reality Television: Oddities of Culture*, ed. by Alison F. Slade et al. (Lanham, MD: Lexington Books, 2014), 143–61.

6. Rebecca M. Curnalia. "Frugal Reality TV During the Great Recession: A Qualitative Content Analysis of TLC's *Extreme Couponing*," in *Reality Television: Oddities of Culture*, ed. by Alison F. Slade et al. (Lanham, MD: Lexington Books, 2014), 101–21.

7. Hernandez, "'I was Born this Way.'"
8. Cavalcante, "You Better 'Reneckognize'!"
9. Cavalcante, "You Better 'Reneckognize'!" 42.
10. Cavalcante,"You Better 'Reneckognize'!" 43.
11. Cavalcante, "You Better 'Reneckognize'!" 44, 48.
12. Nicole B. Cox, "Bravo's *The Real Houswives*: Living the (Capitalist) American Dream?" in *Reality Television: Oddities of Culture*, ed. by Alison F. Slade et al. (Lanham, MD: Lexington Books, 2014), 77–99, 87.
13. Cox, "Bravo's *The Real Houswives*,'" 87.
14. Donald Horton, R. Richard Wohl. "Mass Communication and Parasocial Interaction." *Psychiatry* 19, no. 3 (1956): 215–29.
15. Horton and Wohl, "Mass Communication and Parasocial Interaction."
16. Horton and Wohl, "Mass Communication and Parasocial Interaction."
17. Horton and Wohl, "Mass Communication and Parasocial Interaction."
18. Philip J. Auter, "Psychometric: TV That Talks Back: An Experimental Validation of a Parasocial Interaction Scale." *Journal of Broadcasting & Electronic Media* 36, no. 2, (1992): 173–81. Mark R. Levy. "Watching TV News as Para-Social Interaction." *Journal of Broadcasting* 23, no. 1 (1979): 69–80. Elizabeth M. Perse and Rebecca Rubin. "Attribution in Social and Parasocial Relationships." *Communication Research* 16, no. 1 (1989): 59–77; Alan M. Rubin and Elizabeth E. Perse. "Audience Activity and Soap Opera Involvement: A Uses and Effects Investigation." *Human Communication Research* 14, no. 2 (1987): 246–68.
19. Rick Houlberg. "Local Television News Audience and the Para-Social Interaction." *Journal of Broadcasting* 28, no. 4 (1984): 423–42. Elizabeth M. Perse. "Media Involvement in Local News Effects." *Journal of Broadcasting & Electronic Media* 34, no. 1 (1990): 17–36.
20. Adam C. Earnheardt, and Paul M. Haridakis. "An Examination of Fan-Athlete Interaction: Fandom, Paraosical Interaction, and Identification." *Ohio Communication Journal* 47 (2009): 27–53.
21. Lawrence. A. Wenner. "Political News on Television: A Reconsideration of Audience Orientations." *Western Journal of Speech Communication* 47, no. 4 (1983): 380–95.
22. Cynthia A. Hoffner and Elizabeth L. Cohen. "Responses to Obsessive Compulsive Disorder on *Monk* Among Series Fans: Parasocial Relations, Presumed Media Influence, and Behavioral Outcomes." *Journal of Broadcasting & Electronic Media* 56, no. 4 (2012): 650–68.
23. Neil Alperstein. "Imaginary Social Relationships with Celebrities Appearing in Television Commercials." *Journal of Broadcasting & Electronic Media* 35, no. 1 (1991): 43–58.
24. Holger Schramm and Tilo Harmann. "The PSI-Process Scales. A New Measure to Assess the Intensity and Breadth of Parasocial Processes." *Communications* 33, no. 4 (2008): 385–401.
25. Schramm and Harmann, "The PSI-Process Scales. A New Measure to Assess the Intensity and Breadth of Parasocial Processes."
26. Schramm and Harmann, "The PSI-Process Scales. A New Measure to Assess the Intensity and Breadth of Parasocial Processes," 386.
27. Jan-Erik Nordlund. "Media Interaction." *Communication Research* 5, no. 2 (1978): 150–75.
28. Schramm and Harmann, "The PSI-Process Scales. A New Measure to Assess the Intensity and Breadth of Parasocial Processes," 387.
29. Schramm and Harmann, "The PSI-Process Scales. A New Measure to Assess the Intensity and Breadth of Parasocial Processes."
30. Schramm and Harmann, "The PSI-Process Scales. A New Measure to Assess the Intensity and Breadth of Parasocial Processes."
31. Schramm and Harmann, "The PSI-Process Scales. A New Measure to Assess the Intensity and Breadth of Parasocial Processes."
32. Jonathan Cohen. "Parasocial Breakups: Measuring Individual Differences in Responses to the Dissolution of Parasocial Relationships." *Mass Communication & Society* 6, no 2 (2003): 191–202.
33. Schramm and Harmann, "The PSI-Process Scales. A New Measure to Assess the Intensity and Breadth of Parasocial Processes."

34. Perse. "Media Involvement in Local News Effects."
35. Nick Wingfield and Brian Stelter. (2011), "How Netflix Lost 800,000 Members, and Good Will," *New York Times*, October 24, 2011, accessed on Sept. 9, 2014, http://www.nytimes.com/2011/10/25/technology/.
36. Wingfield and Stelter, "How Netflix Lost 800,000 Members, and Good Will."
37. Horton and Wohl, "Mass Communication and Parasocial Interaction."
38. Joe Satran. "Paula Deen Aims For Comeback with 'Uncensored' New Digital Network." *Huffington Post*, September 24, 2014, accessed on September 27, 2014, http://www.huffingtonpost.com/2014/09/24/paula-deen-network_n_5876080.html?1411595012.

REFERENCES

Alperstein, Neil. "Imaginary Social Relationships with Celebrities Appearing in Television Commercials." *Journal of Broadcasting & Electronic Media* 35, no. 1 (1991): 43–58.
Auter, Philip J. "Psychometric: TV That Talks Back: An Experimental Validation of a Parasocial Interaction Scale." *Journal of Broadcasting & Electronic Media* 36, no. 2 (1992): 173–81.
Cavalcante, Andre. "You Better 'Reneckognize'!: Deploying the Discourses of Realness, Social Defiance, and Happiness to Defend Here Comes Honey Boo Boo on Facebook." In *Reality Television: Oddities of Culture*, edited by Alison F. Slade, Amber J. Narro, and Burton P. Buchanan, 39–58. Lanham, MD: Lexington Books, 2014.
Cox, Nicole B. "Bravo's 'The Real Houswives': Living the (Capitalist) American Dream?" In *Reality Television: Oddities of Culture*, edited by Alison F. Slade, Amber J. Narro, and Burton P. Buchanan, 77–99. Lanham, MD: Lexington Books, 2014.
Curnalia, Rebecca M. "Frugal Reality TV During the Great Recession: A Qualitative Content Analysis of TLC's Extreme Couponing." In *Reality Television: Oddities of Culture*, edited by Alison F. Slade, Amber J. Narro, and Burton P. Buchanan, 101–21. Lanham, MD: Lexington Books, 2014.
Earnheardt, Adam C. and Paul M. Haridakis. "An Examination of Fan-Athlete Interaction: Fandom, Parasocial Interaction, and Identification." *Ohio Communication Journal* 47 (2009): 27–53.
Gennis, Sadie. "A Timeline of Paula Deen's Downfall," *TV Guide*, June 27, 2013, accessed, June 30, 2014, http://www.tvguide.com/news/paula-deen-scandal-timeline-1067274.aspx.
Hernandez, Leandra H. "'I Was Born This Way': The Performance and Production of Southern Masculinity in A&E's Duck Dynasty." In *Reality Television: Oddities of Culture*, edited by Alison F. Slade, Amber J. Narro, and Burton P. Buchanan, 21–37. Lanham, MD: Lexington Books, 2014.
Hoffner, Cynthia A. and Elizabeth L. Cohen. "Responses to Obsessive Compulsive Disorder on Monk Among Series Fans: Parasocial Relations, Presumed Media Influence, and Behavioral Outcomes." *Journal of Broadcasting & Electronic Media* 56, no. 4 (2012): 650–68.
Horton, Donald and R. Richard Wohl. "Mass Communication and Parasocial Interaction." Psychiatry 19, no. 3 (1956): 215–29.
Houlberg, Rick. "Local Television News Audience and the Para-Social Interaction." *Journal of Broadcasting* 28, no. 4 (1984): 423–42.
Levy, Mark R. "Watching TV News as Para-Social Interaction." *Journal of Broadcasting* 23, no. 1 (1979): 69–80.
Morris, Pamela L. and Charissa, K. Niedzwiecki. "Odd or Ordinary: Social Comparisons Between Real and Reality TV Families." In *Reality Television: Oddities of Culture*, edited by Alison F. Slade, Amber J. Narro, and Burton P. Buchanan, 143–61. Lanham, MD: Lexington Books, 2014.
Nordlund, Jan-Erik. "Media Interaction." *Communication Research* 5, no 2 (1978): 150–75.
Perse, Elizabeth. "Media Involvement in Local News Effects." *Journal of Broadcasting & Electronic Media* 34, no. 1 (1990): 17–36.
Perse, Elizabeth and Rebecca Rubin. "Attribution in Social and Parasocial Relationships." *Communication Research* 16, no. 1 (1989): 59–77.

Rubin, Alan and Elizabeth E. Perse. "Audience Activity and Soap Opera Involvement: A Uses and Effects Investigation." *Human Communication Research* 14, no 2 (1987): 246–68.

Schramm, Holger and Tilo Harmann. "The PSI-Process Scales. A New Measure to Assess the Intensity and Breadth of Parasocial Processes." *Communications* 33, no. 4 (2008): 385–401.

Wenner, Lawrence A. "Political News on Television: A Reconsideration of Audience Orientations." *Western Journal of Speech Communication* 47, no. 4 (1983): 380–95.

Wingfield, Nick and Brian Stelter. (2011). "How Netflix Lost 800,000 Members, and Good Will." *New York Times*, October 24, 2011, accessed on Sept. 10, 2014. http://www.nytimes.com/2011/10/25/technology/.

Chapter Fifteen

Fans Can Be Journalists Too

A Look at Fan Interaction with HBO's The Newsroom

Julia E. Largent and Jason Roy Burnett

"I'm going to single-handedly fix the Internet!" exclaims Will McAvoy, a fictional news anchor on the popular HBO show *The Newsroom*.[1] Created by Aaron Sorkin, the political drama series focuses on a newsroom that strives to change the way the American people get their news, and fans have reacted with fake Twitter accounts for characters with real, current news. An abundance of research has been conducted on Twitter, Facebook, and various other social media sites in relation to the various uses people have for these platforms, including reading and spreading news.[2] However, these platforms are also utilized for alternative reasons; some individuals use them to further a fandom and carry on fictional voices in realistic ways. Sometimes this performance is so seamless that the fan is completely folded into the voice of the character he or she is portraying. So, when a voice moves from the stage to the screen and then into real life, does that voice somehow lose its legitimacy, even if the persona that performs the voice is fictional?

The characters of *The Newsroom* exist in several ways, beyond that of the actors who portray them on HBO. Each and every character has a Twitter account, and each one posts and interacts with all those who follow their accounts. Given the proliferation of "characters" on various shows posting on social media, this wouldn't be unusual, save for the fact the accounts of *The Newsroom* characters are not sponsored by nor controlled by HBO, and that the "real world" identity of the various voices behind these accounts remain anonymous and undisclosed. In this unusual dynamic is a powerful and tangible example of performative identity: the theory that identity is not fixed and stable, but fluid and shifting, moving from one expression to another as the context changes. In these Twitter accounts, the characters of *The*

Newsroom actually exist, even if they are "fictional" and are voiced by out-side individuals.

In researching the intersections of performativity and the Twitter accounts of *The Newsroom* characters, this research utilized a modified rhetorical analysis, searching the content of the tweets for the narrative elements that are indicative of identity expression. These included things like voice, perso-na, tone, and structure. After establishing a baseline of each character's per-formance, the research then examined the Twitter narratives for evidence of code switching or other breaks in character. In doing so, the researchers remained aware that these Twitter performances represent a single facet of any given identity. For example, Will McAvoy on Twitter overall remains a journalist, and performs through that facet of his persona; relational and interpersonal performances were largely confined to the show. Finally, this data allowed the contrasting of a given character's presentation on each stage. This multilayered analysis allowed us to argue for identity as a contex-tual performance, made real by the consistency of the portrayal.

This chapter will provide a look into various Twitter accounts of fictional characters from *The Newsroom* such as Will McAvoy, Maggie Jordan, and MacKenzie MacHale. Drawing from an aspect of identity theory, this chapter will also provide some insight as to why fans are willingly penning a fiction-al Twitter account, some for characters who are not even major players in the show. Beginning with an introduction to Twitter, social media, and the tele-vision show itself, *The Newsroom* fan involvement on Twitter is viewed through the lens of performative identity.

THE VARIETIES OF TWITTER EXPERIENCE

With more than 271 million monthly active users, Twitter is a micro-blog-ging site which allows users to post updates in 140 characters or less.[3] Their mission, "To give everyone the power to create and share ideas and informa-tion instantly, without barriers," allows users to discuss an infinite list of topics, create social movements, and spread news about a particular event.[4] In June 2009, protests erupted in Iran after the national presidential election. Tehran authorities had shut down universities, blocked cellphone transmis-sions and access to social networking sites, and later blocked text-messaging services.[5] With social media sites blocked, many people posted on Twitter new IP addresses to access the Internet, with several social networking sites helping those in Iran access these sites in a different manner. During the second week of the protest, videos were surfacing on Facebook and Twitter, as well as making news within America.

Social media played an important role to the Iranian protests, and had so much of an impact the U.S. State Department had asked Twitter to postpone a scheduled maintenance on its site to prevent any sort of issue for the Iranian protesters to remain connected to the outside world.[6] Commenting on this action, Mark Pfeifle, a former national-security advisor, stated, "Without Twitter, the people of Iran would not have felt empowered and confident to stand up for freedom and democracy."[7] Ben Parr states social media helped Iranians communicate with each other and the outside world, and it helped those in the outside world better understand, as well as communicate with Iranians.[8] Because of social media, thousands of first-hand accounts from people on the streets of these protests are recorded.

Aside from protests, fans are also using Twitter to discuss their favorite moments of a show or movie, often live-tweeting.[9] Live-tweeting during a show, which is when a user tweets thoughts, questions, or quotes while watching or participating in a television show or performance, is one way for a fan to interact and show affection or distaste for a particular event in a series or movie. Many popular shows or events, including Netflix's *Orange Is the New Black*, ABC's *Once Upon a Time*, BBC's *Doctor Who*, the Olympics, the World Cup, and NBC's *The Voice* all create official Twitter hashtags such as #TheVoice or #OITNB (*Orange Is the New Black*).[10] Hashtags (#) congregate all tweets containing the hashtag in one timeline allowing for fans to see what other fans are saying. NBC's *The Voice* used hashtags to "save" contestants who were voted off the show: "Previously, only coaches on *The Voice* had the power to make a save. Now, using Twitter as a real-time public platform in conjunction with a live TV broadcast, the power is yours. You'll have just as much say as @xtina, @blakeshelton, @adamlevine, and @CeeloGreen."[11] Fans simply tweeted #VoiceSave plus the contestants name at a specific time during the show. The contestants with the most saves were kept on the show. In the case of *The Newsroom*, fans are giving a voice to fictional characters from the show as a way to show their interest and carry on the spirit of the series to current events.

REPRESENTING PRESENT TENSE
IN *THE NEWSROOM*

The Newsroom is not the first political drama Aaron Sorkin has created, who is best known for *A Few Good Men*, *The West Wing*, and *The Social Network*.[12] Sorkin's films and series had an impact on some individuals who chose to go into the profession he portrayed with many individuals claiming their inspiration for their career came from watching some of Sorkin's films and series.[13] However, with *The Newsroom* being a show based in the recent past about real current events in a fictional newsroom, some individuals thought Sorkin was telling the journalism industry how to do its job. This was not the intent, and Sorkin made a public apology, stating

I think that there's been a terrible misunderstanding. I did not set the show in
the recent past in order to show the pros how it should have been done. That
was and remains the furthest thing from my mind. I set the show in the recent
past because I didn't want to make up fake news. It was going to be weird if
the world that these people were living in did not in any way resemble the
world that you were living in. I wasn't trying to and I'm not capable of
teaching a professional journalist a lesson. That wasn't my intent and it's never
my intent to teach you a lesson or try to persuade you or anything.[14]

Interestingly, because of the real news aspect, what came out of the show
was a myriad of fans who took the idea of a news team portraying the news
as facts and continuing this onto social media, specifically Twitter.

The Newsroom covers true current event stories that happened a year ago,
ranging from the death of bin Laden to the execution of Troy Davis. Echoing
the mistakes of CNN in 1998 and their Operation Tailwind broadcast, in
season 2, the news team "discovers" a war crime, which they call Operation
Genoa. This story is one of a very few times the team discusses a fictional
story. Starring Jeff Daniels, Emily Mortimer, and Olivia Munn, the cast
creates a humorous, yet thought-provoking look into risks journalists and
news anchors make when reporting the news, including the chasing of a story
that may not be true.

In the 2012 series' premier episode, "We Just Decided To," McAvoy, in
an auditorium in front of students at Northwestern University, gives a deep,
emotional speech about how America is no longer the greatest country in the
world.[15] After this speech, McAvoy returns to work to find a majority of his
staff has left his show and switched to a new show slated for a later timeslot.
Because of this staff change, Atlantis Cable News, or ACN (the fictional
network), is forced to hire a new team, including McAvoy's ex-girlfriend,
MacKenzie McHale. McHale convinces McAvoy to change the way he approaches the news, changing to a delivery of the news with integrity. From
this episode on, the news staff approaches the news with a new lens: providing the news with integrity and truth, not the news that is easy and lovable,
with McAvoy saying, "people should know what they're screaming about."
After covering the story about the BP Gulf of Mexico oil spill in a way all
other news networks did not, McAvoy and his staff leave the newsroom with
a new outlook on what their job means.

INTERACTIVE ENTERTAINMENT:
FAN INVOLVEMENT IN *THE NEWSROOM*

When discussing the sharing of news on various social media networks,
Arianna Huffington, editor and founder of the *Huffington Post*, states, "It's
not just about consuming content, but sharing it, passing it on, and adding to

it."[16] That is precisely what people use the Internet for; whether that is email or social media. Many of us get our news from social media; according to a November 2013 research study by Pew Research Center, 30 percent of Americans are getting their news on Facebook while 8 percent are getting their news on Twitter.[17] This study also found those who consume their news on Twitter rather than on Facebook are younger, more mobile, and more educated than those who consume their news on Facebook. Fans of *The Newsroom* are filling a somewhat-fictional and somewhat-nonfictional gap on Twitter. By creating Twitter accounts for almost every character (major and minor—including the janitor who is never mentioned and a psychologist who is only in a few episodes) from the show, they are providing real-life current information about what is going in the world today.

Not only are the accounts reporting news, the characters are also interacting with each other much like they would during an episode. They make jokes and poke at each other (in the character's voice and demeanor), and comment on vacations the character goes on. Not much is known about the fans behind these Twitter accounts due to the choice to remain anonymous, only allowing interviews if their identity is kept secret. In an interview with Poynter, the fan behind Will McAvoy (@WillMcAvoyACN), who is also the voice behind the Twitter account of President Bartlet from *The West Wing*, stated, "Part of the illusion of the account would be ruined if my actual name and identity were revealed. It stops being 'some guy tweeting as Will McAvoy' and it just is me tweeting, which takes away some of the charm of the account."[18] The voice behind @WillMcAvoyACN stated he started this project as a creative writing exercise, and due to Twitter's character limit (140 characters) makes @WillMcAvoyACN be even more creative with some of his tweets, sharing

> I do think Twitter has made me a better writer because when you give yourself constraints it requires you to be stronger. When you only have so many characters to use, you have to think more about phrasing, about effective use of words. You might use one word when you would usually 10, you might find the adjective that works slightly better than the more common one you would usually use. . . . Some of my best tweets are the ones that were originally too long—because I had to revise them multiple times.[19]

@WillMcAvoyACN has had difficulties in personifying McAvoy, but enjoys the challenge.[20]

There are several accounts for the same character on the show, such as Will McAvoy or Neal Sampat, but some are more active than others. What is even more fascinating is the anonymity between the characters. @WillMcAvoyACN does not know who is behind the several other character accounts on Twitter. The fans bounce off each other as the characters do in the show, without the awareness of who the voice is behind the other account.

@WillMcAvoyACN interacts daily with fans, arguing and commenting on comments made about political events and ideas, often arguing with the fans. @WillMcAvoyACN has a mission,

> My job is to try and find snippets of [McAvoy's] character that, as a whole, show someone with something to say. The theme of the McAvoy account is holding people accountable for the things they say—and encouraging people to demand that the real media do the same thing. Considering the number of followers I have from the real media, I think it's something they want to do as well. Or maybe they just enjoy a program about their profession. [21]

The act of arguing with someone who is real, yet not real is a surreal act, one that feels trivial and important all in the same realm. These interactions can help impact an identity, whether that is the fan behind these Twitter accounts, or if it is the fan that is being interacted with.

INTRODUCING PERFORMATIVE IDENTITY

As people navigate the sensory and information overload that so defines our experience of the "real," a phenomenon in and of itself worthy of discussion, especially in the context of globalization, it can be difficult to keep in mind that so much of what is seen and encountered is, at its core, a performance. Performance has a number of lexical definitions, ranging from "to attain or accomplish" to a description in computer science about the workload capacity of a given system, all adapted to the term since the sixteenth century. But as people consider the dynamic playing out between *The Newsroom* and anonymous individuals on Twitter, it is fair to ask about the notions concerning performance; is it something done or watched, whether at work or on the big screen, or does it speak to something more, an interaction between people and contexts?

In many ways it is tempting to assume performance, in this discussion, as being no more or less complicated than a group of devoted fans who have taken it upon themselves to get into character and continue the performance begun by the cast on HBO. This would meet with little argument throughout fandom culture. But this assumption has to overlook or ignore certain peculiarities and nuances to the performances given by members of *The Newsroom* in this specific venue of social media. These individuals are not merely playing a role: they actively perform an identity. Within their chosen context, everyone on the show, from Will McAvoy to the Janitor, does more than act out a scripted part, but interacts with thousands of people every day and in a manner consistent with their visual representations on cable. In every meaningful way that one can say he or she exists in digital space and is able to portray his or her identity (personhood, sense of self, or even a simple collection of what you like and dislike), so can Will McAvoy. He and his colleagues are real.

This may at first glance seem disconcerting; after all, it implies that a "real" (or authentic) expression of any given identity in cyberspace isn't authentic at all. The identity molds itself to the context. By extension, that means I might be doing the same thing, and is it possible that this contextual expression of the self goes well beyond the digital world? This is a very rough explanation of performative identity, a perspective on the narrative self arguing that our identity is a function of language expressed in relationship. We cannot exist outside relationship, and relationships require we utilize the communicative tools of our culture (language, gesture, nonverbals, and so on) to enable and engage in discourse. This cultural dimension means that to comprehend any relational exchange, that exchange must employ a common reference; therefore, the "performance" of identity brings with it "a history of relationships, manifesting them, expressing them."[22] Performative identity argues that our stories (the narratives we build to share with each other and describe who we are through our experiences) are neither singular nor stable, but "multiple, fragmentary, unfinished, always changing."[23] We perform ourselves, moment to moment, person-to-person, context to context, and each performance is an identity negotiated by all the sociocultural contexts in which that particular relational exchange takes place. Identity, then, is less who we are and more what we are doing where with whom.

In other theories of identity (the psychosocial, the intersubjective, the storied resource, and the dialogic), a sense of personal psychology, of a separate individual, forms part of the theory to some degree.[24] There is no sense of this; in fact, it is rejected outright in performativity. Our identities exist, are expressed, and determined only in discourse, which is a social act,

> Relational constructionism clearly extends and deepens the social orientation. However, this does not necessarily eliminate psychological explanations so much as beg for a reworking of what is meant by psychology. The challenge here is that of bracketing the mind-world dualism that has long plagued philosophy and obfuscated psychological study.[25]

In their classic work on social construction, Berger and Luckmann point out "If theories of identity are always embedded in the more comprehensive theories about reality, this must be understood in terms of the logic underlying the latter."[26] That is to say understanding our identities is a function of understanding our greater realities, and it is arguable that are realities in the globalized, twenty-first-century world are complex, interrelated, multiple, and unstable. Ken Gergen observes, "In the last year alone it is estimated that 8 trillion text messages were sent via cell phones. Everywhere in motion are meanings being shaped and reshaped on virtually every issue of importance to our lives—government, education, religion, family, work, leisure, the economy, love, appearance, and so on."[27] Because so much is moving around us and through us so fast, our realities reveal that older, modernist notions of who we are cannot withstand the lived experience of what we are when we are in the different contexts of our multifaceted worlds.

This brings us at last to the postmodern foundations of performativity. Generally, postmodern thought is skeptical, if not hostile, to meta-narratives (any view that argues for one specific thing as being the ultimate foundation of everything in society and self). It also understands power as negative or oppressive, and that the self has an inner essence that is not influenced by power relations. This is contrasted with arguments supporting multiple, contextual, and fluid theories describing intersecting realities; power as a ubiquitous force that is both repressive and constitutive; and the self as unstable, produced by power exchanges (discourse) and never existing beyond power or history.

Central to the idea of performative identity is the confluence of discourse and power, especially as articulated in the works of Michel Foucault. For Foucault, discourses are more than a simple exchange of meaningful and systematically organized statements between individuals. The meaning, found as much in the behaviors accompanying the expression as in the language itself, reveals the relationship between language and power in the constitutive role of power in radical social construction.[28] The power of language shapes reality, including and especially the reality of the self: identity. Therefore, the power of the discursive act might constrain an individual in one sense while enabling the individual in another. Given the context for each particular discourse is constantly fluid, changing (even in multiple discourses with the same person), the power expressed through the discourse and its constitutive impacts also change, ultimately ending with different discourses involving different identities of the participating individuals.[29]

Working after Foucault, Jean Baudrillard explores the substitution of signs of the real for the real.[30] This precession of simulacra (of the copy without an original) determines the real. This does not culminate in the "despairing of meaning" but in layers of improvisation, be it of meaning, nonmeaning, or multiple concurrent meanings resulting in mutual destruction. Multiple meanings reflect the mercurial nature of identities within an individual, and all that is destroyed is the identity associated with a particular performance; the curtain coming down, so to speak. Gergen and Gergen also work with in performativity: "In contrast, from a social constructionist perspective, narratives are discursive actions. They derive their significance from the way in which they are employed within relationships."[31] A discourse cannot exist outside of relationships, just as one cannot perform before an empty theater. As these relational interactions are such an intrinsic part of lived experience, the power embedded in them and expressed by them is unavoidable.

Finally, even if one accepts that identity may be fluid, perhaps even influenced in the moment by discursive acts, performativity is certainly not required by that acceptance. But it is the very shift in how the discourse is engaged that evidences performativity. People choose how to tell the same story depending upon who may be listening. Cultural critic Hayden White describes this as emplotment, a conscious choice in the construction of narrative (specifically historical narrative), in that the historian "emplots" or emphasizes certain elements of the narrative at the expense of others, transforming a historical account into a particular type of story readers will recognize.[32] So if my story has five pieces (abcde) and in one discourse I emphasize it this way: abcDe, my story is comedic. But if in another context I emphasize like this: aBcde, it becomes tragic. One elicits laughter and camaraderie, the other sympathy and support. The elements of the story have not changed, but the relational processes surrounding performance of the story have, and reveal the multiple identities of the storyteller. With all this in mind, this research can fully turn its attention to *The Newsroom*, to the characters existing in multiple spaces, and the storytellers performing their identities across and within these various discourses.

AND NOW FROM ATLANTIS CABLE NEWS (ACN)

Fans voicing characters in various media outlets is nothing new. Social media, particularly Twitter, is a unique avenue as the messages are sent in real time. Also, Twitter allows an individual an impressive amount of creative freedom in establishing an online persona. Performance, however, is not an exercise of free will, but a function of what we believe the audience will accept. Like it or not, when we tweet, we are projecting a facet of ourselves that has been altered according to our beliefs about those people watching us. The whole point of social media platforms, especially Twitter, is to share your facets with a wider audience. To tweet then is to be hyper-aware that what one writes not only will be read but shared, and if done well, will bring more people to his or her account.

The voices of the ACN staff on Twitter perform so convincingly that HBO requested all of the fan accounts include a line on their profiles disclaiming the network's involvement with the accounts.[33] The staff is very cognizant of and determined to maintain personas, enabling real voices for fictional characters that, as @WillMcAvoyACN previously stated, are performing journalistic duties in the real media. Below are four tweets, all pulled from different character Twitter accounts. Maggie Jordan (@MaggieJordanACN) is a member of the Newsroom staff whose loyalty to Will

McAvoy during the first season gets her promoted from an intern to an associate producer. MacKenzie McHale (@MackMcHaleACN) is the ex-girlfriend of McAvoy who returns to fills the Exectutive Producer position after McAvoy's previous Executive Producer, Don Keefer (@DonKeefe-rACN), leaves. Don is a reoccuring character and is also dating Maggie throughout the series (two parts of a love triangle that moves from the show on to Twitter). Lastly, Neal Sampat (@NealSampat_ACN) is the technology and blog writer for McAvoy. The four tweets chosen all resemble four different types of tweets posted on Twitter by the staff.

First, @NealSampat_ACN tweeted on March 5, 2014, "Hearing about a Russian anchor quitting on air is inspiring. May we all have the strength to put our journalistic integrity above all else."[34] @NealSampat_ACN reacts to the quitting of a Russian anchor on air in March 2014 during the Russia-Ukraine debacle. Earlier, Neal is a loud advocate for the Occupy Wall Street movement and strives to get the story heard on a national level. This tweet represents how some of the staff on Twitter embrace what is happening in current events and apply it to how their character, in this case Neal, would react as a real staff member in a real newsroom. Neal's character is enthusiastic, caring deeply for those who are voiceless and for events that need to discussion, evidenced by his activism shared in his tweets.

This real time commentary, in the case of the ACN staff, is distinct and needs to be contrasted with other types of fandom culture. Specifically, whether it's cosplay at Comicon or roleplay in online forums, the act of participating in fandom culture involves dressing the part, but it stops there. A fan may dress like Thor, but should that fan act like Thor in such a way as to convince others that the act reflects an actual persona (and not merely a tongue-in-cheek charade), it would likely cause no small amount of discomfort. The larger fan culture is based on play, not performance. Even more devoted fan gatherings, like Star Trek conventions, where cosplay is elaborate, alien languages are spoken fluently, and the convention floor feels like something pulled from a random series episode, all of that stops and starts throughout the day. (Jason witnessed two Klingons at a convention drop from the alien tongue into the English of the southeastern United States, as apparently Klingon has a difficult time asking whether or not Burger King was acceptable for lunch.) All this to say that, by contrast, Neal interacts with his followers in a voice that is consistently recognizable as Neal, and not someone playing the part. Certainly the voice has developed over time, but it is now just as much Neal's digital embodiment as it was when he opened his account. This continuity-in-context is, as described earlier, a core element of performativity. Ken Gergen's positions are well illustrated here, in that through the social platform Neal performs those facets of his identity relevant to those watching him, interacting with and commenting on the analog world outside cyberspace and thereby solidifying his reality. That Neal is someone else outside Twitter, in this sense, is more than a foregone conclusion; it's impossible to conceive of him not being someone else, which is no different from anyone else.

Next, @MackMcHaleACN tweeted on July 7, 2014, "Hard to get everyone in #newsroom focused after the weekend, especially @WillMcAvoy-ACN who brought sparklers to work-#exasperated."[35] MacKenzie is fun-loving and jocular, bringing her dry British sensibilities to bear on her colleagues. Her tweets, like the one above, evidence the ways the staff play off each other. MacKenzie's cockney terms and slang, brilliantly folded into her tweets, are accessible and humanizing. Her posts play off other staff members in this way, revealing the depth of her personality; strong-headed and determined, and in many ways a comic relief. She often says the wrong things or fixates on one specific part of a story instead of hearing all the details.

MacKenzie is fascinating to follow on Twitter not only because of the continuity just discussed but also these comedic and humanizing elements. Neal is very real, and his voice utterly convincing, but Mackenzie is different. It is easy to tell a joke; it is quite another to be genuinely funny and endearing. Like any great comedic performer, it is not a question of the right material as it is material to which the audience relates. The persona makes a one-liner truly memorable, and to achieve that the audience must accept that persona as a real part of their collective worlds. Here Foucault's arguments concerning language and discourse are expressed in incredibly involved ways that produce an engagement between MacKenzie and her followers that reveals her position, sense of power, and persona in a charmingly unselfconscious manner. MacKenzie doesn't have to try to be MacKenzie (and good thing, as a fair portion of social media seems determined to sniff out frauds). She just is. Her discourse with her followers establishes her parameters and authenticity.

By contrast, @MaggieJordanACN tweeted on July 29, 2014, "BREAKING: CO Supreme Court Orders Boulder to stop issuing SSM licenses while it considers constitutionality of law. bzfd.it/1kljbSw."[36] Maggie disseminated this news at the height of the national debate regarding same-sex marriage (SSM), and she, along with @WillMcAvoy_ACN, often breaks news on Twitter, sometimes before larger network news agencies do. As an associate producer, Maggie remains aware of the happenings in the world around her. She often posts commentaries to different current events, as well as just providing news of which this tweet is an example. The minute a story breaks, Maggie has it tweeted, as would any journalist.

Likewise and finally, @DonKeeferACN tweeted on April 28, 2014, "Thoughts go out to everyone affected in Oklahoma/Arkansas. @ElliotHirschACN will be broadcasting live from the worst hit areas from 10pm ET."[37] Don, much like Maggie, knows what is happening in the world around him, viewing his journalistic responsibilities as sacred. As the executive producer for Elliot Hirsch (another ACN news program), Don knows all that is happening on Elliot's show, and his posts demonstrate his continuing devotion to the news even when his show is not broadcasting. This is not uncommon among the various ACN staff member accounts on Twitter (even MacKenzie's tweet alludes to this same gesture).

The interaction and the continual journalistic tweets create this false illusion that this particular newsroom, with all of these characters, is in fact a real place. Due to Sorkin's use of real current events, he created a unique venture for fans to take the idea the show puts forth and carry it into the real world. These fans are most likely not journalists, just fans who wanted to partake in this illusion and carry on the work of the characters. But for these fans, what does this do to their identity? Do they mentally become a different person when tweeting as @MackMcHaleACN or @DonKeeferACN? Or are they themselves just tweeting as a creative writing exercise? How do the fans interact with these characters? Are they aware they are interacting with John Smith or do they believe they are actually interacting with Will McAvoy?

ANALYZING PERFORMANCE IN *THE NEWSROOM*

In examining the social and cultural histories of the ancient Greek city states (foundational to Western culture as a whole), one discovers that an individual identity is in many ways just as nuanced as an individual identity is often thought of today. To be Greek was everything; Greek culture, the Greek way, the Greek language, in fact the entire Greek civilization was immeasurably superior to any and all others, and today is still widely praised in the Western world for its "superior" contributions.[38] The thought that this may not be so was not even possible to form. You were Greek or non-Greek, no matter where in the world you are from, and the Greeks typically describe non-Greeks by mocking their language, the incoherent babble of all those "Others." This mocking and the repetitious *bar bar bar bar* gave rise to "barbarian."[39] With a moment's reflection, this sort of ethnocentrism might seem familiar.

But being Greek was far more than not being a barbarian. One was also Corinthian, or Athenian, or Spartan, living in the *polis* because (it was understood) that the human being was by nature a "social animal."[40] But even more important than these largely geographical and political distinctions was the presence throughout all ancient Greek culture of competition: to perform. And everything was a competition; in addition to the Olympic Games, oration (public speaking, usually of a political nature), music, singing, fighting, and even poetry were central parts of festivals and public life. The glory or shame earned in such performances would forge an individual's identity in his or her mind and the minds of all Greeks. An extreme example of this is found in Euripides, considered one of the three great dramatists of the age, who presented the classic *Medea* at the Dionysia (a rowdy three-day festival held every year in which playwrights would compete) early on in his career. His play was detested, and he was cursed and booed off the stage; the reaction was so severe, that is, his failure to perform and embody the identity he presented to his audience, that he was still cursed for it thirty years later and up until his death.[41]

This is far from a perfect illustration, but generally speaking, there is in this ancient society a very real sense of meaning to perform a given identity in a given place at a given time. In many ways, this is what Edith Hamilton is referring to when she speaks of the "Greek miracle"; the Greeks were the first to imagine the gods in human form, and by doing so put humanity at the highest place in the universe.[42] This is one reason why Greek athletic contests were done naked, as nothing was as perfect as the human form. Therefore, in whatever identity you perform, in many ways you perform it to be recognized, honored, praised, and even feared for the magnificence of the presentation, much of which took place in language.

In July 2010, author Peggy Orenstein wrote an article for the *New York Times* entitled "I tweet, therefore I am," briefly discussing a cognitive dissonance, an unsettled feeling that she encounters as she reflects on the use of social media and its impact on the sense of self.[43] She admits that she came late to social media (Twitter in particular), and that as she adapted to the medium she found herself constantly on alert, consistently scanning her environment for that next 140 characters. In deciding upon a tweet to describe sitting with her daughter outdoors listening to E. B. White, she has an epiphany of sorts: "Yet the final decision was not really about my own impressions: it was about how I imagined, and wanted, others to react to them. That gave me pause. How much, I began to wonder, was I shaping my Twitter feed, and how much was Twitter shaping me?"[44]

Orenstein goes on to invoke the sociologist Erving Goffman (who argued that all life is performance) and to contrast it with the work of MIT professor Sherry Turkle, who was at the time publishing a book on the rising use of social media and cell phones. Throughout the course of the discussion, it becomes evident that Orenstein, while reflexively commenting on "performance culture" in rather macabre tones, writing

> When every thought is externalized, what becomes of insight? When we reflexively post each feeling, what becomes of reflection? When fans become friends, what happens to intimacy? The risk of the performance culture, of the package itself, is that it erodes the very relationships it purports to create, and alienates us from our own humanity.[45]

She concludes with the admission that these questions are almost enough to get her to cancel her Twitter account . . . that is, if she weren't hooked, and her parting thought (somewhat self-congratulatory) is that she will overcome the performance culture by not tweeting about her next trip outside with her daughter.

Again, this was in 2010, in between then and the time of this composition some four years later, the world has drastically changed. But in Orenstein's reflection is a rather common misconception, and misperception, about performance culture and performativity as is discussed here. It is well-documented that we act differently in different circumstances; put someone in front of the camera, for example, and countless studies demonstrate that there is a very discernible change in individual behavior. We have always put a different part of ourselves forward when we think someone is watching. The reality of human existence, being social in all its expressions, is that someone is always watching, even if it is only us watching ourselves. What happens in social media, in the performance that Orenstein is so troubled by as to not remove herself from, is simply another facet on the face of our interrelatedness. It is an expression of performative identity, of the context and the discourse shaping our involvement in the exchange (something many feminist and identity theorists like Judith Butler explore at length, taking their cue from Foucault).[46] Performative culture, on the other hand, is more like the context that allows us to perform in the first place, and not the world of manufactured representations that Orenstein fears.[47]

How is this relevant to the discussion? Simply, the context and the discourse determines self, all within social relations. The individual at the keyboard might not be named Will McAvoy, but that individual is not the person engaging in discourse on Twitter. It is a sustained and conscious effort, now lasting several seasons of *The Newsroom*, and unlike cosplay, the mask never comes off. These twitter accounts are clear examples of performative identity expressed in the digital age. The third and final season of *The Newsroom* aired fall of 2014, and, as of printing, these fan accounts are still being used and are still educating their followers, including one character who died in the last season. With the show off air, these fans are still taking Will McAvoy's vision and turning it into a reality.

NOTES

1. Aaron Sorkin, *The Newsroom*, HBO Studios, 2012.

2. Jesse Holcomb, Jeffrey Gottfried, and Amy Mitchell, *News Use Across Social Media Platforms*. November 14, 2013. http://www.journalism.org/2013/11/14/news-use-across-social-media-platforms/.

3. Twitter, *About Twitter, Inc.* July 30, 2014. https://about.twitter.com/company.

4. Ibid.

5. Robert F. Worth and Nazila Fathi, *Protests Flare in Tehran as Opposition Disputes Vote.* June 13, 2009. http://www.nytimes.com/2009/06/14/world/middleeast/14iran.html?pagewanted=all.

6. Paul Levinson, *New New Media* (Boston: Allyn & Bacon, 2009); Malcolm Gladwell, "Small change." October 4, 2010. http://www.newyorker.com/reporting/2010/10/04/101004fa_fact_gladwell.

7. Gladwell, *Small Change*, para. 7.

8. Ben Parr, "#IranElection crisis: A Social Media Timeline." June 21, 2009. http://mashable.com/2009/06/21/iran-election-timeline/.

9. Tim Highfield, Stephen Harrington, and Axel Bruns. "Twitter as a Technology for Audiencing and Fandom." *Information, Communication & Society* 16, no. 3 (April 2013): 315–39.

10. *The Voice.* Television series. Directed by Alan Carter. Produced by NBC. 2011; *Doctor Who* (The New Series). Directed by Steven Moffat. Produced by BBC. 2005; *Once Upon a Time*. Television Series. Directed by Mark Mylod. Produced by ABC. 2011; *Orange Is the New Black.* Television Series. Directed by Jenji Kohan. Produced by Netflix. 2013.

11. Fred Graver, *Tweet to "Save" Artists on NBC's The Voice.* https://blog.twitter.com/2013/tweet-to-save-contenders-on-nbc-the-voice.

12. *The West Wing.* Television series. Directed by Aaron Sorkin. Produced by NBC. 1999; Aaron Sorkin, *A Few Good Men.* Film. Produced by Columbia Pictures. 1992; *The Social Network.* Film. Directed by Aaron Sorkin. Produced by Sony. 2010.

13. Juli Weiner, *How Aaron Sorkin's West Wing Inspired a Legion of Lyman Wannabes.* April 2012. http://www.vanityfair.com/politics/2012/04/aaron-sorkin-west-wing.

14. Margaret Eby, "Aaron Sorkin Apologizes for 'The Newsroom': 'I Think There's Been a Terrible Misunderstanding.'" April 22, 2014. http://www.nydailynews.com/entertainment/tv-movies/aaron-sorkin-apologizes-reporters-newsroom-article-1.1764796.

15. *The Newsroom.* "We Just Decided To." (24/June/2012), written by Greg Mottola.

16. C. Stec. "5 Quotes That Explain How Effective Social Media Marketing Works." April 12, 2014. http://www.impactbnd.com/blog/5-quotes-that-explain-how-effective-social-media-marketing-works .

17. Jesse Holcomb, Jeffrey Gottfried, and Amy Mitchell, *News Use Across Social Media Platforms.*

18. M. J. Tenore, "The Real Writer Behind the Fictional Will McAvoy's Twitter Account." Para 7. http://www.poynter.org/latest-news/top-stories/187822/the-real-writer-behind-the-fictional-will-mcavoys-twitter-account/.

19. Ibid., para 14–15.

20. Ibid.

21. Ibid.

22. Brett Smith and Andrew Sparks, (2008). "Contrasting Perspectives on Narrative Selves and Identities: An Invitation to Dialogue." *Qualitative Research*, vol. 8: no. 5, 23.

23. Ibid., 24.

24. Ibid.

25. Ken Gergen and Mary Gergen, "Narrative Tensions: Perilous and Productive." *Narrative Inquiry* 21, no. 1 (2011): 379, doi: 10.1075/nl.21.2.17ger.

26. Peter Berger and Thomas Luckman, (1966). *The Social Construction of Reality* (New York: Doubleday) 175.

27. Ken Gergen, (2012). "From Reflecting to Making: Psychology in a World of Change." *European Journal of Psychology*, vol. 8, no. 4: 512.

28. Michel Foucault, (1980). *Power/Knowledge* (New York: Vintage).

29. Michel Foucault, (1984). *The Foucault Reader*, Ed. Paul Rabinow, (New York: Pantheon Books).

30. Jean Baudrillard, (1994). *Simulacra and Simulation*, (Ann Arbor: University of Michigan Press), 40.

31. Mary Gergen and Ken Gergen, (2006). "Narratives in Action." *Narrative Inquiry*, vol. 16, no.1: 118.

32. Hayden White, (1978). *Tropics of Discourse* (Baltimore: John Hopkins University Press).
33. Tenore, 2012.
34. Neal Sampat, Twitter post, March 4, 2014, 11:51p.m., https://twitter.com/NealSampat_ACN.
35. MacKenzie McHale, Twitter post, July 7, 2014, 3:47p.m., https://twitter.com/MackMcHaleACN.
36. Maggie Jordan, Twitter post, July 29, 2014, 5:49p.m., https://twitter.com/MaggieJordanACN.
37. Don Keefer, Twitter post, April 28, 2014, 2:37p.m., https://twitter.com/DonKeeferACN.
38. Edith Hamilton, (1998). *Mythology* (Boston: Bay Back Books).
39. Thomas Cahill, (2003). *Sailing the Wine-Dark Sea* (New York: Random House).
40. I. F. Stone, (1989). *The Trial of Socrates* (New York: Anchor Books).
41. Cahill, 2003.
42. Hamilton, 1998.
43. Peggy Orenstein, "I Tweet, Therefore I Am." July 30, 2010. http://www.nytimes.com/2010/08/01/magazine/01wwln-lede-t.html?_r=3& .
44. Ibid., para. 4.
45. Ibid., para. 7.
46. Foucault, 1984.
47. Ibid.

REFERENCES

Baudrillard, J. 1994. *Simulacra and Simulation.* Ann Arbor: University of Michigan Press.
Berger, P. and Luckman, T. 1966. *The Social Construction of Reality.* New York: Doubleday.
Cahill, T. 2003. *Sailing the Wine-Dark Sea.* New York: Random House.
Doctor Who (The New Series). Directed by Steven Moffat. Produced by BBC. 2005.
Eby, M. "Aaron Sorkin Apologizes for 'The Newsroom': 'I Think There's Been a Terrible Misunderstanding.'" April 22, 2014. http://www.nydailynews.com/entertainment/tv-movies/aaron-sorkin-apologizes-reporters-newsroom-article-1.1764796.
Ellison, N. B., C. Steinfield, and C. Lampe. 2007. "The Benefits of Facebook 'Friends': Social Capital and College Students' Use of Online Social Network Sites." *Journal of Computer-Mediated Communication* 12, no. 4 (July): 1143–68.
Foucault, M. 1984. *The Foucault Reader.* Ed. Paul Rabinow. New York: Pantheon Books.
———. 1980. *Power/Knowledge.* New York: Vintage.
Gergen, K. 2012. "From Reflecting to Making: Psychology in a World of Change," *European Journal of Psychology*, vol. 8, no. 4: 511–14.
Gergen K. and Gergen, M. 2011. "Narrative Tensions: Perilous and Productive," *Narrative Inquiry* 21, no. 1: 374–81, doi: 10.1075/nl.21.2.17ger.
———. 2006. "Narratives in Action," *Narrative Inquiry*, vol. 16, no. 1: 112–21.
Gladwell, M. "Small Change." October 4, 2010. http://www.newyorker.com/reporting/2010/10/04/101004fa_fact_gladwell.
Graver, F. "Tweet to 'Save' Artists on NBC's *The Voice*." https://blog.twitter.com/2013/tweet-to-save-contenders-on-nbc-the-voice.
Hamilton, E. 1998. *Mythology.* Boston: Bay Back Books.
Highfield, T., S. Harrington, and A. Bruns. 2013. "Twitter as a Technology for Audiencing and Fandom." *Information, Communication & Society* 16, no. 3 (April): 315–39.
Holcomb, J., J. Gottfried, and A. Mitchell. "News Use Across Social Media Platforms." November 14, 2013. http://www.journalism.org/2013/11/14/news-use-across-social-media-platforms/.
Levinson, P. *New New Media.* Boston: Allyn & Bacon, 2009.
The Newsroom. Television series. Directed by Aaron Sorkin. HBO, 2012.
Once Upon a Time. Television series. Directed by Mark Mylod. Produced by ABC. 2011.

Orange Is the New Black. Television series. Directed by Jenji Kohan. Produced by Netflix. 2013.

Orenstein, P. "I Tweet, Therefore I Am." July 30, 2010. http://www.nytimes.com/2010/08/01/magazine/01wwln-lede-t.html?_r=3&.

Parr, B. "#IranElection Crisis: A Social Media Timeline." June 21, 2009. http://mashable.com/2009/06/21/iran-election-timeline/.

Smith, A. "Six New Facts about Facebook." February 3, 2014. http://www.pewresearch.org/fact-tank/2014/02/03/6-new-facts-about-facebook/.

Smith, B. and Sparks, A. (2008). "Contrasting Perspectives on Narrative Selves and Identities: An Invitation to Dialogue." *Qualitative Research*, 8, no. 5: 5–35.

The Social Network. Film. Directed by Aaron Sorkin. Produced by Sony. 2010.

Sorkin, Aaron. *A Few Good Men.* Film. Produced by Columbia Pictures. 1992.

Stec, C. "Five Quotes That Explain How Effective Social Media Marketing Works." April 12, 2014. http://www.impactbnd.com/blog/5-quotes-that-explain-how-effective-social-media-marketing-works.

Stone, I. F. (1989). *The Trial of Socrates.* New York: Anchor Books.

Tenore, M. J. "The Real Writer Behind the Fictional Will McAvoy's Twitter account." September 14, 2012. http://www.poynter.org/latest-news/top-stories/187822/the-real-writer-behind-the-fictional-will-mcavoys-twitter-account/.

The Voice. Television series. Directed by Alan Carter. Produced by NBC. 2011.

Twitter. "About Twitter, Inc." 2014. https://about.twitter.com/company (accessed July 30, 2014).

Weiner, J. "How Aaron Sorkin's West Wing Inspired a Legion of Lyman Wannabes." April 2012. http://www.vanityfair.com/politics/2012/04/aaron-sorkin-west-wing.

The West Wing. Television series. Directed by Aaron Sorkin. Produced by NBC. 1999.

White, H. 1978. *Tropics of Discourse.* Baltimore: The John Hopkins University Press.

Worth, R. F., and N. Fathi. *Protests Flare in Tehran as Opposition Disputes Vote.* June 13, 2009. http://www.nytimes.com/2009/06/14/world/middleeast/14iran.html?pagewanted=all.

Chapter Sixteen

It's Bigger on the Inside

Fandom, Social Media, and Doctor Who

Krystal Fogle

The British science fiction TV program, *Doctor Who*, kicked off more than fifty years ago aiming to teach children about history and moral values. The show, which began airing in the United States in the 1980s, has developed a strong fanbase internationally, especially since the advent of the Internet.

Online fandom, relationships, and fan groups surrounding fictional works are particularly ripe for study due to fan involvement in the co-creation of texts. Fans engage with the original text through, among other forms of expression, fan fiction, and artistry. These new texts add to the original, allowing fans to explore options for plot and provide commentary on characters and story line and the larger sociopolitical context surrounding the fictional text.

Pinterest, in particular, provides an arena in which fans can pull together content from many different places, supplying commentary on both the show and the state of the fandom. In particular, the *Doctor Who* fandom has taken to Pinterest to bond over ancillary content.

Doctor Who chronicles the adventures of an alien, of the race of Time Lords, who goes by the name "The Doctor." The Doctor travels time and space looking for planets in need of protection. This chapter seeks to examine the nature of fandom, how *Doctor Who* fandom has progressed, what fandom looks like online, and how these fandoms build group identity through the use of rhetorical visions, or fantasy themes.

FANDOM

Fandom as a form of group interaction is somewhat unique. Media culture scholar Kaarina Nikunen defines fandom as "an affective and active relationship towards the media: fans collect, interpret, circulate and re-write media texts."[1] People who identify themselves as "fans"—people who voluntarily participate in activities related to a given artifact—choose to band together with other self-identified fans to construct a community based around liking a particular artifact. This community rehashes events from the original artifact and often creates new content, such as memes and "fan fiction," stories based on the events portrayed in the artifact. These groups are self-selected rather than chance-based; they form because people first choose to identify a certain way—as fans of a particular object—then choose to link with other fans. These self-selected groups indicate quite a bit about the members, but they also continue to reshape the participants within the fandom.

Throughout history, people have naturally grouped together. Group characteristics are often easy to analyze, as groups have historically formed based on characteristics such as proximity, and utility. Similarities were abundant, and homogeneity was the norm. But with the advent of the Internet, globalization became the new norm, and differences reign over similarities. Groups are now self-selected, formed for interest and desire instead of proximity or necessity. Thus, the unifying characteristics of groups are vastly different, and their characteristics will be varied. Psychologists Joel Nadler and Geraldine Hannon argue that self-selected groups are ripe for study because they provide new insight into motivations for grouping.[2] The most obvious, and globalized, group networks are those formed through affinity for a particular object. To put it clearly: fandoms provide a rich area of study because they are self-selected groups and thus have a different motivation for grouping together than traditional "convenience-based" groups. Furthermore, fandoms are increasingly globalized and therefore provide insight into how people from varied background can interact and unite over a shared value system.

When the organization or group is no longer perceived in a positive light, group norms must be modified. In particular, when the group is removed from the original artifact, such as in fandom, the governing body—the object of fan adoration—must step in to dictate new group norms. Fandoms in particular are susceptible to detrimental fan behavior, which can reflect poorly on the team. Sport and culture researchers John Hughson and Emma Poulton report on England's efforts to redefine the perception of sports fandom for football (soccer), explaining that officials used news reporting to claim that the majority of fans were acting appropriately.[3] Because group members have a desire to adhere to group norms, portraying the unruly group actors as the minority applied pressure to those members to begin acting according to group norms. Group members have that fundamental need for belongingness, and so they desire to act according to larger group norms so as not to be disposed of by the group. Of course, these unruly actors may or may not have truly been in the minority, however, despite labeled as deviants, which often provides enough negative connotation to increase the likelihood of good behavior. Management researcher Daniel Feldman argues that group "norms are likely to be enforced if they express the central values of the group and clarify what is distinctive about the group's identity."[4] For fandoms, the distinctiveness of the group originates in the purpose of the relationship: affinity for a certain television show. Thus, to maintain membership, group members desire adherence to normal behavior expectations and help to maintain that which is distinctive about the group. They will partake of both typical behavior and also of "inside jokes." These behaviors provide much insight into what group members as a whole believe. Thus, academic interest is peaked by fandoms.

Fan communities are a unique forum for minority voices. Media and theater scholar Abigail Derecho claims that "since at least the early seventeenth century [fan fiction] has been a compelling choice of genre for writers who belong to 'cultures of the subordinate,' including women . . . and ethnic minorities."[5] Film director Julio Garcia Espinosa, in an essay comparing modern television to art, claims that "popular art has always been created by the least learned sector of society, yet this 'uncultured' sector has managed to conserve profoundly cultured characteristics of art. One of the most important of these is the fact that the creators are at the same time the spectators and vice versa."[6] Media scholar Deborah Kaplan agrees, arguing that "the community that produces and consumes fan fiction is virtually the same as the community of fan critics [those people who critique and examine the television show]."[7]

Of course, in the realm of television, academics' interests may not overlap with the typical viewers' interests, leading to a gap in true knowledge about how viewers actually experience television. Film and television expert Matt Hills explains that "academia, as a 'taste community,' focuses on specific types of television, resulting in a critical concentration on a highly limited range of texts. What can be dubbed 'invisible television' falls outside the bright lights of this attention, residing instead in the penumbra of relative academic disinterest."[8] There is a call, then, for analysis of television and fan situations that overlap both areas: scholarly interest and general interest. As I will discuss in the following section, *Doctor Who* is an ideal artifact to study because it combines both scholarly interests and general interests. Television analyst James Hibbard reports that *Doctor Who's* fiftieth anniversary episode broke television network "BBC America's all-time ratings record."[9] Television is a valuable area for study as values are shaped by television,[10] and values shape how people both perceive reality and how people act within that real world environment.[11] *Doctor Who* contains certain values that will influence how viewers interpret the world.

The *Doctor Who* Fandom

Doctor Who fandom has a long history, dating back to the 1980s in the United States, when PBS began airing episodes of the long-running British show.[12] Of course, *Doctor Who* began long before that in the United Kingdom, which means that this particular fandom has a long and varied past.

Fan-created items cultivate spaces to communicate a deeper understanding of how viewers interpret the story. The original text of an episode provides what is known as "canon:" that which is set in stone for everyone discussing the universe of *Doctor Who*. Cultural theorist Catherine Driscoll argues that "the web of canon drawn from a source text is the primary reality against which fan fiction is written . . . and thus is the only way of accounting for ideology . . . but every claim about canon nevertheless raises the specter of its opposite."[13] Driscoll is arguing that the very nature of canon allows fans to imagine a world in which the story line evolved in a different direction. Perhaps this character did not die; perhaps those characters fell in love. Thus, fan creations containing these rebellious story markers indicate areas in which fans may be dissatisfied with the story taking place in the canon. Cultural researcher Ika Willis expounds on this point, claiming that "fan fiction, then, is generated first of all by a practice of reading which . . . *reorients* a canonical text, opening its fictional world onto a set of demands determined by the individual reader and her [sic] knowledge of the (fictional and nonfictional) world(s)."[14] This is important because, as critical fan scholar Mafalda Stasi claims, "fandom requires that there is a common base understanding: shared culture. The story is important but so are cultural references. It must have a basis, be grounded in something shared and commonly understood."[15] Thus, to belong to a fandom, one must have a grasp of the plot, of the culture, and accompanying references from which other members of the fandom are drawing. Willis agrees, claiming that "all texts, of course, depend for their legibility on intertextual and extratextual knowledge."[16] The *Doctor Who* fandom, with a vast history, contains a great number of plot lines that could be altered for the sake of fan experimentation and exploration. The ways in which fans both adhere to and stray from the original text provides information about their consumption of the show itself and their common understanding of the show and surrounding culture. To engage with these erroneous plot spin-offs, fans must have a firm grasp of the plot of the show and must also understand any cultural references. Understanding will manifest in content created and shared online.

Fandom Online

All in all, fandom interactions that take place online are increasingly important for individuals, societies, and academics. Cultural and media researchers
Megan M. Wood and Linda Baughman argue that social media allows "a
specific form of participatory culture behavior that involves an active, collective audience that collaborates on a new, more intense level of fandom to
create a deeper, more fulfilling experience from an existing produced media
text."[17] In other words, interacting online actually provides a deeper sense of
cohesion for fandom members than simply interacting face-to-face, or simply, for example, viewing the television show solo. In the digital community,
there is a deeper sense of connection and, indeed, a larger magnitude of
connection, as digital interactions can span continents. Thus, digital fandom
increases group members' sense of belongingness. Furthermore, the advent
of social media has allowed fandom to flourish as interactions online supplement the consumption of texts about the plotline. Episodes only air weekly,
with hiatuses occurring between seasons, so fandom-centric texts such as
"fan art compensate for deficiencies and gaps in the marketplace."[18] Fans
who are eagerly awaiting the next episode seek out fans online or in person to
analyze previous plot points and to predict what may come next.

Digital fandom also allows for competition between fan groups. Bertha
Chin, fan studies expert, argues that social media and the Internet allow fans
to create fan fiction that will ultimately aid in demonstrating how their fandom is superior to other fandoms.[19] This competition both cultivates fan
bonding within the fandom and solidifies the relationship between this fandom and the outside world—an important distinction as belongingness is a
fundamental concern for human beings. Humans crave not only the sense of
"I belong to this group," but also, ultimately, the sense of "I belong to this
group *over* that group." Fan creations additionally construct and define the
boundaries between one fandom and another, assisting individuals in clas-
sifying their position in the larger, globalized world. Indeed, one theory,
reinforced by film scholar Robert Stam, posits that "television . . . is founded
upon the pleasure of looking (scopophilia) and the pleasure of hearing . . .
and [television] allows us to see without being seen and hear without being
heard . . . our privileged position triggers a fictitious sense of superiority."[20]
Thus, fans inherently feel that they are in a privileged position, as onlooker.
This facilitates their investment in the show and fosters the desire to create
ancillary content.

Fandom studies have traditionally focused on the response individuals
have to the original airing of a television episode. This field certainly illustrates the value of understanding viewers' habits; however, the advent of
social media has irrevocably altered how viewers experience television.
Now, fans can communicate with other members via the Internet long after

an episode has originally aired, and fans can create and share original content—such as fan fiction and memes—around the world, leading to an ongoing discussion about the show in question. Communication scholar Rebecca Williams argues that "fandom will not necessarily end in the post-object period," which she defines as any time after the original episode (object) has aired, and that "attachment to ancillary textual materials might continue . . . and people may continue to self-identify as fans of objects, persisting in [engaging with the material] or discussing the fan objects with fellow fans."[21] Now, the area of study with the most relevance is fan interactions after—sometimes long after—an episode has aired.

Still, examining fandom on the Internet can only go so far, as Western societies will be overrepresented. Intercultural communication scholars Guo-Ming Chen and Xiaodong Dai argue that "it is unfortunate that non-Western countries have fewer chances to articulate their opinions and define themselves than their Western counterparts. Moreover, in transforming their traditions, non-Westerners face a perplexing dilemma: to incorporate modern Western ideas denotes both empowerment and marginalization."[22] Furthermore, women in these underrepresented countries will be additionally underrepresented. New media scholar Robert Shuter argues this point, claiming that "there is ample evidence that women in developing countries like India have less access to a variety of information and communication technologies (ICTs)—which has been referred to as the gender divide."[23] This underrepresentation may skew the way scholars understand fandom and analyze the online communication of fans in general. Some societies will inadvertently be privileged over others. Men in developing countries have significantly more access to the Internet, decreasing understanding of fan motivations across genders. Over time, as access expands, the field of fandom will undoubtedly change, indicating a need for continued study in this area.

Online communication elicits transformation: as people self-disclose and connect with various others, their "selves" become simultaneously more defined and more amorphous, as they are shaped to fit the community. Debbie Rodan, Lynsey Uridge, and Lelia Green, media and culture scholars, explain "the fact that online community is generally expressed in written form means that communications in this context are amenable to a process starting with self-disclosure, moving to reflection, and resulting in transformation over time."[24] Thus, people who self-disclose and engage in fandom interactions are actually participating in collaborative identity construction. This identity construction influences one's own sense of self, but also one's identity as a group member. *Doctor Who*, with its vast history, is particularly useful for identity construction.

DOCTOR WHO

Watched worldwide and dearly beloved by many, *Doctor Who* was dreamed up by Sydney Newman, then head of drama at the BBC, in the 1960s.[25] Newman's concept took a time-and-space traveling alien, belonging to the species called Time Lords, around the galaxy, aiming to teach children about history and moral values. However, some at the BBC were concerned a science-fiction piece could not captivate audiences, as evidenced by the debates that took place among BBC brass.[26] Much of the argument questioned whether the decidedly British television company would be able to sustain a story arc for science fiction, at the time a mostly American genre.[27] This argument now seems humorous, considering the nearly thirty years during which *Doctor Who* has reigned on screen. After enjoying a period of popularity, *Doctor Who* experienced an unfortunate decline in ratings in the 1980s and was subsequently removed from programming.[28] In the early 2000s, however, Russell T. Davies decided to resurrect his beloved show.[29] Through Davies's guidance, *Doctor Who* achieved remarkable success, expanding its audience among British subjects as well as widening a viewing base internationally, particularly in the United States. Davies remained with the show until 2008, at which time he chose to leave the program, bequeathing writers' rights to Steven Moffat,[30] who ultimately expanded the show into what media critic Jace Lacob calls "a global phenomenon."[31]

The main character, answering to the name "The Doctor," traverses the universe and millennia, looking for adventure but inevitably ending up where he is needed most. His time-machine/spaceship, the Tardis (Time And Relative Dimension In Space) takes the shape of a blue police box.[32] Although intended to blend in to whatever surroundings the Doctor may find himself in, the Tardis's shape-shifting technology[33] became stuck on this shape, and it is now an iconic image for fans to use to signal their membership. The Doctor often locates a companion to travel with him, who is usually young and female. Some are involved with the Doctor romantically, but many serve to keep the Doctor company and assist him as he saves the world. Throughout these adventures, the Doctor and his companion encounter many alien races and must rescue humanity, and many other species at various times, from imminent annihilation.

Doctor Who Fandom Online

The BBC utilizes social media to perpetuate fan interest in *Doctor Who*. Fans in the States can follow *Doctor Who* on BBCA (BBC America) for information about upcoming episodes and behind the scenes photos.[34] *Doctor Who* also has an official Twitter page, Doctor Who Official, with updates regarding actors, writers, episodes, history, and rumors surrounding the show.[35]

The BBC has a Facebook page, allowing fans to interact with the brand and allowing the BBC to post teaser trailers for the show and to share fan-made art and memes.[36] *Doctor Who Magazine* also runs a Twitter page; although the magazine is not run by the BBC, it is the official magazine of *Doctor Who* and partners with the BBC to obtain information, interviews, and inside reports.[37] These online faces of the show represent an additional access point for fans to garner an understanding of the plot and context of the show as well as offering opportunities for fans to interact with and alter their perceptions of the show.

Fans of *Doctor Who* have also taken to social media to express their interest in the show and to interact with other fans. For example, Twitter user @Barbarian57 says, "I took Zimbio's 'Doctor Who' quiz and I'm Rory Williams!"[38] This quiz and tweet not only allow @Barbarian57 to interact with the characters of *Doctor Who* but to invite further discussion of the characters and @Barbarian57's own life with other Twitter users and fans. @lovejoydiver tweeted, "In your dreams, they'll still be there. The Doctor and Amy Pond . . . and the days that never came. Live well. Love Rory. Bye-bye, Pond," in an effort to memorialize favorite—now gone—characters.[39] Although these characters left the show in the previous season, @lovejoydiver continues tweeting about them. Her tweets can lead to several assumptions. First, she could be tweeting because she still feels a connection to these characters and mourns their departure. Second, she could be rewatching the episodes these characters inhabit, prompting her to tweet quotes she believes her Twitter followers will recognize. Third, she may be behind on the episodes and tweeting in real-time for her as she views the episodes in order. Regardless of which circumstance is true, all of them mean that the relationship viewers have with characters of *Doctor Who* not only exist past the air date but also bleed over into the online realm. These relationships are formed around understandings of both the story line and one's culture.

FANTASY THEMES

All groups are developed around a common understanding, as group members strive to understand and co-create identity. Groups formed around television shows are no different. These groups begin with their members' common understanding of the show's plot and characters, as well as their interpretation of the show's meaning. Communication scholar Lisa Waite claims that "audiences frequently shape their own connotation of an artifact, and this interpretation may vary from the rhetor's message."[40] These connotations, however, generally fit into certain preconceived frameworks of understanding. Ernest Bormann, renowned communication scholar and developer of symbolic convergence theory, argues that "in symbolic convergence, the term *fantasy* means the creative and imaginative shared interpretation of events that fulfills a group's psychological or rhetorical need to make sense of its experience and anticipate its future . . . the fantasy theme is the pun, figure, or analogy that characterizes [a significant] event, or it is a narrative that tells the story in terms of specific characters going through a particular line of action."[41] These themes become norms for a group's communication, providing reference points for the group developing a common understanding. Bormann clarifies that fantasy should not connote what most people think of when reading "fantasy" literature—fairies, talking animals, and epic adventures—but rather that "*fantasy* within this theory is the way that communities of people create their social reality [and should be understood to be] the imaginative and creative interpretation of events that fulfills a psychological or rhetorical need."[42]

These fantasy themes exist below the surface of an individual's perceptions of the world; he or she would not be able to define a certain fantasy theme but would simply know that a story or event "makes sense" according to the framework of the fantasy theme. Essentially, as a person views television, he or she interprets events according to certain themes that exist within the society the individual belongs to. He or she is unaware of these themes, yet compares television to those themes nonetheless. As Bormann puts it, "most people have been watching television news for so long that they have learned how to participate in its dramatizations until it has become second nature to them"—their interpretations operate beneath the surface of their consciousness.[43] When individuals come to a common understanding of what they are viewing, psychologist Darin Arsenault asserts that "the meanings that members share with each other create convergence, or understanding, between members" of a society.[44] Arsenault argues that fantasy themes are necessary as members cannot truly communicate without those shared meanings: "in other words, other members must be able to 'latch on' to what the arguer is promoting and relate this argument to personal experience."[45] Television allows for mass permeation of fantasy themes which in turn allows individuals, then larger societies, to make sense of the world—ultimately leading to groups acting as a coherent whole to function in a world of chaos. How? Bormann explains that individuals involved in communication acts containing fantasy themes "will attribute motives, purposes, and causes to the people in the story and will fit the events into a meaningful sequence of events. Fantasies always provide an organized artistic explanation of happenings and thus create a social reality which makes sense out of the blooming buzzing confusion of the experience."[46] Certainly, then, television is instrumental in helping people understand the world they make up and are simultaneously shaped by. These fantasy themes, and any communicative act within a group, are bound by culture. Belonging to a group requires a certain level of cultural understanding, as well as an understanding of the rules of the chosen medium through which fans engage.

PINTEREST

Pinterest, the social medium most closely associated with preteens and housewives, is increasingly a host for fans to congregate and aggregate their fan knowledge. Originally designed to aggregate favorite content from around the Web as a means of categorizing do-it-yourself projects and home decorating ideas, other members of the online community morphed Pinterest into an entirely different type of community. This medium has taken off probably due to the very nature of fandom; as media expert Deborah Kaplan puts it, "a large part of the fannish experience lies in analyzing the source texts of fandom."[47] Pinterest allows fans to see created content from other pins, while allowing people to pin—save for future reference—content from various Web sites. These pins lead to conversations about both the original artifact and the pin itself. This analysis manifests in conversations and fan-created content that aids in a deeper understanding of the original text—what fans construct or remember from episodes is indicative of the social state of the world fans live in. Fans pin content from the Internet that is related to the show in an effort to relive favorite moments or to take part in inside jokes, leading to a deeper investment in the fan community. Pinterest is unlike any other form of social media, and it is an underresearched area of communication. It is still evolving to adapt to user needs and wants, which indicates a rich area of communication and research.

METHODOLOGY

This chapter seeks to answer two essential questions: (1) How do members of a fan group utilize social media to connect with the text and with each other, to reimagine the text, and to integrate the text into their day-to-day lives? And (2) How does fandom, specifically fan groups communicating via the Internet, allow fans to interact with the text of a show? To answer these questions, fan content containing visual material that is produced or perpetuated online, and found on Pinterest was examined. To find these pins, the "Geek" category of Pinterest, the category under which most fan content is pinned, was selected. The *Doctor Who* pins were categorized, keeping only those directly related to the show or, which were obviously fan content. Pins that had a caption, indicating that the pinner was invested in that pin, were specifically searched. The selected pins were the first ones to appear under the search criteria, which necessarily left many pins unstudied. The selected pins were chosen as a convenience sample to represent all *Doctor Who* pins on Pinterest. After collecting pins, patterns among the pins were analyzed to see what categories fans were most drawn to in pinning content from the web. Most pins fell into four categories: crossover pins, inside jokes, products, and pins which indicated that *Doctor Who* had a wider reach into fans' lives than as a simple television show, which will be labeled as lifestyle pins.

ARTIFACT DESCRIPTION

Fans are able to pin content from all over the Web. Some of these pins contain content created by fans to memorialize, parody, or expound upon the show. Other pins contain content from various Web sites—content that was created by others and simply "pinned" to a Pinterest user's virtual bulletin board. Both types of pins were utilized by fans to deepen the fandom experience.

Crossover Pins

One increasingly popular type of pin is the crossover pin. These pins take two different artifacts—other television shows, popular movies and songs, and so forth—and combines aspects of each one to either compare them or to construct a joke that only other fans will understand. Currently, the Disney movie *Frozen* is a popular crossover medium, as evidenced by Cassidy Bosso and James Porter's pins.[48] Bosso's pin depicts Anna, the main character of *Frozen,* singing her popular lyric, "Do you wanna build a snowman?" with one of the *Doctor Who* villains—creepy snowmen—featured underneath, with the words, "No . . . No, I do not" adamantly placed across them.[49] Bosso's comment on the pin says, "It is two of my favorite things put into one picture!" In contrast, Porter's pin features a particularly emotive scene when the Doctor's companion and love, Rose, is sucked into an alternate dimension and then gets locked into that dimension with no return. The scene depicts the Doctor and Rose both pressed against a wall—the "wall" symbolizing the divide between them—listening for the other. Porter's pin superimposes the words: "Rose? Do you wanna build a snowman?"[50] These lyrics are further utilized in a pin from Tessa Gautreau, portraying little Anna, the protagonist of *Frozen,* peering into the Tardis, the Doctor's mode of transportation. Gautreaux's caption parodies the lyrics by proclaiming, "Doctor? Do you wanna fight a Dalek? C'mon let's go and play! I never see you anymore, come out the door . . . [sic]."[51] Another Disney crossover is found in Maggie Spreier's pin, featuring lyrics to the popular *Mulan*[52] song, "Be a Man," but with a *Doctor Who* twist.[53] Popular song, "Call Me Maybe," by Carly Rae Jepson,[54] is spoofed in Tara Hodges's pin that features the character River Song with the words, "Hey I just met you and this is crazy but you're my husband and I'm their baby [sic]."[55] These pins indicate that a particular scene from a movie or lyrics from a song have caused the individual to remember a scene or quote from *Doctor Who* (or vice versa). Prior to Pinterest and the Internet, fans would only be able to share these revelations with people they personally came into contact with. Pinterest allows fans across the globe to share, enjoy, and co-create these memes and memories.

Inside Joke Pins

Pinterest is also used to revisit moments from the show that made an impression on people, or that were humorous. These inside jokes from the show itself provide a feeling of belongingness for the fans. Megan Henson's pin captures a scene from *Doctor Who* that features Clara, the Doctor's latest companion, discussing the Tardis with the Doctor over instant video communication.[56] Clara asks, "When you say 'mobile phone' why do you point at that blue box? [sic]," to which the Doctor replies, "Because it's a surprisingly accurate description!" This post was pulled from Tumblr, as many fan pins are, and demonstrates the joy fans get from remembering humorous moments from the show.[57] This pin's caption from Tumblr declares, "THIS IS THE BEST JOKE THAT HAS EVER BEEN MADE ON DOCTOR WHO AND IF THEY NEVER MAKE IT AGAIN I WILL CRY I SWEAR."[58] Because it was then pinned from Tumblr, this moment was one that was poignant to multiple people, including (at least) the Tumblr user and the original pinner. Another instance of *Doctor Who* fans utilizing inside jokes comes from jaira phillips' [sic] pin, which depicts a news headline claiming, "Scientists Discover That Mars Is Full of Water."[59] Underneath, an image of the Doctor looking concerned immediately points all fans back to a specific episode[60] in which people living on Mars became infected by drinking the water.[61] Jaira's comment reinforces this point: "Don't drink it." Moments such as this allow fans to remember moments from the show and to apply instances from the show to real-life situations. Pinterest is a powerful tool for aggregating content from all over the Internet—such as the Mars story, or the Tumblr post—for fans to keep in one place to be accessed and shared repeatedly. These inside jokes construct identity for members—"I understand this joke, therefore I belong in this community."

Product Pins

Doctor Who bleeds over into all areas of life for fans. Pinterest is just one vehicle that demonstrates this phenomenon and allows critical examination of how users utilize and refresh *Doctor Who*'s content. Obviously, many products featuring *Doctor Who* content are made with commercial interest primarily in mind. However, fans also take these products as ways to identify themselves as fans to other fans, increasing the bonds of community. Rachel

Priest pinned a coffee maker in the shape of a Dalek, a popular *Doctor Who* villain, with the words, "CAFFENINATE"[62] superimposed above it.[63] Rachel's comment proclaims, "Oh my gosh, do they ACTUALLY make this ???? They totally should if they don't!!! [sic]," to which another user, I Urobouros, adds, "For this i would learn to drink coffee, so cool [sic]." Jen Garn, through etsy.com (an online store for handmade products), located and pinned Swarovski crystal *Doctor Who* police box/Tardis peep-toe heels, commenting, "Oh my. I need to do this."[64] These are just a few examples of the countless *Doctor Who* products made, many of which are pinned by fans.

Another product, though of an entirely different form, is tattoos. Buzz-Feed pinned a link to its article "50 Fantastic 'Doctor Who' Tattoos,"[65] featuring body ink as varied as words in Galifreyan,[66] the Doctor's face, popular villains, and famous scenes.[67] These fans went so far as to permanently portray their fanhood through inking it under their skin. The collection posted by Buzzfeed demonstrates that this is a trend; there were enough people who had publically shared images of their *Doctor Who* tattoos for Buzzfeed to compose this list. Body art is a particularly interesting medium of expressing fandom—it is far more permanent that purchasing a T-shirt. Being able to share tattoos on Pinterest guarantees a wider reach than simply showing off the artwork to close friends an individual might see face-to-face. All of these products aid in community-building efforts, as they take the fan object and display one's affiliation for other fans to identify like fans. The products function as nametags, or membership cards, for the fandom. Here is where fandom online and fandom offline collide, as people find products through social media, but display them for those in their immediate surroundings to see. Although people may originally seek out social media outlets for fandom group construction, products bridge the gap between the two worlds.

Lifestyle Pins

Fans who enjoy *Doctor Who* art yet are not quite committed enough to permanently display it on their skin choose to share their artwork and that of others through social media, including Pinterest. Alicia Orozco's pin depicts a drawing of many of the Doctor's companions throughout time spilling out of the Tardis as the Doctors[68] sit by watching the activity take place.[69] Other fans are so enamored with the show that they cannot keep from bringing their children into the fandom at an early age. Amanda Bryson's pin depicts two young toddlers embracing, one dressed as the Doctor, the other, the Tardis.[70] Fans also utilize GIFs[71] from the show to convey other meanings. One example comes from Maggie Rust's pin. The text of the pin urges the viewer to "Google 'Scotland's National Animal.' GOOGLE IT."[72] Underneath the text is a gif of the tenth Doctor's face reacting in shock.[73] This use of the *Doctor Who* gifs in other contexts demonstrates the extent to which *Doctor Who* has saturated the lives of fans. They reach immediately for a reaction shot of the Doctor's face to express their own emotion, which indicates that fans are deeply entrenched in their love of the show.

ANALYSIS

Having one's identity validated is an important aspect of group belongingness. Pinterest, and all social media, allow fans to find others who share interests, reaching beyond geographical bounds. Sharing beliefs and interests, and having those values reinforced by other people, perpetuates group membership. In fandom, particularly in the *Doctor Who* fandom on Pinterest, fantasy themes are clearly evident. Fantasy themes indicate that the story "holds together." In order for pins to be re-pinned or understood, there must be a base knowledge of the story line. In a broader sense, to understand the story line of the show, viewers must have a grasp of the language used by the show as well as any historical events which are referenced in the show—both real-world events and events from the history portrayed on-screen. Additionally, fans come to view the Doctor and his companions as the heroes of the story. In order for the story to make sense and for fans to interact on a common level, this heroic nature must be understood by all group members. Crossover pins indicate that fans have a firm enough grasp on the show that they can apply quotes and events to other stories, which may or may not be very similar to *Doctor Who*. Pins were understood by other fans indicating that fans have a common shared understanding of certain aspects of the show as well as a common, shared understanding of other fictional worlds. Further, these crossovers were never intended or foreseen by the creators of either artifact.

Thus, fans are now able to co-create with the rhetors of shows and movies. Fans are entering into the worlds of the fictional characters and finding and creating new connections between these realms. The Internet allows fans to disseminate these connections to other fans around the world, increasing fan connectivity and interaction with the text, as well as altering how fans view the artifacts.

Inside jokes indicate much of the same—for a joke to be humorous, all parties must agree on the laws, which govern the world in which they live, be they natural or social laws. Within fandom, there are two worlds fans must understand—the world in which they live *and* the world in which they inhabit as they partake of *Doctor Who*. Additionally, Pinterest provides another realm of common understanding, as fans interacting via Pinterest must abide by the rules governing the world of Pinterest. Thus, there is a tri-cultural understanding that takes place among the fans on Pinterest. Inside jokes clearly transcend the barriers of these realms.

Products made to represent aspects of the *Doctor Who* universe indicate, again, that fans understand both worlds. Additionally, products indicate that fans are willing to integrate items from that universe into their day-to-day lives. The pins that have little to do with *Doctor Who* itself but utilize aspects of the show further indicate that fans are interacting with the text in a much deeper way than simply viewing the show once. Fans now have ample time and resources to engage with the text, comparing it to the real world, altering perceptions of the story and characters, and predicting future events.

CONCLUSION

Fandom provides a unique group experience for members. Fans group together based on a common interest in a particular artifact. However, group membership invites increased engagement with the text, leading to altered perceptions of the universe contained within the text. This is clearly seen in the ways *Doctor Who* fans engage with one another and with the story's canon on Pinterest. While fans clearly have a common base understanding of the world of *Doctor Who*, the propagation of memes and inside jokes indicates that fans are open to reimagining the world their characters inhabit. Fandom requires this openness. However, for the story to continue to hold together, fans must maintain an understanding of canon, or what the rhetors have decreed the story line to be. Yet, to continue to interact with other fans online, individuals must remain up-to-date as exploration of the story evolves. This duality of fan experience cultivates and creates fan culture, increasing investment in the group and in the show.

NOTES

1. Kaarina Nikunen, "The Intermedial Practises of Fandom," *NORDICOM Review* 28, no. 2 (2007): 123.

2. Joel T. Nadler and Geraldine Y. Hannon, "Self-Selected Social Identification Measure (SSIM): A Survey Assessing Identity Based on Group Membership," *North American Journal of Psychology*, 15, no. 3 (2013): 443.

3. John Hughson and Emma Poulton. "'This Is England': Sanitized Fandom and the National Soccer Team," *Soccer & Society* 9. no. 4 (2008): 512.

4. Daniel C. Feldman, "The Development and Enforcement of Group Norms," *Academy of Management Review* 9, no. 1 (1984): 49. doi: 10.5465/AMR.1984.4277934.

5. Abigail Derecho, "Archontic Literature: A Definition, a History, and Several Theories of Fan Fiction," in *Fan Fiction and Fan Communities in the Age of the Internet*, eds. Karen Hellekson and Kristina Busse (Jefferson, NC: McFarland & Company, Inc., 2006).

6. Julio Garcia Espinosa, "For an Imperfect Cinema," in *Film and Theory*, eds. Robert Stam and Toby Miller (Malden, MA: Blackwell Publishing Ltd., 2000), 291.

7. Deborah Kaplan, "Construction of Fan Fiction Character Through Narrative," in *Fan Fiction and Fan Communities in the Age of the Internet*, eds. Karen Hellekson and Kristina Busse (Jefferson, NC: McFarland & Company, Inc., 2006), 136.

8. Matt Hills. "When Television Doesn't Overflow 'Beyond the Box': The Invisibility of Momentary Fandom." *Critical Studies in Television* 5, no. 1 (2010): 97.

9. James Hibberd, "'Doctor Who' Ratings Break BBC America Record," *Entertainment Weekly*, November 25, 2013, accessed April 15, 2014, http://insidetv.ew.com/2013/11/25/doctor-who-ratings-day/.

10. L. J. Shrum, James E. Burroughs, and Aric Rindfleisch, "Television's Cultivation of Material Values," *Journal of Consumer Research* 32, no. 3 (2005): 473–79.

11. N. Feather, Lydia Woodyatt, and Ian McKee, "Predicting Support for Social Action: How Values, Justice-Related Variables, Discrete Emotions, and Outcome Expectations Influence Support for the Stolen Generations," *Motivation & Emotion* 36, no. 4 (2012).

12. Francesca Coppa, "A Brief History of Media Fandom," in *Fan Fiction and Fan Communities in the Age of the Internet*, eds. Karen Hellekson and Kristina Busse (Jefferson, NC: McFarland & Company, Inc., 2006), 51.

13. Catherine Driscoll, "One True Pairing: The Romance of Pornography and the Pornography of Romance," in *Fan Fiction and Fan Communities in the Age of the Internet*, eds. Karen Hellekson and Kristina Busse (Jefferson, NC: McFarland & Company, Inc., 2006), 88.

14. Because most fan fiction is written by females, Willis's usage of "her" instead of "his or her" is accurate. Ika Willis, "Keeping Promises to Queer Children: Making Space (for Mary Sue) at Hogwarts," in *Fan Fiction and Fan Communities in the Age of the Internet*, eds. Karen Hellekson and Kristina Busse (Jefferson, NC: McFarland & Company, Inc., 2006), 155.

15. Mafalda Stasi, "The Toy Soldiers from Leeds: The Slash Palimpsest," in *Fan Fiction and Fan Communities in the Age of the Internet*, eds. Karen Hellekson and Kristina Busse (Jefferson, NC: McFarland & Company, Inc., 2006), 123.

16. Ika Willis, "Keeping Promises to Queer Children: Making Space (for Mary Sue) at Hogwarts," in *Fan Fiction and Fan Communities in the Age of the Internet*, eds. Karen Hellekson and Kristina Busse (Jefferson, NC: McFarland & Company, Inc., 2006), 157.

17. Megan M. Wood and Linda Baughman. "Glee Fandom and Twitter: Something New, or More of the Same Old Thing?" *Communication Studies* 63, no. 3 (2012): 331. doi:10.1080/10510974.2012.674618.

18. Coppa, "Media Fandom," 42.

19. Bertha Chin. "Locating Anti-Fandom in Extratextual Mash-Ups," *M/C Journal* 16, no. 4 (2013): 6.

20. Robert Stam, "Television News and Its Spectator," in *Film and Theory*, eds. Robert Stam and Toby Miller (Malden, MA: Blackwell Publishing Ltd., 2000), 364.

21. Rebecca Williams. "'This Is the Night TV Died': Television Post-Object Fandom and the Demise of The West Wing," *Popular Communication* 9, no. 4 (2011): 268–69. doi:10.1080/15405702.2011.605311.

22. Guo-Ming Chen and Xiaodong Dai, "New Media and Asymmetry in Cultural Identity Negotiation," in *New Media and Intercultural Communication: Identity, Community and Politics*, eds. Pauline Hope Cheong, Judith N. Martin, and Leah P. Macfadyen (New York: Peter Lang Publishing, 2012), 130.

23. Italics original. Robert Shutter, "When Indian Women Text Message: Culture, Identity, and Emerging Interpersonal Norms of New Media," in *New Media and Intercultural Communication: Identity, Community and Politics*, eds. Pauline Hope Cheong, Judith N. Martin, and Leah P. Macfadyen (New York: Peter Lang Publishing, 2012), 210.

24. Debbie Rodan, Lynsey Uridge, and Lelia Green, "Negotiating a New Identity Online and Off-Line: The HeartNET Experience," in *New Media and Intercultural Communication: Identity, Community and Politics*, eds. Pauline Hope Cheong, Judith N. Martin, and Leah P. Macfadyen (New York: Peter Lang Publishing, 2012), 150.

25. Marc Horne, "How Doctor Who Nearly Became the Time Lady," *The Telegraph*, October 10, 2010, accessed May 8, 2014,http://www.telegraph.co.uk/culture/tvandradio/doctor-who/8052694/How-Doctor-Who-nearly-became-the-Time-Lady.html.

26. "Science Fiction," *BBC Archive*, March 1962, accessed May 8, 2014, http://www.bbc.co.uk/archive/doctorwho/6400.shtml.

27. Ibid.

28. Ibid.

29. Tony Smith, "*Doctor Who* to Return to TV," *The Register*, September 26, 2003, accessed May 8, 2014, http://www.theregister.co.uk/2003/09/26/doctor_who_to_return/.

30. Ben Dowell, "Moffat Named Doctor Who Supremo," *The Guardian*, May 20, 2008, accessed May 8, 2014, http://www.theguardian.com/media/2008/may/20/bbc.television2.

31. Jace Lacob, "Doctor Who's Global Takeover," *The Daily Beast*, August 22, 2011, accessed May 8, 2014, http://www.thedailybeast.com/articles/2011/08/22/steven-moffat-interview-doctor-who-season-7-amy-pond-river-song.html.

32. These boxes were popular in Britain in the 1960s, when *Doctor Who* began its on-air adventures. Immanuel Burton, "A Brief History of the Police Box," November 2006, accessed May 8, 2014, http://www.policeboxes.com/pboxhist.htm.

33. Known as the Chameleon Circuit. For this topic, Wikipedia, as a fan-constructed and maintained entity, seems a fitting source of information: http://en.wikipedia.org/wiki/TARDIS.

34. @DoctorWho_BBCA, Twitter page, accessed April 20, 2014, https://twitter.com/DoctorWho_BBCA.

35. @bbcdoctorwho, Twitter page, accessed April 20, 2014, https://twitter.com/bbcdoctorwho.

36. Doctor Who's Facebook page, accessed April 20, 2014, https://www.facebook.com/DoctorWho.

37. @DWMtweets, Twitter page, accessed April 20, 2014, https://twitter.com/DWMtweets.

38. Barbara Leonard, Twitter post, April 15, 2014, 8:46 p.m., accessed April 15, 2014, https://www.facebook.com/DoctorWho.

39. Caroline, Twitter post, April 15, 2014, 7:50 p.m., accessed April 15, 2014, https://twitter.com/lovejoydiver.

40. Lisa Waite, "Rock and Roll! Using Classic Rock as a Guide to Fantasy-Theme Analysis," *Communication Teacher* 22, no. 1 (208): 10.

41. Ernest G. Bormann, "Symbolic Convergence Theory," in *Small Group Communication: Theory & Practice*, eds. Randy Y. Hirokawa, Robert S. Cathcart, Larry A. Samovar, and Linda D. Henman (Los Angeles: Roxbury Publishing, 2003), 41–42.

42. Ernest G. Bormann, "The Symbolic Convergence Theory of Communication: Applications and Implications for Teachers and Consultants," *Journal of Applied Communication Research* 10, no. 1 (1982): 52.

43. Ernest G. Bormann, "A Fantasy Theme Analysis of the Television Coverage of the Hostage Release and the Reagan Inaugural," *Quarterly Journal of Speech* 68, no. 2 (1982): 133.

44. Arsenault, "Rhetorical Vision," 59–60.

45. Arsenault, "Rhetorical Vision of the Independent and Sovereign Nation of Hawai'i: A Fantasy Theme Analysis," 60.

46. Bormann, "A Fantasy Theme Analysis of the Television Coverage of the Hostage Release and the Reagan Inaugural," 134.

47. Kaplan, "Fan Fiction," 135.

48. Cassidy Bosso, Pinterest post, "The Fandom Made Me Do It" board, accessed April 20, 2014, http://www.pinterest.com/pin/429038301973554282/; James Porter, Pinterest post, "Doctor Who" board, accessed April 20, 2014, http://www.pinterest.com/pin/561050066051185592/.

49. The popular lyrics originated in Disney's musical *Frozen*, directed by Chris Buck and Jennifer Lee (2013: Burbank, CA: Walt Disney Animation Studios).

50. James Porter, Pinterest post, "Doctor Who" board, accessed April 24, 2014, http://www.pinterest.com/pin/561050066051185592/.

51. Tessa Gautreaux, Pinterest post, "Things That Are Not Books [sic]" board, accessed April 27, 2014, http://www.pinterest.com/pin/182255116144956660/.

52. *Mulan*, directed by Tony Bancroft and Barry Cook, (Burbank, CA: Walt Disney Feature Animation, 1998).

53. Maggie Spreier, Pinterest post, "Books, Movies & TV" board, accessed April 27, 2014, http://www.pinterest.com/pin/186547609539316153/.

54. Carly Rae Jepsen, Tavish Crowe, Josh Ramsay, *Call Me Maybe*, Carly Rae Jepsen, 2011 by Schoolboy Records, MP3.

55. Tara Hodges, Pinterest post, "Doctor Who LOVES" board, accessed April 27, 2014, http://www.pinterest.com/pin/89227636341417656/.

56. Megan Henson, Pinterest post, "Doctor Who" board, accessed April 30, 2014, http://www.pinterest.com/pin/271904896227235813/.

57. Silabor, *Where You Tend a Rose, a Thistle Cannot Grow*, Tumblr post, accessed April 30, 2014, http://silabor.tumblr.com/post/71760126939/dohctor-this-is-the-best-joke-that-has-ever [sic].

58. Quite a bit can be said about fans on Tumblr, but there is sadly little space here to discuss what should really be its own research paper. Instead, I will provide some brief information. First, fans on Tumblr are highly emotional creatures who identify deeply with the characters they cherish. Second, they often seem to be speaking an entirely different language, with strange (or no) grammatical sense and punctuation. This you can observe in the odd quote, in all caps, on page 306.

59. This story was reported by many sources, such as by Alex Knapp, "NASA Finds Clues That There's Flowing Water on Mars," *Forbes*, February 10, 2014, accessed April 30, 2014, http://www.forbes.com/sites/alexknapp/2014/02/10/nasa-finds-clues-that-theres-flowing-water-on-mars/.

60. Russell T. Davies and Phil Ford, "The Waters of Mars," *Doctor Who*, season 4, episode 16, directed by Graeme Harper, aired November 15, 2009.

61. jaira phillips, Pinterest post, "Nerdness [sic]" board, accessed April 27, 2013, http://www.pinterest.com/pin/309129961895297164/.

62. The Daleks' typical saying is, "Exterminate." Fans enjoy riffing on the popular phrase.

63. Rachel Priest, Pinterest post, "Doctor Who Stuff I Want" board, accessed April 30, 2014, http://www.pinterest.com/pin/355291858074891586/.

64. Jen Garn, Pinterest Post, "Doctor Who" board, accessed September 6, 2014, http://www.pinterest.com/pin/47498971044376384/.

65. A popular news source for young adults, featuring both genuine news and popular interest articles. Rosa Pasquarella, "50 Fantastic 'Doctor Who' Tattoos," *BuzzFeed Community*, November 17, 2013, accessed April 30, 2014, http://www.buzzfeed.com/rosap/50-fantastic-doctor-who-tattoos-p98.

66. The language spoken by the Doctor and others of his race, the Time Lords, because they are from the planet Galifrey.

67. BuzzFeed, Pinterest post, "Tattoos" board, accessed April 30, 2014, http://www.pinterest.com/pin/137500594846785493/.

68. Because the Doctor regenerates, there are many iterations of his person throughout the show. This artwork features Doctors numbered 9, 10, and 11.

69. Alicia Orozco, Pinterest post, "Doctor Who [sic]" board, accessed April 30, 2014, http://www.pinterest.com/pin/167970261076871617/.

70. Amanda Bryson, Pinterest post, "#10 & 11" board, accessed April 30, 2014, http://www.pinterest.com/pin/74872412528736951/.

71. Animated images, usually drawn from popular television shows or movies.

72. If one follows the advice of the pin, one will discover that Scotland's National Animal is a unicorn.

73. Maggie Rust, Pinterest post, "For the Inner Crazy Fan Girl in Me (no shocks there!)" board, accessed April 30, 2014, http://www.pinterest.com/pin/139048707217885443/.

REFERENCES

@bbcdoctorwho. Twitter page. Accessed April 20, 2014, https://twitter.com/bbcdoctorwho.

@DoctorWho_BBCA. Twitter page. Accessed April 20, 2014, https://twitter.com/DoctorWho_BBCA.

@DWMtweets. Twitter page. Accessed April 20, 2014, https://twitter.com/DWMtweets.

Arsenault, Darin J. "Rhetorical Vision of the Independent and Sovereign Nation of Hawai'i: A Fantasy Theme Analysis." *Journal of Critical Postmodern Organization Science* 3, no. 2 (2005): 57–73.

BBC Archive. "Science Fiction." Accessed May 8, 2014. http://www.bbc.co.uk/archive/doctorwho/6400.shtml.

Bormann, Ernest G. "Symbolic Convergence Theory." In *Small Group Communication: Theory & Practice*, edited by Randy Y. Hirokawa, Robert S. Cathcart, Larry A. Samovar, and Linda D. Henman. Los Angeles: Roxbury Publishing, 2003.

———. "A Fantasy Theme Analysis of the Television Coverage of the Hostage Release and the Reagan Inaugural." *Quarterly Journal of Speech* 68, no. 2 (1982): 133–45.

———. "The Symbolic Convergence Theory of Communication: Applications and Implications for Teachers and Consultants." *Journal of Applied Communication Research* 10, no. 1 (1982): 50–62.

Bosso, Cassidy. Pinterest post, "The Fandom Made Me Do It" board. Accessed April 20, 2014, http://www.pinterest.com/pin/429038301973554282/.

Bryson, Amanda. Pinterest post, "#10 & 11" board. Accessed April 30, 2014, http://www.pinterest.com/pin/74872412528736951/.

Burton, Immanuel. "A Brief History of the Police Box," November 2006. Accessed May 8, 2014, http://www.policeboxes.com/pboxhist.htm.

BuzzFeed. Pinterest post, "Tattoos" board. Accessed April 30, 2014, http://www.pinterest.com/pin/137500594846785493/.

Caroline. Twitter post, April 15, 2014, 7:50 p.m. Accessed April 15, 2014, https://twitter.com/lovejoydiver.

Chen, Guo-Ming and Dai, Xiaodong. "New Media and Asymmetry in Cultural Identity Negotiation." In *New Media and Intercultural Communication: Identity, Community and Politics*, edited by Pauline Hope Cheong, Judith N. Martin, and Leah P. Macfadyen. New York: Peter Lang Publishing, 2012.

Chin, Bertha. "Locating Anti-Fandom in Extratextual Mash-Ups." *M/C Journal* 16, no. 4 (2013): 6.

Coppa, Francesca. "A Brief History of Media Fandom." In *Fan Fiction and Fan Communities in the Age of the Internet*, edited by Karen Hellekson and Kristina Busse, 41–60. Jefferson, NC: McFarland & Company, Inc., 2006.

Derecho, Abigail. "Archontic Literature: A Definition, a History, and Several Theories of Fan Fiction." In *Fan Fiction and Fan Communities in the Age of the Internet*, edited by Karen Hellekson and Kristina Busse, 61–78. Jefferson, NC: McFarland & Company, Inc., 2006.

Doctor Who's Facebook page. Accessed April 20, 2014, https://www.facebook.com/DoctorWho.

Dowell, Ben. "Moffat Named Doctor Who Supremo." *The Guardian*, May 20, 2008. Accessed May 8, 2014, http://www.theguardian.com/media/2008/may/20/bbc.television2.

Driscoll, Catherine. "One True Pairing: The Romance of Pornography and the Pornography of Romance." In *Fan Fiction and Fan Communities in the Age of the Internet*, edited by Karen Hellekson and Kristina Busse, 79–96. Jefferson, NC: McFarland & Company, Inc., 2006.

Espinosa, Julio Garcia. "For an Imperfect Cinema." In *Film and Theory*, edited by Robert Stam and Toby Miller, 287–97. Malden, MA: Blackwell Publishing Ltd., 2000.

Feldman, Daniel C. "The Development and Enforcement of Group Norms." *Academy of Management Review* 9, no. 1 (1984): 47–53. doi: 10.5465/AMR.1984.4277934.

Gautreaux, Tessa. Pinterest post, "Things That Are Not Books [sic]" board. Accessed April 27, 2014, http://www.pinterest.com/pin/182255116144956660/.

Henson, Megan. Pinterest post, "Doctor Who" board. Accessed April 30, 2014, http://www.pinterest.com/pin/271904896227235813/.

Hibberd, James. "'Doctor Who' Ratings Break BBC America Record." *Entertainment Weekly*, November 25, 2013. Accessed April 15, 2014. http://insidetv.ew.com/2013/11/25/doctor-who-ratings-day/.

Hills, Matt. "When Television Doesn't Overflow 'Beyond the Box': The Invisibility of Momentary Fandom." *Critical Studies in Television* 5, no. 1 (2010): 97–110.

Hodges, Tara. Pinterest post, "Doctor Who LOVES" board. Accessed April 27, 2014, http://www.pinterest.com/pin/89227636341417656/.

Home, Marc. "How Doctor Who Nearly Became the Time Lady." *The Telegraph*, October 10, 2010. Accessed May 8, 2014, http://www.telegraph.co.uk/culture/tvandradio/doctor-who/8052694/How-Doctor-Who-nearly-became-the-Time-Lady.html.

Hughson, John, and Poulton, Emma. "'This Is England': Sanitized Fandom and the National Soccer Team." *Soccer & Society* 9, no. 4 (2008): 509–19.

Kaplan, Deborah. "Construction of Fan Fiction Character Through Narrative." In *Fan Fiction and Fan Communities in the Age of the Internet*, edited by Karen Hellekson and Kristina Busse, 134–52. Jefferson, NC: McFarland & Company, Inc., 2006.

Lacob, Jace. "Doctor Who's Global Takeover." *The Daily Beast*, August 22, 2011. Accessed May 8, 2014, http://www.thedailybeast.com/articles/2011/08/22/steven-moffat-interview-doctor-who-season-7-amy-pond-river-song.html.

Leonard, Barbara. Twitter post, April 15, 2014, 8:46 p.m. Accessed April 15, 2014, https://www.facebook.com/DoctorWho.

Nadler, Joel T., and Hannon, Gerladine Y. "Self-Selected Social Identification Measure (SSIM): A Survey Assessing Identity Based on Group Membership." *North American Journal of Psychology* 15, no. 3 (2013): 425–46.

Nikunen, Kaarina. "The Intermedial Practises of Fandom." *NORDICOM Review* 28, no. 2 (2007): 111–28.

Orozco, Alicia. Pinterest post, "Doctor Who [sic]" board. Accessed April 30, 2014, http://www.pinterest.com/pin/167970261076871617/.

Pasquarella, Rosa. "50 Fantastic 'Doctor Who' Tattoos." BuzzFeed Community, November 17, 2013. Accessed April 30, 2014, http://www.buzzfeed.com/rosap/50-fantastic-doctor-who-tattoos-p98.

Phillips, Jaira. Pinterest post, "Nerdness [sic]" board. Accessed April 27, 2013, http://www.pinterest.com/pin/309129961895297164/.

Porter, James. Pinterest post, "Doctor Who" board. Accessed April 20, 2014, http://www.pinterest.com/pin/561050066051185592/.

Priest, Rachel. Pinterest post, "Doctor Who Stuff I Want" board. Accessed April 30, 2014, http://www.pinterest.com/pin/355291858074891586/.

Rodan, Debbie, Uridge, Lynsey, and Green, Lelia. "Negotiating a New Identity Online and Off-Line: The HeartNET Experience." In *New Media and Intercultural Communication: Identity, Community and Politics*, edited by Pauline Hope Cheong, Judith N. Martin, and Leah P. Macfadyen, 139–54. New York: Peter Lang Publishing, 2012.

Rust, Maggie. Pinterest post, "For the Inner Crazy Fan Girl in Me (no shocks there!)" board. Accessed April 30, 2014, http://www.pinterest.com/pin/139048707217885443/.

Shutter, Robert. "When Indian Women Text Message: Culture, Identity, and Emerging Interpersonal Norms of New Media." In *New Media and Intercultural Communication: Identity, Community and Politics*, edited by Pauline Hope Cheong, Judith N. Martin, and Leah P. Macfadyen, 209–20. New York: Peter Lang Publishing, 2012.

Silabor. Where You Tend a Rose, a Thistle Cannot Grow. Tumblr post, accessed April 30, 2014, http://silabor.tumblr.com/post/71760126939/dohctor-this-is-the-best-joke-that-has-ever.

Smith, Tony. "Doctor Who to Return to TV." *The Register*, September 26, 2003. Accessed May 8, 2014, http://www.theregister.co.uk/2003/09/26/doctor_who_to_return/.

Spreier, Maggie. Pinterest post, "Books, Movies & TV" board. Accessed April 27, 2014, http://www.pinterest.com/pin/186547609539316153/.

Stam, Robert. "Television News and Its Spectator." In *Film and Theory*, edited by Robert Stam and Toby Miller, 361–80. Malden, MA: Blackwell Publishing Ltd., 2000.

Stasi, Mafalda. "The Toy Soldiers from Leeds: The Slash Palimpsest." In *Fan Fiction and Fan Communities in the Age of the Internet*, edited by Karen Hellekson and Kristina Busse, 115–33. Jefferson, NC: McFarland & Company, Inc., 2006.

Waite, Lisa. "Rock and Roll! Using Classic Rock as a Guide to Fantasy-Theme Analysis." *Communication Teacher* 22, no. 1 (208): 10–13.

Williams, Rebecca. "'This Is the Night TV Died': Television Post-Object Fandom and the Demise of The West Wing." *Popular Communication* 9, no. 4 (2011): 266–79. doi:10.1080/15405702.2011.605311.

Willis, Ika. "Keeping Promises to Queer Children: Making Space (for Mary Sue) at Hogwarts." In *Fan Fiction and Fan Communities in the Age of the Internet*, edited by Karen Hellekson and Kristina Busse, 153–70. Jefferson, NC: McFarland & Company, Inc., 2006.

Wood, Megan M., and Baughman, Linda. "Glee Fandom and Twitter: Something New, or More of the Same Old Thing?" *Communication Studies* 63, no. 3 (2012): 328–44. doi:10.1080/10510974.2012.674618.

Chapter Seventeen

Television Inspired Cosplay and Social Media

Laura Kane and William Loges

Every weekend somewhere in the world there is a convention hall filled with fans of television shows, video games, movies, anime, and manga. Many of these individuals choose to express their love through dressing as their favorite characters. These individuals, known as cosplayers, spend weeks to months creating and obtaining perfect replicas of the outfits worn by characters like Ned Stark from *Game of Thrones*, Captain Malcolm Reynolds from *Firefly*, and any of the twelve incarnations of The Doctor from *Doctor Who*. "The word cosplay is a portmanteau word that combines both elements costume and play together. While play represents that it is one sort of performing activity, costume implies that people need particular outfits and accessories to be a specific character."[1] Costumed fans gather at conventions to socialize, take photographs, attend meet and greets with actors, participate in costume contents, and purchase memorabilia from their favorite shows.

The interaction of cosplayers does not end at the convention. Hundreds of thousands of cosplayers gather together online through social media platforms like Facebook, Twitter, Tumblr, and DeviantArt. Cosplayers share photos of their creations, swap convention stories, ask other cosplayers questions on how to construct specific parts of costumes, and provide tutorials to help others. Platforms like Facebook provide a space for individuals to form specific groups dedicated to cosplaying specific television series such as *Game of Thrones*, *Star Trek*, *Sherlock*, and *Doctor Who*. In these spaces, individuals share their fandom with others and participate in weekly episode discussion, view costume analyses and upload photos of their own costumes. The relationships that social media provide between fans and producers can be understood through Stuart Hall's encoding/decoding framework.[2] Hall introduced this framework in the historical context of network television in the United Kingdom in the early 1970s, but the strength of the approach is such that it provides insight into the current technological environment.

Hall described the relationship between media producers and audiences in terms of (1) a *technical infrastructure* that delimits what can be done by both producers and audiences, (2) *relations of production* that describe who has access to and controls the available technologies, and (3) *frameworks of knowledge* that producers and audiences use to makes sense of the content they produce and consume.[3] The television environment that Hall was describing was such that the technical infrastructure available to producers of TV was much more sophisticated than that of the TV audience. The relations of production also favored producers, since the ordinary audience member could not be expected to produce a TV show, let alone distribute a show on a mass scale. Hall argued that as a result of the imbalances in technology and control of technology, the frameworks of knowledge of those who produced television were unduly influential on audiences. Audiences were left to "decode" the content they encountered, but not to respond to it in kind.

With many television networks like HBO now sharing space on Facebook with their audience, connections to fans is deeper than has ever been possible. Cosplayers who wish to recreate the costumes from *Game of Thrones* can now follow the show's embroidery designer on Facebook and view in depth discussion and tutorials about how the actual show costumes were created. The *Doctor Who* Tumblr page frequently shares the fan makings of its audience, which includes costumes and craft items. This close interaction between television media and its fans allows the opportunity for a fan to feel more connected to their chosen series. With additional information about the characters, construction of costumes, and behind-the-scenes information usually only provided online, cosplayers can feel more confident about their project of choice and an enhanced connection to their fandom.

In this chapter we bring to light the connection between the ways that television producers use social media to interact with fans that express their love of their series through fan works. A case study of the *Game of Thrones* embroiderer Michelle Carragher will provide an example of producers of content connecting to fans, more specifically cosplayers. Social media platforms like Facebook and YouTube have allowed producers to interact on the same level as fans, allowing a back-and-forth relationship whereby producers post inside information or behind-the-scenes detail of the show, and fans can come back and share how they have appropriated and used this information. The relationships between producers and fans can be understood through Hall's encoding/decoding framework, but only if new relations of production are recognized on the decoders' side of the model.

THE STUDY OF FANDOM AND FANDOM PRACTICES

According to Henry Jenkins, belonging to a show's fandom means not only consuming the show as it's provided by producers but taking a step further to read into the show, analyze the characters' motives, create new stories based on the provided world, and reconstruct the content to one's desire.[4] For those in fandom, being a fan is more than just taking the material at face value, it's immersing oneself in the experience.

Current studies of fandom trace their origins back to the works of de Certeau, Fiske, and Jenkins, and were concerned primarily with coming to an understanding of the concept of fandom and breaking popular stereotypes of fans.[5] Henry Jenkins is credited with bringing the study of fandom into the forefront of academic thought with his seminal work *Textual Poachers.*[6] Jenkins adapted de Certeau's analogy describing readers of text as "poachers" to fans, who embrace content and bend it to their whim.[7] These early studies of fandom occurred around the early 1990s in the time of VCRs instead of DVRs, AOL chat rooms instead of Facebook pages and before it was cool to be a fan.[8] Fandom communities were seen as niche, and odd and suffered from stereotypes perpetuated through popular media outlets.[9] This "First Wave" of fandom studies attempted to break through these stereotypes and explain what drew fans to particular content and how their fan communities provided support and a sense of security.[10] Jenkins stressed the idea that fans "lack direct access to the means of commercial cultural production and have only the most limited resources with which to influence entertainment industry's decisions."[11] Fans coped with this "powerlessness" by taking personal hold of the media and creating new stories through fan art and fan fiction.[12]

In Hall's terms, Jenkins was describing an attempt by fans to find a relation of production that would allow them to produce content that they could share on a large (mass) scale, but still be traced to them as authors. The author/fans didn't necessarily demand money in return for their labor. The opportunity to share the work they produced—clearly inspired by someone else's copyrighted work—was sufficient to the fan community. In Hall's terms, this is a relation of production on the part of the audience that conventional television never allowed. In the environment Jenkins describes, the audience considers TV shows as raw material, not finished products. And yet the fans could not distribute the content they created.

The "Second Wave" of fandom studies came to terms with the idea that fans were a part of consumer society and were embraced by mainstream media outlets.[13] Studies celebrated fandom, investigated collection practices, social make up, and definitions of fan inclusion.[14] The stereotypes of the fan as being obsessive, anti-social, and out of touch with reality was breaking away. This wave recognizes the increased production capability of fans.

What were once obvious home movies were increasingly approaching professional standards of production. What was missing was professional standards of distribution (and criticism).

The third and current wave of fandom studies focuses on fandom in the new world of technology, focusing on how fans interact with the fan objects themselves and to "capture fundamental insights into modern life."[15] Life in 2014 means an almost constant connection to the Internet via a multitude of devices. "Blackberries, iPods, PSPs laptops, PDAs, and cell phones all bring fan objects with their users to the subway, the street, and even the classroom."[16] Fandom today relies heavily on the Internet, with larger and larger numbers of individuals taking part in fan communities than ever before. "Online fans express their attachment to television narratives by creating or visiting Web sites associated with a program, and/or by interacting with other fans who share a common zest for the same TV series."[17]

Studies of fandom frequently discuss the concept of fan works with the most popular examples being fan-created art, fan fiction, and discussion boards.[18] These texts usually make a passing mention of the fan works that are the focus of this chapter: the recreation of a costume worn by a character in a given television show, movie, or other text. This type of fan creation is known as cosplay (costume play). Contemporary cosplay takes advantage of technologies, relations of production, and frameworks of knowledge that enhance the experience of cosplayers in ways that Hall's original model in fact contemplates, but couldn't describe in the early 1970s.

TV CREATORS AND CONNECTING TO FANS

Today's fan activities that used to take place in convention halls, through small in-person gatherings, and email lists now take place largely online in large web communities. The massive Cosplay.com Web site features a forum, photo gallery, and wig store and has over 359,000 members.[19] Tumblr, a Mecca for fandom, has over 197 million users sharing content. The fan fiction site An Archive of Our Own, has 370,000 users and features fan fiction in 15,770 different fandoms.[20] These Web sites allow a place for fans to gather and share their passion and enthusiasm for the things they love. These Web sites provide a technological infrastructure to the audience that Hall could not have reasonably anticipated. The biggest change is not in production and recording, although those have increased dramatically. The biggest change in the technological infrastructure Hall described is in the distribution of content, which concerns the relations of production.

Fans are now interacting among producers, actors, writers, and television executives (i.e., those who influence the relations of production). Online fan sites are largely open communities that accommodate "lurkers," who can read posts without ever making a post themselves. It is not uncommon to have actors or writers scouring fan sites to get an inside look at what their viewers think about the program. Individuals are hired to act as a TV show's social media liaison—taking care of Twitter feeds, posting to Facebook pages and updating the Web site. The most common uses of these communication platforms are to promote and endorse the show to as wide an audience as possible.[21]

"Companies can achieve intimacy between customers and their branded offerings by displaying to the customer that he/she is an important partner."[22] This can easily be done by acknowledging and sharing fan-created content related to the brand. Many television networks and producers of shows with large fan bases such as *Game of Thrones*, *Doctor Who*, and *Sherlock* acknowledge that their fans interact with their show on an emotional and personal level through fan works. These networks have used social media in the past to highlight and feature fan work not only as a way to say "thank you" to the fans, but also as a way to strengthen the bond between the fan and the brand.[23] For example, in 2011, BBC America held a contest called "Where's the TARDIS?" to promote season 6 of *Doctor Who* where fans of the show were asked to submit photos and stories about their home made version of the iconic time machine. Other fans could vote on their favorites and the producers of the show picked a winner among the most popular entries.[24] The winner was showcased on the BBC America Facebook page and shown on the BBC America channel in between shows.

Showcasing the cosplay costumes created by fans has become a way for networks to acknowledge the dedication of fans and to promote fan works in a positive light. Gone are the days when people only dressed up from *Star Trek* and dressing up was considered bizarre and the epitome of nerdom. Today, a TV show or movie obtaining a cult-like following full of fans who create new and imaginative works (including cosplay) is often a goal that indicates success.[25] Fans are able to take advantage of a new technological infrastructure and new relations of production to force program producers to contemplate (if not adopt) a new framework of knowledge when new episodes are produced (or old episodes are interpreted).

COSPLAY AS A FANDOM OUTLET

Popular works about fandom practices such as Jenkins' *Textual Poachers* and *Fans, Bloggers and Gamers* focus on fan fiction, fan art, and fan video-making as expressions of fan enthusiasm for a popular text, and only mention cosplaying in passing.[26] Cosplaying is mentioned in these works as one of many ways that fans express their love of a series, but it has not been analyzed to the extent that other types of fan work has been. Studies about cosplay have alluded to the idea that cosplayers participate in the hobby in order to express their love for a character or a series.[27] Wearing the costume of a character and performing as him or her for a day strengthens the connection the fan feels for their character or series.[28]

The play aspect of cosplay usually occurs at anime, video game, or science fiction conventions. "Conventions offer the perfect setting for open and public celebration of fandom by hosting various events related to fan interests in a short amount of time."[29] It is generally understood that cosplayers choose to participate in the hobby as a means of expressing their love for a character, enjoying the creation process, and escaping from mundane life.[30] Some cosplayers choose to just wear their costumes in social settings, posing for pictures every now and then, and attending convention events like usual, while other cosplayers choose to adapt the persona of their character, compete in cosplay competitions, and partake in large or elaborate photoshoots in and around the convention.

Participation in the cosplay community does not end when the convention is over. Many cosplayers join and take part in online cosplay Web communities where participants can exchange ideas for costumes, provide photos of their work, offer constructive criticism, and talk about their favorite series. New and experienced cosplayers alike gather and share ideas, photos, and experiences with creating costumes on social media Web sites such as Facebook, YouTube, and Tumblr. These communities bring together cosplayers of particular types of texts, including anime, video games, and television shows, specific shows like *Game of Thrones,* and popular franchises like Disney.[31] This allows relations of production from the message-creation side that Hall knew, but also production from the audience that Hall could not reasonably describe or anticipate. In the early 1970s, not only was production capability in the hands of a tiny amount of professionals, but distribution capability was even more reserved to an even more tiny number of people.[32] The relations of production have somewhat balanced between professionals and fans (but they still favor pros); it is in distribution that the most disruption has occurred.

A growing trend in cosplay groups is the creation of a personal cosplay fan page on Facebook, a page dedicated to the activities of a single cosplayer. Cosplayers showcase their work by uploading photos and statuses about their current projects. There are dozens of Facebook cosplay pages for individual cosplayers that now have an almost celebrity-like status in the cosplay community. Famous cosplayers often have small Web stores where fans can purchase prints of their most popular cosplay photos. Other types of merchandise include patterns for costumes, books detailing how to build props, and calendars filled with the cosplayer wearing a dozen different costumes. Many of the most popular cosplayers, such as Yaya Han and Kamui, are often invited as special guests to conventions and are given autograph booths and chances to host and judge major competitions.[33]

While cosplayers preach that the hobby is about fun and carefree dressing up, many cosplayers take dressing as a character very seriously, stating the importance of authenticity and accuracy to the original character.[34] Accuracy and authenticity is subjectively defined for each cosplayer, but many believe that it is most important to recreate the outfit of the character as closely as possible. This authentic representation means having the correct shoes, outfit, accessories and hair (which usually means wearing wigs). For others, having a similar body shape or facial features to the original character is essential to maintaining accuracy and authenticity to the original. Not being able to live up to the sometimes-impossible standards of looking like a TV actor or a cartoon character whose body cannot be emulated by a healthy human being, many individuals simply ignore this hurdle, and cosplay the character anyway. In the creation of a costume, cosplayers are engaged in relations of production that produce costumes that they can distribute widely online, or display to others at conventions.

As much as the cosplay community claims that there are no rules to cosplaying, as soon as a cosplay photo is uploaded and shared on the Internet there is the potential for negative and unwarranted feedback. If the costume is not complete, accurate, or tailored well, participants on Web sites like 4chan.org/CGL will be happy to point it out. Not the same body type of the original character? Don't even bother! Wearing a costume of a character that happens to show cleavage or a lot of leg? You're a slut! These types of comments are not uncommon, and only a sliver of how elitist and strict individuals can be about cosplay standards. Those cosplayers with more exposure in the community (like Yaya Han) have made a point to try and stop negative comments and bad attitudes in the cosplay community.[35]

Almost as important as the actual wearing of the costume is the creation of the ensemble. Participating in cosplay requires re-creating the costume of a specific character, and most of the time these garments cannot be found at your local clothing retail store. For many individuals, the costumes of their favorite characters have to be made and constructed from scratch. While there is no data to indicate exactly what percentage of cosplayers make their own costumes, there is a large enough foundation of cosplayers that make their own costumes that there are tutorials and guides online detailing how to construct virtually any type of costume. Cosplayers who choose to make their own costumes extensively research sewing, prop making, wig styling, and armor making in order to faithfully re-create the look of the original character. Many of these participants start off with no previous experience with these creation methods and through their desire to dress as their favorite character push themselves to learn new skills.[36]

Now, more than ever, it is easy for cosplayers to gain access to high quality materials, wigs, and fabrics to use for their costumes. Thermoplastic materials like Worbla and Wonderflex make it easy to create huge armored costumes. Specialty wig fibers make it possible to style wigs like real hair. Advances in home sewing machines make it possible to create custom embroidery at the touch of a button. As much as the methods of creation have improved, so have the ways that individuals record their hard work. Cosplayers now have access to advanced photography and photo editing software, allowing them to create high quality images of themselves in costume that rival Hollywood production screen shots. HD video recording software makes it easy to film cosplay videos, with individuals able to create their own Web series borrowing their favorite characters from their favorite shows. All of this represents democratization of production technology. Fans of *Nightwing*, *Doctor Who*, and *Castlevania* have come together to make their own versions of the popular stories, following a short episode format and posting directly to YouTube channels.[37] This democratized distribution of content is more revolutionary than production. People have been making home movies for decades; it was only very recently that those movies could be shared with everyone online.

In the early 1970s, Stuart Hall could account for TV producers' technological infrastructure, relations of production, and frameworks of knowledge in a model that privileged producers and characterized TV viewers in a dependent position.[38] In 2014 we find ourselves in a technological environment where many TV viewers (better described now as TV *users*) have a relation of production that was literally impossible when Hall wrote his landmark article. The model Hall provided still offers a fruitful way of interpreting the changing relationship between cosplayers and TV producers. To fully understand this new relationship, one must recognize that the imbalance of technological infrastructures has narrowed considerably. That has resulted in an adjustment in the relations of production (particularly such that viewers or users can distribute content on a scale the approaches that of professionals). Thus, the frameworks of knowledge Hall described become poignant.

For many cosplayers, obtaining the perfect material or accessory may mean re-creating it from scratch or finding the original item used in the production. When screen-accurate fabrics, accessories, or props are discovered by one individual (such as the exact same boots worn by the Doctor from *Doctor Who*, or the manufacturer of Sherlock Holmes's iconic wool coat) this information is shared with other cosplayers in the community. It is easier (and usually much cheaper) to find boots or jackets that look similar to the items worn on screen, but many cosplayers will go out of their way to re-create the outfit down to the exact items used by the costume crew of a certain production. This could often put the cosplayer out thousands of dollars for screen-accurate pieces, not to mention having to enter bidding wars for one-of-a-kind items. For those who do purchase screen-accurate pieces for their costumes, having these same items used in the production of the TV show creates a more authentic cosplaying experience. This sentiment is similar to the way historical re-enactors feel about using period-accurate clothing, behavior, and weaponry.[39]

Obtaining wardrobe items from TV shows and movies where the costume department utilized modern designer garments and lower budget accessories is much easier than it is for fantasy or period series. The fantasy series *Game of Thrones* is a prime example of this. Costume designer Michelle Clapton has mentioned in interviews that every costume on the show is made from scratch, including the costumes for the extras. When cosplayers choose to re-create costumes from the series, the majority of them rely on fabrics and materials that mimic the materials used in the show, going for a similar appearance rather than screen accuracy.[40]

Exhibitions showcasing the exquisite detail work of the costumes from television shows and movies allow hopeful cosplayers to see these garments up close and personal, being able to note the types of closures, seaming detail, and color used on screen. These events offer an opportunity for a television show's audience to obtain a behind-the-scene experience into the costumes of the show. It is not unusual for several attendees to photograph the costumes on display up close with high resolution cameras, offering extremely detailed photographs of weathering, beading, fabric structure, and layering that otherwise may not have come across on the screen. While all this detail is usually unique to individual pieces, cosplayers now have access to extraordinary details of costumes that boost the potential of a cosplayer to create an authentic costume recreation. This opportunity is wonderfully promoted by Michelle Carragher, the embroiderer of many of the most iconic costumes from the *Game of Thrones* television series.

MICHELLE CARRAGHER EMBROIDERY

With four seasons and over 7.1 million people tuning into its 2014 season finale, HBO's television series *Game of Thrones* has become the network's most popular and successful show ever.[41] The series is based on George R. R. Martin's *A Song of Ice and Fire* series and takes place in the fantasy world of Westeros. The story is full of political drama, power struggles, magic, violence, war, and intrigue. With a massive budget, the production team utilizes exotic locations such as Croatia, Ireland, Malta, and Iceland for backdrops, and creates extensively detailed sets, costumes, and effects that rival movie productions. The costumes of the series, designed by Michelle Clapton, have been awarded and nominated for multiple creative arts Emmy awards. The costumes of the show are richly detailed, with each costume meticulously created from scratch to reflect the development of the character. Clapton works closely with Michelle Carragher, a textile embellisher and artist on many of the costumes for the show.

Michele Carragher is a United Kingdom-based embroiderer who studied at the London College of Fashion and has worked restoring historical arti-facts with the Textile Conservation. Her work includes costume embellish-ment for the 2005 mini-series *Elizabeth I* and assistant maker for costumes from 2007's ITV adaptation of *Mansfield Park*.[42] Carragher's work is ex-tremely detailed, incorporating smocking, beading, needlework, and felting techniques to create highly dramatic motifs that are directly sewn onto the garment or onto silk organza to be applied separately afterward.

Carragher's work is most prominent on several principal characters in *Game of Thrones* who wear some of the most iconic costumes in the show. The female characters Daenerys Targaryen, Cersei Lannister, Sansa Stark, and Margaery Tyrell have been outfitted in elaborately detailed costumes in several important scenes. These garments usually incorporate a narrative within the detail of the costume itself. For example, the wedding dress of the character Sansa Stark into the Lannister family featured an overlapping band around the torso embellished with images reflecting the overtaking of a dire-wolf (the Starks' sigil) by powerful lions (the Lannister sigil).[43] The intricate detailing was only seen at a distance onscreen and likely went entirely unno-ticed by the average viewer. After the episode aired, Carragher uploaded close-up shots of the costume embroidery on her Web site and Facebook page, along with a tutorial on how she constructs her embroidery pieces from start to finish.

Here some of the producers' control over the elements of message-crea-tion that Hall described as relations of production is offered to viewers (who, in the cosplay community become users) of television. The frameworks of knowledge of the viewers who choose to become users become competitive with the producers' original intent, but also conform to the producers' under-standing of the characters and their costume. When fans acquire production technology, control its use, and distribute their productions, fans reply to television in ways that Hall could not anticipate, but in ways that his model does in fact describe.

Perhaps Carragher's most influential contribution for *Game of Thrones* cosplayers is her inclusion of a tutorial for the dragonscale smocking she used throughout the costumes of the character Daenerys Targaryen. Daener-ys is one of *Game of Thrones'* most cosplayed characters, and has several iconic blue scaled and beaded tunics she wears through seasons 3 and 4 of the show.[44] Several versions of the dragonscale tunics exist, with the cluster-ing and spread of the dragonscale embroidery extending further and further throughout the garment as the character develops through the season. Casual viewers may not notice that the scaling increases throughout the season.

When Carragher uploaded her dragonscale tutorial on her Web site cosplayers were able to re-create the detail work on their own costumes. Cosplayers could add authenticity to their costume re-creations because they had access to the same techniques used on screen.

Fans reacted overwhelmingly positively to Carragher's tutorials. Cosplayers documented their own experience using the tutorials, and shared the links among their respective friends and fellow cosplayers.[45] Media began to take note of the popularity of Carragher's postings, and interviews and features swarmed in. Carragher's photos were featured on BuzzFeed, Italian Vogue, Hollywood Reporter, and Fashionably Geek.[46] A common theme throughout these articles is the fact that without the photos Carragher posted, much of the intricate work would go unnoticed by television viewers.

Carragher further endorses cosplayer fans of *Game of Thrones* by being a member of *Game of Thrones* cosplay groups on Facebook. She "likes" posts and shares the photos of cosplayers that used her technique on her own personal Facebook page. Carragher is bridging the gap between producer and fan by supporting and celebrating fans who choose to re-create costumes as part of their fandom experience. This type of relationship between costume designer and costume recreators would likely never have existed without the pervasive use and features of social media. Keeping up with Carragher's work is as easy as clicking a button, and Carragher can show support for *Game of Thrones* cosplayers just as easily.

THE FUTURE OF THE ROLE OF SOCIAL MEDIA IN CONNECTING FANS TO TV SHOWS

Producers have taken notice as the methods of production for cosplay costume becomes more advanced. However, "[t]he fact that the companies embrace some forms of fan productivity doesn't mean all fans are equally welcome."[47] Creating fan works and costumes for personal use is usually ok with producers, but take it any further and the hand of the law comes down. Entertainment giant Marvel has sent cease and desist letters to those attempting to produce and sell 3D printed Iron Man suits, and sellers on Etsy selling handmade versions of Jayne Cobb's hat from *Firefly* have been asked to remove their listings due to copyright infringement.[48] When there's a conflict between the relations of production of commercial creators and the interests of fans inspired by commercial creators, the interests of commerce will often prevail, at least in strictly legal terms.

Television fan pages for shows like *Sherlock* or *Doctor Who* that celebrate fan works also only do so if it does not interfere with the canon of the show. For example, with the current BBC show *Sherlock*, there are thousands of images online of cosplayers dressed as the modern Holmes and Watson, but with their relationship taken to a level not shown on the screen.[49] The *Sherlock* Facebook page is unlikely to share pictures of cosplayers dressed as their two heroes in a lip-locked embrace. While they may not share these types of fan works through their own promotional page, this does not mean that they do not support fans bending the media to their will.

Many Facebook pages for major television shows post interactive Web sites and links to videos that allow audience members an inside look at the show. Providing additional background and behind-the-scenes information otherwise not available until DVD release allows fans to delve further into the worlds created in the shows. Fans can use this background information to enhance their own fan works. For example, a 360° 3D rendering of the whole of 221B Baker Street could help fan fiction writers stage scenes in their work and describe parts of the set not seen in the show aired on TV. Posting this kind of information for fans creates excitement within the audience and can inspire fans to continue to create their own fan works.

This potential creates a relationship between fans, audiences, and producers that was unimaginable in the early 1970s. Nonetheless, Hall provides a valuable distinction between the technological infrastructure available to producers and users (audiences), the relations of production that describe who controls that technology, and the frameworks of knowledge that producers and users bring to the content they share. There is a power dynamic in media use that Hall described. The technology has changed, and thus the foundations of Hall's encoding/decoding model might need new attention. The behavior of fan communities provides a window into that change in power. It remains to be seen how far the balance will tip between users and producers.

New media have offered audiences a very different relationship to media producers, still bounded by the concepts Hall described. The technical infrastructure still favors producers on purely state-of-the-art electronics, but the gap between the technology available to professionals and consumers has shrunk considerably (evidence by the term *prosumer*, which describes technology that consumers can use to produce professional grade content). Relations of production have adjusted to this new technological environment, and much of our analysis stems from this change. Fans' ability to not merely consume professional productions but to respond to them through media is not contemplated in Hall's original description of the producer/audience relationship. Producers' responses to fans' action has to account for the increased technical savvy of fans. The frameworks of knowledge of producers and audiences thus can confront one another on grounds more equal than the situation Hall described. They may still be different, but they are more likely to be communicated mutually than in the media environment Hall described in the early 1970s.

The most significant change in the technology and relations of production is the ability to distribute content. People have been writing, taking photographs, making home movies, and (later) videos for as long as those basic technologies were available. What has distinguished ordinary people from "the media" is the ability to distribute what they created. As the gap between producers and audiences has narrowed both in technological and distribution terms, a new relationship has emerged. Producers have to be more sensitive to the work their audiences might do in response to an original production. Audiences can communicate with one another on a larger and more immediate scale than before, giving them individual voices and collective power that Hall knew the TV audience did not have.

The cosplay community has used the new media in a number of ways that can be understood through Hall's framework. They have used the new technological infrastructure to enhance their knowledge of the costumes they seek to re-create and to share what they've learned. They have used their control over new media to create and maintain online identities on personal Web sites and social media in order to share their costumes and create a network of communication not controlled by the producers. They have used their individual frameworks of knowledge to evaluate one another's work, not just to react to the original production. While each of these changes is fraught with its own complications for the cosplay community, taken together, they represent a significant change in the media environment in which cosplay occurs.

NOTES

1. Kanzhi Wang, "Cosplay in China: Popular Culture and Youth Community" (Master thesis, Lund University, 2010), 18, http://lup.lub.lu.se/student-papers/record/1698210/file/1698215.pdf.

2. Stuart Hall, "Encoding/decoding," in *Culture, Media, Language,* ed. Stuart Hall and D. Hobson (London: Hutchinson, 1980), 128–38.

3. Ibid.

4. Henry Jenkins, *Textual Poachers: Television Fans and Participatory Culture,* 2nd ed. (New York: Routledge, 2012).

5. Jonathan Gray, Cornel Sandvoss, and C. Lee Harrington, *Fandom: Identities and Communities in a Mediated World,* 1st ed. (New York: New York University Press, 2007).

6. Ibid.; Jenkins, *Textual Poachers.*

7. Jenkins, *Textual Poachers.*

8. Gray, Sandvoss, and Harrington, *Fandom.*

9. Jenkins, *Textual Poachers.*

10. Gray, Sandvoss, and Harrington, *Fandom.*

11. Jenkins, *Textual Poachers,* 26.

12. Jenkins, *Textual Poachers.*

13. Gray, Sandvoss, and Harrington, *Fandom.*

14. John L. Sullivan, *Media Audiences: Effects, Users, Institutions, and Power* (Thousand Oaks, CA: Sage, 2012).

15. Gray, Sandvoss, and Harrington, *Fandom,* 9.

16. Ibid., 8.

17. Victor Costello and Barbara Moore, "Cultural Outlaws: An Examination of Audience Activity and Online Television Fandom," *Television & New Media* 8, no. 2 (May 1, 2007): 127, doi:10.1177/1527476406299112.

18. Gray, Sandvoss, and Harrington, *Fandom*; Sullivan, *Media Audiences*; Henry Jenkins et al., *Spreadable Media: Creating Value and Meaning in a Networked Culture* (New York; London: New York University Press, 2013); Jenkins, *Textual Poachers*.

19. "Cosplay.com—The World's Largest Cosplay Community," accessed July 31, 2014, http://www.cosplay.com/.

20. "Home | Archive of Our Own," accessed July 31, 2014, https://archiveofourown.org/.

21. Anna M. Turri, Karen H. Smith, and Elyria Kemp, "Developing Affective Brand Commitment through Social Media," *Journal of Electronic Commerce Research* 14, no. 3 (2013): 201–14.

22. Ibid., 210.

23. Turri, Smith, and Kemp, "Developing Affective Brand Commitment through Social Media."

24. *Doctor Who: "Where's the TARDIS?"* Winner, accessed June 30, 2014, http://www.bbcamerica.com/doctor-who/videos/wheres-the-tardis-winner/.

25. Jenkins et al., *Spreadable Media*.

26. Jenkins, *Textual Poachers*; Henry Jenkins, *Fans, Bloggers, and Gamers: Media Consumers in a Digital Age* (New York: New York University Press, 2006).

27. Osmud Rahman, Liu Wing-sun, and Brittany Hei-man Cheung, "'Cosplay': Imaginative Self and Performing Identity," *Fashion Theory: The Journal of Dress, Body & Culture* 16, no. 3 (September 1, 2012): 317–42, doi: 10.2752/175174112X13340749707204; Marjorie Cohee Manifold, "What Art Educators Can Learn from the Fan-Based Artmaking of Adolescents and Young Adults," *Studies in Art Education* 50, no. 3 (2009): 257–71.

28. Rahman, Wing-sun, and Cheung, "'Cosplay.'"

29. Jayme Rebecca Taylor, "Convention Cosplay: Subversive Potential in Anime Fandom" (University of British Columbia, 2009), 17, https://circle.ubc.ca/bitstream/id/18022/ubc_2009_spring_taylor_jayme.pdf&page=3.

30. Henrik Bonnichsen, "Cosplay-Creating or Playing Identities?: An Analysis of the Role of Cosplay in the Minds of Its Fans," 2011, http://www.diva-portal.org/smash/record.jsf?pid=diva2:424833; Ashley Lotecki, "Cosplay Culture: The Development of Interactive and Living Art Through Play," 2012; Natasha Nesic, "No, Really: What Is Cosplay?" 2013, https://ida.mtholyoke.edu/xmlui/handle/10166/3217; Rahman, Wing-sun, and Cheung, "'Cosplay.'"

31. "Disney Cosplayers," Facebook Group, *Disney Cosplayers*, (September 23, 2014), https://www.facebook.com/login.php?next=https%3A%2F%2Fwww.facebook.com%2Fgroups%2Fdisneycosplayers%2F; "Game of Thrones Costuming," Facebook Group, *Game of Thrones Costuming*, (September 23, 2014), https://www.facebook.com/login.php?next=https%3A%2F%2Fwww.facebook.com%2Fgroups%2FGameofThronesCostuming%2F.

32. Hall, "Encoding/decoding."

33. "Yaya Han," accessed July 31, 2014, http://yayahan.com/; "Kamui Cosplay," *Kamui Cosplay*, accessed July 31, 2014, http://www.kamuicosplay.com/.

34. Taylor, "Convention Cosplay"; Lotecki, "Cosplay Culture."

35. Yaya Han, "Some Thoughts About Cosplay," Personal Blog, *Yayahan.com*, (February 24, 2014), http://www.yayahan.com/news-events/some-thoughts-about-cosplay.

36. Allison DeBlasio and Joey Marsocci, *1,000 Incredible Costume and Cosplay Ideas: A Showcase of Creative Characters from Anime, Manga, Video Games, Movies, Comics, and More* (Beverly, MA: Quarry Books, 2013).

37. *Castlevania: Hymn of Blood [Live Action Fan Series]—Castlevania: Hymn of Blood [Live-Action Fan Series]-Episode 1*, 2012, http://www.youtube.com/watch?v=voEKE G0Y3as&feature=youtube_gdata_player; *Doctor Who: AUFS "Fire and Ice"* (Trailer), 2009, http://www.youtube.com/watch?v=TUBCxJ6lRP8&feature=youtube_gdata_player; *Nightwing: The Series—Trailer (Fan Film)*, 2014, http://www.youtube.com/watch?v=EhHoUGITFEo&feature=youtube_gdata_player.

38. Hall, "Encoding/decoding."

39. *Doctor Who: AUFS "Fire and Ice" (Trailer)*, 2009, http://www.youtube.com/watch?v= TUBCxJ6lRP8&feature=youtube_gdata_player; *Nightwing: The Series—Trailer (Fan Film)*, 2014, http://www.youtube.com/watch?v=EhHoUGlTFEo&feature=youtube_gdata_player; *Castlevania: Hymn of Blood [Live Action Fan Series]—Castlevania: Hymn of Blood [Live-Action Fan Series]—Episode 1*, 2012, http://www.youtube.com/watch?v=voEKEG0Y3as& feature=youtube_gdata_player. Mitchell D. Strauss, "A Framework for Assessing Military Dress Authenticity in Civil War Reenacting," *Clothing and Textiles Research Journal* 19, no. 4 (September 1, 2001): 145–57, doi:10.1177/0887302X0101900401; Stephanie K. Decker, "Being Period: An Examination of Bridging Discourse in a Historical Reenactment Group," *Journal of Contemporary Ethnography* 39, no. 3 (June 1, 2010): 273–96, doi: 10.1177/ 0891241609341541.

40. "Game of Thrones Costuming."

41. "'Game of Thrones' Season 4 Finale Draws 7.1 Million Viewers," *NY Daily News*, accessed July 29, 2014, http://www.nydailynews.com/entertainment/tv/game-thrones-season-4-finale-draws-7-1m-viewers-article-1.1834464.

42. "Costume Embroidery & Illustration by Michele Carragher for Film & TV—About," accessed July 29, 2014, http://www.michelecarragherembroidery.com/About(2829561).htm.

43. "Costume Embroidery & Illustration by Michele Carragher for Film & TV—Sansa's Wedding Dress Gallery," accessed July 31, 2014, http://michelecarragherembroidery.com/ Sansas-Wedding-Dress-Gallery(2880131).htm.

44. "Game of Thrones Costuming."

45. Ibid.; "Costume Embroidery & Illustration by Michele Carragher for Film & TV— About."

46. Elizabeth Snead, "'Game of Thrones' Costume Designers Reveal the Secrets Stitched into the Actresses' Gowns—Hollywood Reporter," *The Hollywood Reporter*, accessed September 28, 2014, http://www.hollywoodreporter.com/news/game-thrones-costume-designers-reveal-705873; "Game of Thrones Costuming"; "I Costumi Di Game of Thrones—Vogue.it," accessed September 28, 2014, http://www.vogue.it/people-are-talking-about/vogue-arts/2014/ 06/game-of-thrones-costumi; "These Close-Ups Of 'Game Of Thrones' Fashion Will Take Your Breath Away," *BuzzFeed*, accessed September 28, 2014, http://www.buzzfeed.com/ donnad/these-close-ups-of-game-of-thrones-fashion-will-take-your-br; Geek Girl Diva on July 25 and 2013, "We Never Truly Realized How Gorgeous Game of Thrones Costumes Are Until Now," *Fashionably Geek*, accessed September 28, 2014, http://fashionablygeek.com/ handmade/we-never-truly-realized-how-gorgeous-game-of-thrones-costumes-are-until-now/.

47. Jenkins, *Textual Poachers*, XXVI.

48. "Marvel Drops the Lawsuit Hammer on Iron Man Cosplay Factory," *Lazygamer: The Worlds Best Video Game News*, accessed July 31, 2014, http://www.lazygamer.net/24/marvel-drops-the-lawsuit-hammer-on-iron-man-cosplay-factory/; Robo Panda, "Serenity Now: 'Firefly' Fans Making Jayne Hats Get Cease And Desist Orders (Plus A Cosplay Gallery)," *UPROXX*, April 11, 2013, http://www.uproxx.com/gammasquad/2013/04/firefly-jayne-hats-cosplay/.

49. "Cosplay.com," accessed October 14, 2014, http://www.cosplay.com/results.php?id= 13217580&sort=; "Browsing Photography on deviantART," accessed October 14, 2014, http:// www.deviantart.com/photography/?order=9&q=sherlock.

REFERENCES

Geek Girl Diva on July 2013. "We Never Truly Realized How Gorgeous Game of Thrones Costumes Are Until Now." *Fashionably Geek*. Accessed September 28, 2014. http:// fashionablygeek.com/handmade/we-never-truly-realized-how-gorgeous-game-of-thrones-costumes-are-until-now/.

Bonnichsen, Henrik. "Cosplay-Creating or Playing Identities?: An Analysis of the Role of Cosplay in the Minds of Its Fans," 2011. http://www.diva-portal.org/smash/record.jsf?pid= diva2:424833.

"Browsing Photography on deviantART." Accessed October 14, 2014. http://www.deviantart. com/photography/?order=9&q=sherlock.

Castlevania: Hymn of Blood [Live Action Fan Series]—Castlevania: Hymn of Blood [Live-Action Fan Series]—Episode 1, 2012. http://www.youtube.com/watch?v=voEKEG0Y3as& feature=youtube_gdata_player.

"Cosplay.com." Accessed October 14, 2014. http://www.cosplay.com/results.php?id= 13217580&sort=.

"Cosplay.com—The World's Largest Cosplay Community." Accessed July 31, 2014. http:// www.cosplay.com/.

Costello, Victor, and Barbara Moore. "Cultural Outlaws An Examination of Audience Activity and Online Television Fandom." *Television & New Media* 8, no. 2 (May 1, 2007): 124–43. doi: 10.1177/1527476406299112.

"Costume Embroidery & Illustration by Michele Carragher for Film & TV—About." Accessed July 29, 2014. http://www.michelecarragherembroidery.com/About(2829561).htm.

"Costume Embroidery & Illustration by Michele Carragher for Film & TV—Sansa's Wedding Dress Gallery." Accessed July 31, 2014. http://michelecarragherembroidery.com/Sansas-Wedding-Dress-Gallery(2880131).htm.

DeBlasio, Allison, and Joey Marsocci. *1,000 Incredible Costume and Cosplay Ideas: A Show-case of Creative Characters from Anime, Manga, Video Games, Movies, Comics, and More*. Beverly, MA: Quarry Books, 2013.

Decker, Stephanie K. "Being Period: An Examination of Bridging Discourse in a Historical Reenactment Group." *Journal of Contemporary Ethnography* 39, no. 3 (2010): 273–96. doi: 10.1177/0891241609341541.

"Disney Cosplayers." Facebook Group. *Disney Cosplayers*, September 23, 2014. https://www. facebook.com/login.php?next=https%3A%2F%2Fwww.facebook.com%2Fgroups%2Fdis neycosplayers%2F.

Doctor Who: AUFS "Fire and Ice" (Trailer), 2009. http://www.youtube.com/watch?v= TUBCxJ6lRP8&feature=youtube_gdata_player.

Doctor Who: "Where's the TARDIS?" Winner. Accessed June 30, 2014. http://www. bbcamerica.com/doctor-who/videos/wheres-the-tardis-winner/.

"Game of Thrones Costuming." Facebook Group. *Game of Thrones Costuming*, September 23, 2014. https://www.facebook.com/login.php?next=https%3A%2F%2Fwww.facebook.com% 2Fgroups%2FGameofThronesCostuming%2F.

"Game of Thrones' Season 4 Finale Draws 7.1 Million Viewers." *NY Daily News*. Accessed July 29, 2014. http://www.nydailynews.com/entertainment/tv/game-thrones-season-4-finale-draws-7-1m-viewers-article-1.1834464.

Gray, Jonathan, Cornel Sandvoss, and C. Lee Harrington. *Fandom: Identities and Communities in a Mediated World*. 1st ed. New York: New York University Press, 2007.

Hall, Stuart. "Encoding/decoding." In *Culture, Media, Language*, edited by Stuart Hall and D. Hobson, 128–38. London: Hutchinson, 1980.

Han, Yaya. "Some Thoughts About Cosplay." Personal Blog. *Yayahan.com*, February 24, 2014. http://www.yayahan.com/news-events/some-thoughts-about-cosplay.

"Home | Archive of Our Own." Accessed July 31, 2014. https://archiveofourown.org/.

"I Costumi Di Game of Thrones—Vogue.it." Accessed September 28, 2014. http://www. vogue.it/people-are-talking-about/vogue-arts/2014/06/game-of-thrones-costumi.

Jenkins, Henry. *Textual Poachers: Television Fans and Participatory Culture*. 2nd ed. New York: Routledge, 2012.

———. *Fans, Bloggers, and Gamers: Media Consumers in a Digital Age*. New York: New York University Press, 2006.

Jenkins, Henry, Sam Ford, and Joshua Green. *Spreadable Media: Creating Value and Meaning in a Networked Culture*. New York; London: New York University Press, 2013.

"Kamui Cosplay." *Kamui Cosplay*. Accessed July 31, 2014. http://www.kamuicosplay.com/.

Lotecki, Ashley. "Cosplay Culture: The Development of Interactive and Living Art through Play." (diss., Ryerson University, 2012).

Manifold, Marjorie Cohee. "What Art Educators Can Learn from the Fan-Based Artmaking of Adolescents and Young Adults." *Studies in Art Education* 50, no. 3 (2009): 257–71.

"Marvel Drops the Lawsuit Hammer on Iron Man Cosplay Factory." *Lazygamer: The Worlds Best Video Game News*. Accessed July 31, 2014. http://www.lazygamer.net/24/marvel-drops-the-lawsuit-hammer-on-iron-man-cosplay-factory/.

Nesic, Natasha. "No, Really: What Is Cosplay?" 2013. https://ida.mtholyoke.edu/xmlui/handle/10166/3217.

Nightwing: The Series—Trailer (Fan Film), 2014. http://www.youtube.com/watch?v=EhHoUGlTFEo&feature=youtube_gdata_player.

Panda, Robo. "Serenity Now: 'Firefly' Fans Making Jayne Hats Get Cease And Desist Orders (Plus A Cosplay Gallery)." *UPROXX*, April 11, 2013. http://www.uproxx.com/gammasquad/2013/04/firefly-jayne-hats-cosplay/.

Rahman, Osmud, Liu Wing-sun, and Brittany Hei-man Cheung. "'Cosplay': Imaginative Self and Performing Identity." *Fashion Theory: The Journal of Dress, Body & Culture* 16, no. 3 (September 1, 2012): 317–42. doi: 10.2752/175174112X13340749707204.

Snead, Elizabeth. "'Game of Thrones' Costume Designers Reveal the Secrets Stitched into the Actresses' Gowns—Hollywood Reporter." *The Hollywood Reporter*. Accessed September 28, 2014. http://www.hollywoodreporter.com/news/game-thrones-costume-designers-reveal-705873.

Strauss, Mitchell D. "A Framework for Assessing Military Dress Authenticity in Civil War Reenacting." *Clothing and Textiles Research Journal* 19, no. 4 (September 1, 2001): 145–57. doi: 10.1177/0887302X0101900401.

Sullivan, John L. *Media Audiences: Effects, Users, Institutions, and Power*. Thousand Oaks, CA: Sage, 2012.

Taylor, Jayme Rebecca. "Convention Cosplay: Subversive Potential in Anime Fandom." University of British Columbia, 2009. https://circle.ubc.ca/bitstream/id/18022/ubc_2009_spring_taylor_jayme.pdf&page=3.

"These Close-Ups Of 'Game Of Thrones' Fashion Will Take Your Breath Away." *BuzzFeed*. Accessed September 28, 2014. http://www.buzzfeed.com/donnad/these-close-ups-of-game-of-thrones-fashion-will-take-your-br.

Turri, Anna M., Karen H. Smith, and Elyria Kemp. "Developing Affective Brand Commitment through Social Media." *Journal of Electronic Commerce Research* 14, no. 3 (2013): 201–14.

Wang, Kanzhi. "Cosplay in China: Popular Culture and Youth Community." Master thesis, Lund University, 2010. http://lup.lub.lu.se/student-papers/record/1698210/file/1698215.pdf.

"Yaya Han." Accessed July 31, 2014. http://yayahan.com/.

Chapter Eighteen

Who Killed @TheLauraPalmer?

Twitter as a Performance Space for Twin Peaks *Fan Fiction*

Kathryn L. Lookadoo and Ted M. Dickinson

Fandoms have come a long way in recent years. Society has moved from the conceptualization of fans as the maligned nerds huddling around fan-fiction in the dark corner of the Internet to viewing fans as sources of both economic and creative influence.[1] Media producers and advertisers have begun to see the economic potential in these fans and how their loyalty as consumers can impact their media product and associated branding.[2] Additionally, the participatory nature of fandoms extends beyond purchasing power and demonstrates vast creativity. For instance, fans of the MTV *Teen Wolf* created the "Sterek Campaign" based on their love for two characters, Stiles and Derek, and raised and donated ten thousand dollars to Wolf Haven International.[3] Further, fans continue to breakdown online copyright protection through textual poaching (e.g., writing fan fiction or creating artwork related to a particular piece of media) and connect with other fans based on their mutual love for a piece of media.[4]

Contemporary scholarship has taken notice of the participatory culture of fans and is working to study the many facets of fandoms. In particular, researchers have highlighted what the role of social media platforms like Twitter play in the fan experience. Fandom studies have explored how people comment on their fandom on Twitter while consuming media in real-time,[5] the implications of developing transmedia narratives,[6] and how Twitter affects the traditional fan-celebrity relationship.[7]

The case study presented here focuses on how fans have used Twitter not only to associate with celebrities and other fans, but to continue the narratives of their favorite television shows. In particular, the 2014 *Enter the Lodge* project serves as an online production of fan-fiction for the 1990s television series *Twin Peaks*.

FAN-FICTION AND THE INTERNET

Fan-fiction is a story about characters written by fans of the source material (e.g., books like *Harry Potter* or *Twilight*, games like *Pokémon* or *Halo*, and television series like *Buffy the Vampire Slayer* or *Sherlock*).[8] Fan fiction is most often published on Web sites specifically dedicated to sharing this material such as fanfiction.net. Before social media arose, sites like this served as the primary outlet for fans to gather and creatively express their fan-fiction stories.

Since the early days of social networking sites, fans have used social media to engage in transmedia storytelling. For instance, fans created personal accounts for *Gilmore Girls* characters on MySpace.[9] Twitter offers another space for this type of storytelling. Wood and Baughman[10] note this opportunity in their case study in transmedia convergence in which they examined ten fan-operated *Glee* character role-play Twitter accounts. These accounts interacted with each other online to build upon the show's narrative with short, improvised conversations between characters. This type of story augmentation occurred right after an episode aired on television. These characters would tweet in a style similar to a script, complete with dialogue and descriptions of nonverbal actions marked by asterisks, like: "*Laughs and walks to the counter, paying for the two of them*."[11] The fans running the character accounts did not deviate from their characters' identities, and even communicated with other fans on Twitter as those characters. The Wood and Baughman study highlighted how fans' use of Twitter marks a new form of participatory culture and outlet for fan creativity.

Building upon this area of fan-operated character accounts, Bore and Hickman[12] provide a more in-depth look at the connections and relationships between fans that tweet as fictional characters, using *The West Wing* as an example. In a series of interviews with the people who portray the characters on Twitter, Bore, and Hickman[13] found that the fans operating the accounts did not know each other in real life and only on occasion planned out a Twitter conversation in advance. Additionally, through their study of *The West Wing* character accounts, they note this activity's shared characteristics with online fan activities such as fan fiction writing and fan message boards. However, like Wood and Baughman,[14] the authors cite the improvisation of tweets (defined as "improvised fan simulation") as a clear distinction between these Twitter activities and fan fiction. In improvised fan simulation, characters performed as "normal" Twitter users, meaning that they used

Twitter in a style similar to how a lay person does (e.g., presents thoughts or links, replies to other users, codifies tweets with hashtags).

PARASOCIAL INTERACTION

Scholars recognize the potential impact a media user-figure relationship can have on media enjoyment. Concepts like identification, wishful identification, and parasocial phenomena can affect a person's enjoyment of a piece of media.[15] A parasocial interaction (PSI) is a one-sided, emotional bond between a media user and media persona (e.g., character or celebrity).[16] This interaction can develop into a parasocial relationship (PSR) in which the user feels as if he or she "knows" the persona like a friend.[17] When a television show ends, viewers can experience a parasocial breakup (PSB) with a character or characters.[18] If a breakup occurs when there is a strong parasocial relationship with a character, the event can result in a media user experiencing effects similar to a real-life breakup such as depression, loneliness, and stress.[19]

Past research in this area has examined variables that predict anxiety in PSBs[20] and behavioral reactions to temporary PSBs.[21] One way people deal with their PSBs is to make attempts to maintain their PSRs with the associated media figure. For instance, Sanderson[22] found that people who experienced a parasocial breakup with singers from the boy band New Kids on the Block tried to continue their parasocial relationship by re-listening to the group's music or by thinking about the group often. Although these strategies are specific to nonfictional celebrities, media consumers may use similar strategies like watching reruns of shows, reading fan fiction, or following characters on Twitter to continue their PSR with fictional characters.[23]

Fan-operated character Twitter accounts offer a unique way for media users to reconcile their PSBs and continue their parasocial relationships with characters. First, as Bore and Hickman[24] note, people running these accounts stay in character. This consistency and predictability is a key gratification related to parasocial relationships[25] and allows fans to continue to maintain their PSRs. A part of staying in character is making references to past events that happened in the series or building upon pre-established character traits. Through these references, the account runners may build upon PSRs by rewarding users for their extensive knowledge of the character and thus enhancing the "intimacy" they already feel with the character.[26] Second, like fan fiction, these accounts provide new content for media users. Third, these Twitter accounts offer a way for fans to interact with the character. This type of interaction extends beyond the traditional conception of PSR as "one-sided" that assumes an audience member will never interact with a media figure.

ENTER THE LODGE: A CASE STUDY

The present case study follows this line of research by studying fan-operated character Twitter accounts from the *Enter the Lodge* project, a *Twin Peaks*-based fan-fiction story. However, this fan-fiction project differs from the way other fandoms use Twitter. First, the *Enter the Lodge* project is comprised primarily of scripted tweets, a style that deviates from the improvised fan simulation observed in the prior studies. Second, only two people generated content for the numerous Twitter accounts associated with the project. Conversely, Bore and Hickman[27] found that fans operating the Twitter accounts did not know each other and usually only ran one account each. Third, the format of the *Enter the Lodge* tweets differ from other Twitter fandoms because the story does not read as if the characters are actually tweeting (as in Bore and Hickman's *The West Wing* study)[28] or contain detailed nonverbal actions (as in Wood and Baughman's *Glee* study).[29] Fourth, the *Twin Peaks* canon is unique as the series ceased production over twenty years ago, whereas *Glee* fans tweet right after new episodes still in production air and *The West Wing* was cancelled more recently, in 2006.

About the Series

The pilot episode of *Twin Peaks* debuted on the ABC network on April 8, 1990. The series takes place over approximately one month in 1989, and begins with the discovery of the body of Laura Palmer, the homecoming queen of the titular town in northeastern Washington. After local police note similarities to a previous murder (depicted in the prequel film *Twin Peaks: Fire Walk with Me*), the FBI sends Special Agent Dale Cooper to investigate. Cooper serves as the show's most clearly identifiable protagonist, assisted in his investigation by local law enforcement and other FBI agents. While the investigation is a primary catalyst for the events of the series, creators David Lynch and Mark Frost intended the real focus of the show to be the intertwined lives of the town's other inhabitants.[30] The pilot garnered ABC the highest ratings of the season for a two-hour television movie.[31] The series continued to receive critical acclaim in its first season. The final episode of the season, in which Cooper is shot by an unknown protagonist, achieved a 22 percent audience share, enough for ABC to order a second season.[32]

Of greater effect on the series than an attack on its protagonist, Lynch and Frost were faced with substantial pressure from ABC to reveal the identity of Laura's murderer in season two. In the seventh episode of the season, the identity of Laura's killer was revealed to viewers, with Cooper solving the case two episodes later. Lynch and Frost had not planned for the series after the resolution of Laura's murder; without an active investigation, there would be no justification for keeping Cooper in Twin Peaks. He had increasingly become the show's figurehead in promotional material and elsewhere in popular culture, including a spoof of the series on *Saturday Night Live* with Kyle MacLachlan taking a satirical slant on the role he had made famous.[33]

Lynch and Frost introduced numerous changes to the show to keep the story (and Cooper) afloat. Most of the new characters added to the series had direct relations with Cooper, who was now under investigation for crossing the Canadian border on FBI business without authorization. Notable characters include a new love interest for Cooper portrayed by Heather Graham, Cooper's criminally insane former FBI partner, and a Drug Enforcement Administration agent with the ability to transcend gender boundaries portrayed by David Duchovny years before his breakout role on *The X-Files.*[34] Writers also increased the prominence of the show's supernatural elements, including various spirits possessing major characters, and local Native American legends of two "otherworldly" locations of great power, the White Lodge and the Black Lodge.

Casual fans of the series largely soured on the changes following the resolution of Laura's murder, resulting in a rapid drop-off in ratings and the cancellation of the series by ABC. The series' final two episodes aired on June 10, 1991, ending with a cliffhanger implicating that Cooper's soul was trapped in the Black Lodge, with an evil spirit ultimately responsible for Laura Palmer's murder now inhabiting his body. A year later, Lynch wrote and directed a follow-up theatrical release, depicting the final days of Laura's life and her murder. Whereas some questions about the events leading to her murder were answered, many of the show's loose ends regarding its other characters, including Cooper's fate, remained unresolved.

The series has enjoyed a continued fan following since its cancellation, both in the United States and abroad. Episodes have been rebroadcast on the Bravo cable network, and the show has been released repeatedly on VHS and DVD, with a Blu-Ray release in summer 2014. Japanese travel agencies regularly host trips to shooting locations in northwestern Washington, and annual fan festivals have sold out in recent years.[35] This fan support caused Lynch and Frost to consider returning to the story, announcing in October 2014 that premium cable network Showtime would air a nine-episode third season in 2016.

THE *ENTER THE LODGE* PROJECT

Brothers Emmett and Patrick Furey began developing the idea of extending the story of *Twin Peaks* in 2009, initially as a speculative script so Emmett could secure a job writing for a major television series. After deciding to develop the idea further, initially as an animated series, the brothers elected to use Twitter as the primary performance space for the narrative. Emmett stated that the brothers chose Twitter in part for allowing fans to develop parasocial interaction with the characters:

> Twitter allows us the opportunity to release our content in practically real time, which is something that the show wasn't even able to do, so that offers a lot of interesting new wrinkles for this kind of storytelling. And, of course, Twitter also allows people to interact with the characters and to insert themselves into the story.[36]

In developing *Enter the Lodge*, the Furey brothers understood that creating a fan-fiction narrative on Twitter allowed for fans to have interactions with the characters and become more involved in the story. Due to this awareness that PSIs were a possibility, the Fureys worked to create comprehensive Twitter accounts that reflected characters' personalities whether it was how they spoke, (e.g., Cole's capitalization style), what they discussed (e.g., Cooper's interest in Tibet), or who they followed (e.g., Bill Hayward following other fictional doctors). This depth provided to each character set-up an extensive web of information within the *Twin Peaks* world, thus constructing a way for fans to recover from their parasocial breakups and rekindle their parasocial relationships with the characters.

The first tweet in the story was posted on March 25, 2014, the twenty-fifth "anniversary" of the end of the original series' storyline, a deadline selected by the Fureys both for its symbolism and to force themselves to stop tinkering with the story and release it to the public.[37] Most tweets used to tell the story were scheduled via the SocialOomph platform and posted in "real-time" over the next four weeks; tweets directed to fans were posted using Twitter's Web client.

Method

To understand how Twitter serves as a performance space for fictional narratives, *Enter the Lodge* was selected as a case study because it was, to the best of the authors' knowledge, the only fully scripted fan-fiction story told through Twitter. While other fandoms have fake character accounts on Twitter, these accounts mostly tweet during television broadcasts of the related show or tweet mini-narratives (e.g., less than ten tweets) with other characters. Therefore, other fandom tweets are less orchestrated than *Enter the Lodge*.[38]

Twitter is a free microblogging social network that allows users to communicate in 140 characters or less. Twitter users can share posts ("tweets") which most often include text and but can also contain links to Web sites or pictures. Users can also tag other users in their tweets for easier communication. Aside from tweeting, users also personalize their Twitter profile through a unique username, profile picture, location, and brief biography.

To obtain the tweets related to *Enter the Lodge*, the researchers used Twitter's application programming interface (API) to collect all tweets from all accounts included on the "Twin Peaks Season 3" account list. In total, 5,176 tweets were collected in chronological order from sixty-one *Enter the Lodge* character accounts published on Twitter between March 25 and April 20, 2014. This time period marks the beginning and end of the first "season" of the story line.

Analysis

After analyzing the "third season" of *Twin Peaks*, it was apparent that the creators used specific narrative devices to develop the story on Twitter, made an active effort to include elements that appealed to dedicated fans, and interacted with fans as the story developed.

Constraints of Twitter

Enter the Lodge approaches the limitations of Twitter by constructing a narrative almost entirely through dialogue, while capitalizing on familiar conventions such as the 140-character limit, hashtags, and tagging other accounts (both for fans and for characters). With the understanding that the story line would mainly rely on dialogue, the creators began the "third season" by introducing fans to the new storytelling structure by recreating scenes on Twitter from the show's penultimate episode. This tweeting of familiar scenes allowed the creators to show fans how scenes would unfold on Twitter. For instance, the recreated scenes demonstrated the structures of monologues and conversations between characters, how scene changes worked, and the function of hashtags, in addition to other elements. These recreated scenes eased the transition into the Twitter narrative.

A distinguishing feature of Twitter is its constraint that tweets must be 140 characters or less, which only slightly changes the speaking patterns of characters. This change is noted in how characters deliver monologues and have conversations. In the Twitter story, a character delivers a monologue in a series of tweets. For example, early in the story, Sheriff Harry Truman uses nine tweets in a row to deliver a monologue. Each tweet contains one to two whole sentences and does not cut off mid-sentence, thus making the monologue easier to read and follow.

Conversations between two or more characters employ a different structure than the monologue as multiple characters are tweeting in the same scene. To convey who is involved in the conversation at the beginning of the scene, the characters tag who they are talking to with an "@" sign at the start of their tweet. For instance, a dialogue between Dale Cooper and Annie Blackburn begins as follows,

> Annie Blackburn: @DaleBCooperFBI Dale, there you are! Is it really you?

> Dale Cooper: @DoubleR_Annie Through and through, Annie. As sure as the sight of you brings a smile to my face.

After the first tweets in a conversation, the two characters drop the "@" symbol, which allows for the tweets to have more dialogue and makes the conversation easier to follow. With conversations between three or more characters, the "@" sign is used more frequently. The characters use the symbol to indicate who is in the scene and also note if characters are directing their lines to specific people in the scene. For example, when the Bookhouse Boys, a secret society of Twin Peaks citizens, hold a meeting the characters enter the scene at different times and participate in different conversations occurring in the scene. To lessen this confusion, the characters identify to whom they are speaking at the beginning of the conversation (see Harry Truman's tweet). Also, within the larger conversation, the speaker will also direct particular statements to certain participants (see Ed Hurley's tweet):

> Sheriff Harry Truman: @DepHawk @DaleBCooperFBI @BigEdHurley @BookhouseJames @Joey_Paulson Thanks for coming everyone. We've got a few things to discuss.

> Ed Hurley: @Sheriff_HTruman I'm guessing we're here to discuss the tragedy at the Savings & Loan? How is @Deputy_Andy holding up?

Enter the Lodge also uses tweets to communicate nonverbal elements. Specifically, characters indicate nonverbals through parentheses, capitalization, and hashtags (which will be discussed in a later section). Characters use parentheses as a way to demonstrate whispered or "under-their-breath" utterances, similar to this conversation between Bobby Briggs and Ben Horne:

> Bobby Briggs: @BenJHorne You trying to tell me you're seeing things, Mr. Horne?

> Ben Horne: I am indeed.

> Bobby Briggs: (Well, that's just peachy.)

> Ben Horne: Beg your pardon?

Bobby Briggs: Oh, nothing, sir.

Other characters in the scene further emphasize the nonverbal nature of the parentheses by responding as if they could not hear the statement. Only Agent Cooper seems to have the ability to hear quieter statements and respond to them.

Another form of nonverbal communication is the use of capitalization specific to the character Gordon Cole. In the original series, Cole had a hearing impairment which caused him both to frequently misunderstand what others were saying and shout when speaking. To mirror this character trait on Twitter, Cole tweets in all-caps:

Gordon Cole: @DaleBCooperFBI COOP, SO GLAD TO HEAR YOU'RE SAFE. @SHERIFF_HTRUMAN TOLD US WHAT HAPPENED IN THE FOREST, WE WERE WORRIED SICK

Hashtags are an ubiquitous part of the Twitter experience, and are employed throughout *Enter the Lodge* primarily to connect tweets regarding specific events (such as the #MsTwinPeaks beauty pageant) and locations. #TwinPeaks is consistently used in references to the town, as well as the two other-dimensional locations of power, the #WhiteLodge and #BlackLodge. #Ghostwood refers to a planned housing development that features in a recurring subplot in the original series, and a focal point of conflict in *Enter the Lodge* as the #StopGhostwood movement expands. The most notable departure from this convention comes in the use of a single hashtag as a replacement for a gesture commonly used by one character. Cooper frequently uses #ThumbsUp to show his approval of other characters.

One affordance of Twitter, aside from communication via text, is the presence of a thumbnail picture accompanying each account. The Fureys added pictures to every account created for *Enter the Lodge*. For characters who had existed in the original series, they used screen captures or promotional pictures from the original series run. For the new characters, the Fureys "cast" actors who were familiar to audiences in the early 1990s such as Keith David and Teri Hatcher, using images of the actors from that time.

Usage of Other Social Media and Web sites

The Fureys worked around some of Twitter's restrictions by incorporating other social networking sites into the story. There are 271 tweets from character accounts consisting of FourSquare check-ins at landmarks around Twin Peaks. The primary use of these tweets is to serve as the "establishing shot," the exterior shot of a building used in film and television to delineate the end of a scene while showing the location of the next scene. The first instance of such an establishing tweet is only sixteen tweets into the story as the action moves from the Double R Diner to the Great Northern Lodge.

The second use of FourSquare check-ins is to introduce additional characters into a scene in progress. As Twitter does not have the affordance of showing how a character enters an existing scene, such tweets are used to prevent the startling effect of having a character speak with no prior indication of their presence. An example of this use of FourSquare occurs about eight hours after the beginning of the story, when Cooper and Sheriff Truman arrive at the scene of the Miss Twin Peaks Pageant to secure the winner against a kidnapping threat issued earlier.

Pinterest, a "visual bookmarking" Web site, is also occasionally used by Annie Blackburn as she prepares for her wedding to Cooper. Three tweets from Annie's account link to photographs on Pinterest: a pillow set, a man in a suit (with the caption "Dale would look so handsome in this"), and a picture of a hairstyle Annie is considering for her wedding. This use of Pinterest aligns with Annie's character as Pinterest is a popular Web site among young women. In a recent survey, 70 percent of brides-to-be reported that they used Pinterest to plan their wedding prior to getting engaged.[39]

Outside of social media, the official *Enter the Lodge* Web site (http://www.enterthelodge.com) stores several supplemental documents further fleshing out the story line. These documents include transcriptions of therapy sessions by the town psychiatrist, legal documents, and news reports surrounding the mayoral election that becomes a significant plot point in the second half of the story. Several of the documents even "fast forward" past the timeline of the main story to hint at the futures of some characters, including the return of a character believed to be dead in the story, as well as the ultimate fate of Cooper.

References to Popular Culture

True to the form of the original series, there are very few explicit references to popular culture in the story of *Enter the Lodge*. The only formal reference in the story to other television shows can be seen in a throwaway mention of *Invitation to Love*, a fictitious soap opera frequently seen in the background of the first season of *Twin Peaks*.

However, outside of the formal dialogue that builds the story of *Enter the Lodge*, activity on the character accounts shows a sly awareness both of other fictional universes, and of the real world. After the revelation that the man who raised Donna Hayward is not her biological father, her real father tweets "Donna, did I ever tell you how I met your mother? #HIMYM" a direct reference to the CBS television series *How I Met Your Mother*. This reference is itself an anachronism given that the events of Enter the Lodge take place in 1989. Also, character accounts are following accounts for characters from other fictions: @BillHaywardMD, the account for town doctor Bill Hayward, follows accounts from fictional doctors on *Grey's Anatomy*, *House*, and *Rizzoli & Isles*. Two character accounts follow @WalterBish0p, an account for Dr. Walter Bishop from the Fox television series *Fringe*; the account appears to have been created by the Fureys solely for this purpose.

In several cases, character accounts can be seen interacting with Twitter accounts of actors from the original series. One tweet from @MrDaleBCooper on April 3 addresses both @KatieShow (the official account for Katie Couric's talk show) and @KyleMaclachlan in advance of Maclachlan's appearance on the show. @DoubleR_Annie, the account for Cooper's love interest Annie Blackburn, favorited a tweet by Heather Graham, the actress who portrayed Annie, expressing excitement over an upcoming directing project. No less than three character accounts follow @therealraywise, the Twitter account for the actor who played Laura Palmer's father. Notably, there is only one instance of a character account following that character's original actor: @DoubleR_Shelly following @auntwendythecat, Mädchen Amick's account.

The character with perhaps the strongest "following" ties to the real world is @al_rosenfeld: Albert Rosenfeld, a forensic analyst with the FBI. His account follows several university forensic science department accounts, several official accounts related to the *CSI* television franchise, and a parody account for Dexter Morgan, the serial killer/forensic analyst featured on the Showtime series *Dexter*.

Character Interaction with Fans

While not appearing on the official timeline of tweets on the Enter the Lodge Web site, at least one instance exists of a conversation between one of the characters in *Enter the Lodge* and a fan. On March 28, 2014, the character of Leo Johnson, a formerly adulterous and abusive truck driver who has become a born-again Christian after nearly dying at the hands of Cooper's insane ex-partner, interacts with a fan quoting the Bible:

Leo Johnson: Mysterious ways indeed.

@SouthSideSlopes: @Le0Johnson NT says if your spouse was unfaithful, isn't a believer, & wants a divorce, you should let her go. 1st Cor. 7:15. Let go, Leo

Leo Johnson: @SouthSideSlopes Hmm . . . Wow, you're right, all of this is new to me. You've given me a lot to think about . . .

Three days later, @SouthSideSlopes tweeted about the experience, making a reference to an evil spirit familiar to fans of the show:

@SouthSideSlopes: @EnterTheLodge When @Le0Johnson tweeted me back, it was awesomely terrifying. My friends think I've let BOB into the neighborhood somehow.[40]

Shades of parasocial interaction can also be seen in a April 3, 2014 fan tweet:

@wrobson: Following all of @EnterTheLodge Season 3 #TwinPeaks characters on Twitter makes feel (sic) like I'm part of the town. Don't trust @DaleBCooperFBI[41]

FUTURE DIRECTIONS

Delivering fan-fiction in an interactive setting such as Twitter raises new questions on the nature of parasocial interactions. PSIs are based on the assumption that the audience member will never interact with the other member of the relationship. When fans can direct tweets to characters with whom they have a parasocial relationship, and receive a response, is the relationship still parasocial, or does it more closely resemble a real relationship or something in between? Additionally, if fans still deliver fan fiction through interactive settings after a show goes off the air, how does this continued interaction impact parasocial breakups?

Another avenue for future research hinges on whether and how audience members separate the characters from the creative teams behind them, especially when those teams change. In other words, do fans view @MrDaleB-Cooper as interpreted by the Furey brothers differently than they would the Special Agent Dale Cooper as written by David Lynch and portrayed by Kyle MacLachlan? Or is Cooper just Cooper, whether on television or on Twitter?

NOTES

1. Henry Jenkins,"Buying into American Idol: How We Are Being Sold on Reality Television," in *Introduction to New Media*, ed. Daniel Bernardi and Pauline H. Cheong (Boston: Pearson Learning Solutions, 2009), 174–95.

2. Jennifer Gillan, *Television and New Media: Must-Click TV* (New York: Routledge, 2006).

3. *"Teen Wolf* Fandom, the Sterek Campaign, Provides Food + Medical Costs for 48 Wolves," MTV, accessed July 30, 2014, http://act.mtv.com/posts/teen-wolf-the-sterek-campaign-adopts-wolves-wolf-haven-international/.

4. Karen Hellekson and Kristina Busse, *The Fan Fiction Studies Reader* (Iowa City: University of Iowa Press, 2014).

5. Tim Highfield, Stephen Harrington, and Axel Bruns, "Twitter as a Technology for Audiencing and Fandom: The #Eurovision Phenomenon," *Information, Communication & Society* 16, no. 3 (2013): 315–39, doi: 10.1080/1369118X.2012.756053.

6. Megan M. Wood and Linda Baughman. "*Glee* Fandom and Twitter: Something New, or More of the Same Old Thing?" *Communication Studies* 63, no. 3 (2012): 328–44, doi:10.1080/10510974.2012.674618.

7. Elizabeth Ellcessor. "Tweeting @feliciaday: Online Social Media, Convergence, and Subcultural Stardom." *Cinema Journal* 51, no. 2 (2012): 46–66, doi: 10.1353/cj.2012.0010; Jeffrey W. Kassing and James Sanderson. "Fan-Athlete Interaction and Twitter Tweeting Through the Giro: A Case Study." *International Journal of Sport Communication* 3, no. 1 (2012): 113–28.

8. Hellekson and Busse. *Fan Fiction Studies Reader.*

9. Paul Booth. "Rereading Fandom: MySpace Character Personas and Narrative Identification." *Critical Studies in Media Communication* 25, no. 5 (2008): 514–36.

10. Wood and Baughman. "*Glee* Fandom." 328–44.

11. Ibid., 335.

12. Inger-Lise K. Bore and Jonathan Hickman. "Continuing *The West Wing* in 140 Characters or Less: Improvised Simulation on Twitter." *Journal of Fandom Studies* 1, no. 2 (2013): 219–38, doi: 10.1386/jfs.1.2.219_1.

13. Inger-Lise K. Bore and Jonathan Hickman. "Studying Fan Activities on Twitter: Reflections on Methodological Issues Emerging from a Case Study on *The West Wing* Fandom." *First Monday* 18, no. 9.

14. Wood and Baughman. "*Glee* Fandom." 328–44.

15. Jonathan Cohen. "Mediated Relationships and Media Effects: Parasocial Interaction and Identification," in *The SAGE Handbook of Media Processes and Effects*, ed. Robin L. Nabi and Mary B. Oliver (Thousand Oaks, CA: Sage, 2009), 223–36.

16. Donald Horton and Richard R. Wohl. "Mass Communication and Parasocial Interaction." *Psychiatry* 19 (1956): 215–29.

17. Cohen. "Mediated Relationships and Media Effects: Parasocial Interaction and Identification." *The SAGE Handbook of Media Processes and Effects*, ed. Robin L. Nabi and Mary B. Oliver (Thousand Oaks, CA: Sage, 2009) 223–36.

18. Keren Eyal and Jonathan Cohen. "When Good *Friends* Say Goodbye: A Parasocial Breakup Study." *Journal of Broadcasting & Electronic Media* 50, 502–23.

19. Jonathan Cohen. "Parasocial Breakups: Measuring Individual Differences in Response to the Dissolution of Parasocial Relationships." *Mass Communication & Society* 6, no. 2 (2004): 187–202; Eyal and Cohen. "When Good *Friends* Say Goodbye." 502–23.

20. Eyal and Cohen. "When Good *Friends* Say Goodbye." 502–23.

21. Julie Lather and Emily Moyer-Gusé. "How Do We React When Our Favorite Characters Are Taken Away? An Examination of a Temporary Parasocial Breakup." *Mass Communication and Society* 14 (2011): 196–215, doi: 10.1080/15205431003668603.

22. James Sanderson. "You Are All Loved So Much: Exploring Relational Maintenance Within the Context of Parasocial Relationships." *Journal of Media Psychology* 21, no. 4 (2009) 171–82.

23. Lather and Moyer-Gusé. "How Do We React." 196–215.

24. Bore and Hickman. "Studying Fan Activities."

25. Horton and Wohl. "Mass Communication and Parasocial Interaction." 215–29.

26. Cohen. "Mediated Relationships."

27. Bore and Hickman. "Studying Fan Activities."

28. Ibid.

29. Wood and Baughman. "*Glee* Fandom." 328–44.

30. Graham Fuller. "A Town Like Malice: Maverick Director David Lynch Has Made a Bizarre Soap Opera for American Television." *The Independent*, November 24, 1989.

31. Susan Bickelhaupt. "*Twin Peaks* vs. *Cheers*." *Boston Globe*, April 12, 1990.

32. Bill Carter. "*Twin Peaks* Is Renewed on ABC." *New York Times*, May 22, 1990.

33. Shara Lorea Clark. "Peaks and Pop Culture." In *Fan Phenomena: Twin Peaks*, ed. Marisa C. Hayes and Franck Boulegue (United Kingdom: Intellect Books, 2013).

34. Brian Comfort. "Eccentricity and Masculinity in Twin Peaks." *Gender Forum: An Internet Journal for Gender Studies* 27.

35. "Twin Peaks Fest." retrieved August 1, 2014 from http://www.twinpeaksfest.com.

36. Twin Pie. "Exclusive: Behind *Enter The Lodge* with The Twin Peaks Twitter Fiction Creators," Welcome to Twin Peaks,http://welcometotwinpeaks.com/inspiration/enter-the-lodge-creators-interview/.
37. Ibid.
38. Wood and Baughman, "*Glee* Fandom," 328–44.
39. Taryn Hillin, "If You Had a Wedding Pinterest Board Before Getting Engaged, You're in the Majority," Huffington Post, retrieved August 1, 2014 from http://www.huffingtonpost.com/2014/07/25/wedding-survey_n_5618753.html.
40. Pie, "Exclusive."
41. Ibid.

REFERENCES

Bickelhaupt, Susan. "Twin Peaks vs. Cheers." *Boston Globe*, April 12, 1990.
Booth, Paul. "Rereading Fandom: MySpace Character Personas and Narrative Identification." Critical Studies in Media Communication 25, no. 5, 514–36.
Bore, Inger-Lise K. and Jonathan Hickman. "Continuing *The West Wing* in 140 Characters or Less: Improvised Simulation on Twitter." *Journal of Fandom Studies* 1, no. 2 (2013): 219–38, doi: 10.1386/jfs.1.2.219 1.
———. "Studying Fan Activities on Twitter: Reflections on Methodological Issues Emerging from a Case Study on The West Wing Fandom." *First Monday* 18, no. 9.
Carter, Bill. "Twin Peaks Is Renewed on ABC." *New York Times*, May 22, 1990.
Clark, Shara L. "Peaks and Pop Culture." In *Fan Phenomena: Twin Peaks*, edited by Marisa C. Hayes and Franck Boulegue. United Kingdom: Intellect Books, 2013.
Cohen, Jonathan. "Mediated Relationships and Media Effects: Parasocial Interaction and Identification." In *The SAGE Handbook of Media Processes and Effects*, edited by Robin L. Nabi and Mary B. Oliver, 223–36. Thousand Oaks, CA: Sage, 2009.
———. "Parasocial Breakups: Measuring Individual Differences in Response to the Dissolution of Parasocial Relationships." *Mass Communication & Society* 6, no. 2, (2004) 191–202.
Comfort, Brian. "Eccentricity and Masculinity in Twin Peaks." *Gender Forum: An Internet Journal for Gender Studies* 27.
Ellcessor, Elizabeth. "Tweeting @feliciaday: Online Social Media, Convergence, and Subcultural Stardom." *Cinema Journal* 51, no. 2 (2012): 46–66. doi: 10.1353/cj.2012.0010.
Eyal, Keren and Jonathan Cohen. "When Good Friends Say Goodbye: A Parasocial Breakup Study." *Journal of Broadcasting & Electronic Media* 50, 502–23.
Fuller, Graham. "A Town Like Malice: Maverick Director David Lynch Has Made a Bizarre Soap Opera for American Television." *The Independent*, November 24, 1989.
Gillan, Jennifer. *Television and New Media: Must-Click TV*. New York: Routledge, 2006.
Hellekson, Karen and Kristina Busse. *The Fan Fiction Studies Reader*, Iowa City: University of Iowa Press, 2014.
Highfield, Tim, Stephen Harrington, and Axel Bruns. "Twitter as a Technology for Audiencing and Fandom: The #Eurovision Phenomenon." *Information, Communication & Society* 16, no. 3 (2013): 315–39. doi: 10.1080/1369118X.2012.756053.
Hillin, Taryn. "If You Had a Wedding Pinterest Board Before Getting Engaged, You're in the Majority." *Huffington Post*, retrieved August 1, 2014 from http://www.huffingtonpost.com/2014/07/25/wedding-survey_n_5618753.html.
Horton, Donald and Richard R. Wohl. "Mass Communication and Parasocial Interaction." *Psychiatry* 19, 215–29, 1956.
Jenkins, Henry. "Buying into American Idol: How We Are Being Sold on Reality Television." In *Introduction to New Media*, edited by Daniel Bernardi and Pauline H. Cheong, 174–95. Boston: Pearson Learning Solutions, 2009.
Kassing, Jeffrey W. and James Sanderson. "Fan-Athlete Interaction and Twitter Tweeting Through the Giro: A Case Study." *International Journal of Sport Communication* 3, no. 1 (2012), 113–28.

Lather, Julie and Emily Moyer-Gusé. "How Do We React When Our Favorite Characters Are Taken Away? An Examination of a Temporary Parasocial Breakup." *Mass Communication and Society* 14, (2011) 196–215. doi: 10.1080/15205431003668603.

MTV. "Teen Wolf Fandom, the Sterek Campaign, Provides Food + Medical Costs for 48 Wolves." Retrieved July 30, 2014 from http://act.mtv.com/posts/teen-wolf-the-sterek-campaign-adopts-wolves-wolf-haven-international/.

Pie, Twin. "Exclusive: Behind 'Enter The Lodge' with the Twin Peaks Twitter Fiction Creators." *Welcome to Twin Peaks*, retrieved July 30, 2014, from http://welcometotwinpeaks.com/inspiration/enter-the-lodge-creators-interview/.

Sanderson, James. "You Are All Loved So Much: Exploring Relational Maintenance Within the Context of Parasocial Relationships." *Journal of Media Psychology* 21, no. 4, (2009) 171–82.

"Twin Peaks Fest." Retrieved August 1, 2014 from http://www.twinpeaksfest.com.

Wood, Megan M. and Linda Baughman. "Glee Fandom and Twitter: Something New, or More of the Same Old Thing?" *Communication Studies* 63, no. 3 (2012), 328–44. doi:10.1080/10510974.2012.674618.

Chapter Nineteen

Fifty Years of "The Man from U.N.C.L.E."

How the Ever-Changing Media Sustained and Shaped One of the Oldest Fan Communities

Cynthia W. Walker

FINDING U.N.C.L.E.

In 2007, Martin Fisher, an executive producer of DG Entertainment, was hired by Time/Warner to create a package of extras for a DVD release of the 1960s classic cult television series, *The Man from U.N.C.L.E.* At first, Warner Home Video had planned to release just the first season. If the series sold well, then other seasons might follow.

But soon, Time/Life, a direct marketer of books, music, DVD sets, and other multimedia products became involved. The company had enjoyed great success with *Get Smart: The Complete Collection*, previously released in the United States in November of 2006. The collection sold well and pleased *Get Smart* fans. Eventually, the collection won *Best of Show* and *Best 1960s Series* at the fourth annual *TV DVD Awards* sponsored by *Home Media Magazine* in cooperation with TVShowsOnDVD.com, DEG: The Digital Entertainment Group and *The Hollywood Reporter.*[1]

According to Jeff Peisch, Head of Video at Time/Life, an important element that contributed to the success of *Get Smart: The Complete Collection* was the involvement of the *Get Smart* fan community.[2] Not only did the boxed set feature the 138 episodes uncut and remastered, but also included footage from Don Adam's seventy-fifth birthday party and his later memorial service, along with several other features 　h.　　from material provided by the fans.

351

Peisch believed that *The Man from U.N.C.L.E.*, *Get Smart*'s predecessor and the series which *Get Smart* regularly referenced and spoofed, would be a likely candidate for a similar deluxe release. However, beyond the original 105 U.N.C.L.E. episodes, there was not much ancillary behind-the-scenes footage to be found.

Since Time/Life had enjoyed a good working relationship with *Get Smart* fans, Peisch encouraged Fisher to reach out to fans of *The Man from U.N.C.L.E.* That is, if he could find any. Fisher remembers googling the title and "1,000 sites came up. . . . It was really cool."[3] Not only had he succeeded in locating some fans, he had connected with the entire *U.N.C.L.E.* fan community.

THE MYSTIC CULT OF MILLIONS

In this age of "convergence culture" and participatory audiences,[4] when it seems that every popular culture property, both old and new, has some sort of online following, not surprisingly, even a series as old as *The Man from U.N.C.L.E.* still attracts interested, active, and even devoted, fans. After all, for a few years in the mid-1960s, *U.N.C.L.E.* was a phenomenally popular television show.

Produced by MGM and Arena Productions, *The Man from U.N.C.L.E.* premiered on NBC on September 22, 1964. Initially, the ratings were poor and the series was nearly cancelled early in its first season. A change in time period and cross-country promotional appearances by its stars, Robert Vaughn and David McCallum, combined with college students discovering the series while home for the Christmas holidays helped build a large, enthusiastic and notably youthful audience.[5]

Throughout its second season, comfortably settled in a 10 p.m. Friday timeslot, *The Man from U.N.C.L.E.* was a bona fide hit. The show consistently ranked in the top-twenty programs on U.S. television where it regularly earned a 44 percent share of the audience and beat every television series that CBS and ABC scheduled against it.[6] Eventually the show would be telecast in sixty countries, be nominated for a number of Emmys and other awards, and kick off what *Life* magazine dubbed "The Great TV Spy Scramble."[7] The show spawned a host of espionage-themed programs, both successful and unsuccessful, including, in the latter category, a sister series called *The Girl from U.N.C.L.E.*. Eight feature-length films were made from two-part episodes and profitably released in the United States and Europe.

U.N.C.L.E. was also one of the first television shows to be heavily merchandised, with images of the series' stars and its distinctive logo (a man standing beside a skeletal globe) appearing on hundreds of items, from bubble gum cards to a toy replica of its signature gun, the U.N.C.L.E. Special, to a line of adult clothing that included a trend-setting turtleneck sweater for men.[8] One article in *TV Guide* described David McCallum who played one of the lead actors, Russian agent Illya Kuryakin, as "the greatest thing since peanut butter and jelly."[9] Another article in the ubiquitous magazine declared that *U.N.C.L.E.* was "the mystic cult of millions."[10]

Executive producer Norman Felton, who originally conceived the series, and producer/writer Sam Rolfe who stepped in when Ian Fleming, the creator of James Bond, bowed out early in the development process, had hoped for success but were surprised at the size and intensity of the reaction, particularly from fans. The show received some 10,000 letters a week, more than MGM's all-time fan mail champion, Clark Gable.[11] "They are not just watching the program because they dislike the other programs that are on," Norman Felton reported to Mort Werner, senior vice president for programming and talent for NBC, "or because they just like it. They are watching because they are fans, fanatics if you will. They talk about the program with other fans and go beyond that: they proselytize, they want to convert non-viewers! ... The fans of the program are so enthusiastic because the program is so new, so unique. It offers them a new experience."[12]

Indeed, it did. Pitched and promoted as "James Bond for television," in the end, *The Man from U.N.C.L.E.* offered audiences something quite different than the Bondian formula of "sex, snobbery and sadism."[13] In various ways, both deliberately and accidentally, *U.N.C.L.E.* constructed a world that occasionally blurred the line between reality and fantasy, and then, almost like a modern multiplayer game, invited viewers to play along.

The earliest first season episodes began with a pseudo-documentary prologue in which viewers are given a tour of the headquarters of the fictitious United Network Command for Law and Enforcement (U.N.C.L.E.) in New York. Eventually, they meet Napoleon Solo, Illya Kuryakin, and their craggy spymaster boss, Alexander Waverly (played by Leo G. Carroll) who broke the fourth wall and introduced themselves. During the course of each mission—called "affairs"—the agents would recruit an "Innocent" to help them save the world. These were always average, everyday people—housewives, school teachers, stewardesses, secretaries, librarians, college students, tourists, even children—people very much like those sitting in U.N.C.L.E.'s viewing audience. Fans of other television shows like *Star Trek* needed to invent an "identity figure"—what would become known in media fandom as a "Mary Sue"[14]—to participate in the action aboard the Enterprise. In *U.N.C.L.E.*, this identity figure was canon. *U.N.C.L.E.* audiences could imagine themselves as Innocents, as agents, or as both and often did. MGM's publicity department printed up blue, silver, and gold cards that identified the bearer as an agent of U.N.C.L.E. During the summer of 1965, requests for the cards exceeded 200,000 in England alone.[15]

The fantasy world of *U.N.C.L.E.* bumped up against mundane reality in other ways as well. Sprawling, futuristic office complexes with shiny gunmetal corridors were located behind ordinary tailor shops all over the globe. All one had to do to enter was to locate and pull the secret coat hook. The bad guys, usually represented by the megalomaniac organization called Thrush, were lurking everywhere, hidden within normal-looking establishments like a vacuum cleaner repair shop, a funeral home, an art gallery, a car wash, a suburban drug store, a college town book shop, and the thirteenth floor of a Manhattan high rise. To combat the villains, U.N.C.L.E. agents carried advanced weapons and communication technology disguised as everyday items such as wristwatches and money clips, cigarette cases, and pens.[16]

And finally, at the end of every episode after the villains were vanquished and peace was restored, the end credits included the words: "We wish to thank the United Network Command for Law and Enforcement without whose assistance this program would not be possible." This tongue-in-cheek acknowledgment, created as a way to avoid confusion with the real United Nations, hinted that perhaps, just perhaps, the fictional organization might exist after all. When visitors to the United Nations requested a peek at U.N.C.L.E. headquarters, the amused guides often played along.[17]

Children, teens, and even adults wrote to the United Nations and to the Federal Bureau of Investigation (FBI) asking for information on how to become agents. From the latter, in return, they received an information book on choosing law enforcement as a career accompanied by a letter signed by J. Edgar Hoover informing the recipient that U.N.C.L.E. was an entirely fictitious organization.[18] When they began to write to NBC and MGM for information on how to join U.N.C.L.E., it was obviously time to establish an official fan club.

In 1965, *U.N.C.L.E.*'s *Inner Circle* was created. A fan who sent a letter to the series might receive, along with an autographed color photo of Robert Vaughn or David McCallum, an invitation to become a member. One invitation that survives from 1965 is a small slip of blue paper that bears the warning, "Classified Material—Destroy If Nervous."[19] Membership dues were $2 per year, to be sent to "Central Headquarters and Squadron Leader Mark Whitsett, "a retired, mild-mannered machine-gunner recently appointed to head this super-secret arm of U.N.C.L.E." In return, fans received a welcome letter from the most likely fictitious Whitsett [his true identity has never been discovered] and a membership card, followed by irregular mailings of an inexpensively produced offset *Inner Circle Journal* that included behind the scenes photos, "dossiers" on the stars, "classified bulletins" of upcoming episodes, specifications, and information on the U.N.C.L.E. Special, and an occasional contest to encourage fans "to help promote *The Man from U.N.C.L.E.* in your neighborhood." Members were also informed about any changes in *U.N.C.L.E.'s* schedule, an ongoing concern since the series occupied five time periods during its original run.[20]

One of these fans was a twenty-two-year-old science fiction author with just one commercially published book under his belt named David McDaniel. McDaniel was recruited by Ace editor Terry Carr, along with three other young writers, Buck Coulson, Gene DeWeese, and J. Hunter Holly, to write eight of the twenty-three U.N.C.L.E. tie-in novels that Ace ultimately published. These novels, probably the most popular in the paperback series, featured fannish in-jokes and allusions to fictional characters like Sherlock Holmes and *The Avengers'* John Steed, and real-life figures like Forrest J. Ackerman, the editor of the popular pulp magazine *Famous Monsters of Filmland*. These novels, along with a monthly fiction magazine published by Leo Margulies, also expanded the U.N.C.L.E. universe with added characters and more development of the U.N.C.L.E. and Thrush organizations.[21]

Fans also continued to write letters to the stars and producers connected with the series, particularly Norman Felton. He, in turn, often wrote back personally. Many of these letters are archived in the Special Collections Department of the University of Iowa Library in Iowa City, Iowa.

For three-and-a-half seasons, millions of baby boomers, both male and female, watched *U.N.C.L.E.*, talked *U.N.C.L.E.*; went to *U.N.C.L.E.* movies; watched for special appearances by the stars when they popped up in other television shows and feature films; played *U.N.C.L.E.* in the backyard, and listened to *U.N.C.L.E.* soundtrack records. They bought the toys, the games, the plastic guns, the Barbie-sized dolls, the clothes, the wristwatches, the trading cards, the lunchboxes, and the school book covers. When they ran out of *U.N.C.L.E.*-related merchandise to buy or couldn't save enough allowance to do so, they created their own homemade versions. They also read all the *U.N.C.L.E.* paperbacks, monthly magazines and comics, along with the movie magazines that chronicled the lives of the stars. And, once again, when the fans ran out of material, they wrote their own stories as well. The girls moaned and screamed over McCallum/Illya; the boys tried to imitate the cool suavity of Vaughn/Solo. Wherever the stars went, they were mobbed by crowds and treated like rock stars. Even the Beatles were fans.[22]

And then in the fall of 1968, suddenly and somewhat inexplicably, the bottom dropped out of *U.N.C.L.E.*'s ratings. Changes in Nielsen sampling practices, a breakdown in communication between the producers and the network, rising expenses, confused creative decision, personnel changes, and ultimately, disastrous scheduling decisions, all combined to produce a perfect storm. The ratings plummeted from a high of 55 percent share of the audience in April 1966 to a 21 percent share in October 1967. By January 1968 it was canceled, replaced by *Rowan and Martin's Laugh-In* (1968–1973).[23] In articles that followed *U.N.C.L.E.*'s demise, Vaughn observed that the spy craze had finally run its course. Felton was resigned but philosophical, pointing out that despite the poor Nielsen's overall, *U.N.C.L.E.* was still running strong in some of the larger cities. [24]

THE FANDOM SURVIVES

For their part, the fans were not so willing to give up on their favorite show and they were certainly not ready to move on. When MGM sold off many of the studio's famous props in a huge auction held in May, 1970 in Los Angeles, several fans including Robert Short and Dan Biederman showed up to purchase *U.N.C.L.E.* related items or, in some instances, rescue them from the trash. Today, fans own most of the key items that appeared in the series, including the *U.N.C.L.E.* car, the circular desk from Waverly's office, the coat hook from Del Floria's tailor shop, the communicators, the Thrush sniper scope rifles, and five of the six U.N.C.L.E. Specials. [25] Ace continued to publish *U.N.C.L.E.* paperbacks into the following year, although it declined to publish McDaniel's last *U.N.C.L.E.* novel, *The Final Affair*. That novel, along with J. Hunter Holly's *The Wolves and Lambs Affair* was printed as a fanzine in 1977 and circulated among fans. [26]

In addition, in the summer of 1970, Short, McDaniel, and two other fans, Bill Mills and Don Simpson, established *The Inner Circle II* to be followed by the *Inner Circle III*. Several nonfiction amateur publications began to circulate, aimed at a small coterie of *U.N.C.L.E.* fans. [27] Some *U.N.C.L.E.* fans had also crossed over into *Star Trek* fandom and *U.N.C.L.E.* short stories began to pop up in *Trek* and multimedia fanzines. [28] Finally, a fan-published *U.N.C.L.E.* novel, *The Blue Curtain Affair* appeared in 1975 and the author, Pat Munson, went on to establish the long-running *U.N.C.L.E. HQ* fan club the following year. [29] As more and more *U.N.C.L.E.* fanzines and newsletters appeared, both in the United States and now in the United Kingdom, *U.N.C.L.E.* fans began to find each other at *Trek* and science fiction-themed conventions. In 1983, Sue Cole, Munson's successor, organized Spy Con so that *U.N.C.L.E.* fans would have a con of their own. Spy Cons would continue to be held until 1998, with *U.N.C.L.E. HQ* publishing a regular newsletter until 2007.

What is rather amazing about this pre-Internet period of activity is that not only were *U.N.C.L.E.* fans separated by geographic distances from each other, but they were also separated from their favorite show. Unlike *Star Trek*, which was widely syndicated, [30] concerns about violence on television kept reruns of *The Man from U.N.C.L.E* few and far between. [31] Except for three of the eight theatrical features, which occasionally popped up on late night movie programs, *U.N.C.L.E.* was rarely available for viewing. Indeed, despite repeated attempts to revive it, in the forty-six years since the original series ended, Solo and Kuryakin have been united onscreen only once, in the

TV-movie *The Return of The Man from U.N.C.L.E.: The Fifteen Years Later Affair* which ran on CBS in April 1983.[32]

By comparison, *The Avengers*, which premiered on British television in 1961, returned as *The New Avengers* in the 1970s and as a 1998 theatrical film. *Doctor Who*, originally a 1963 educational program for children, has enjoyed several incarnations and is presently on its twelfth doctor. *Star Trek*, which first appeared two years after *U.N.C.L.E.*, has spawned six television series and twelve theatrical films, two of them based on the rebooted version of the original series. Even the aforementioned *Get Smart* has appeared as two theatrical movies, a TV-movie, and as a weekly series on three different networks.

As a result of *The Return* TV-movie, interest in *U.N.C.L.E.* began to revive again. Survival guides for budding and returning fans appeared in various genre specialty magazines.[33] Membership in U.N.C.L.E. HQ went from sixteen in 1978 to thirty in 1980 to one hundred in 1982.[34] Ironically, considering the problems with antiviolence activists in the past, when *U.N.C.L.E.* did finally return in reruns, it was on the cable Christian Broadcasting Network (CBN). Fans then taped what episodes they could and shared the bootlegged video tapes. In 1991, MGM/UA Home Video finally released twenty-two tapes containing forty-four episodes of the show, still less than half the original run of the series.

In his groundbreaking book, *Textual Poachers*, Henry Jenkins describes media fandoms in terms of five levels of activity: (1) a particular mode of reception that mixes emotional proximity and critical distance; (2) a set of critical and interpretive practices that include filling in gaps and preferred readings; (3) consumer activism; (4) cultural production; and finally (5) offering an alternative social community.[35] *U.N.C.L.E.* fandom functioned on all of these levels even before the series ended, and fans have contributed to nearly every *U.N.C.L.E.* branded project, both amateur and commercially produced, ever since.[36] Because of its dependence on audience devotion, determination, and activity as well as its history stretching back even before *Star Trek*, scholars writing about active, participatory audiences increasingly cite *The Man from U.N.C.L.E.* as the beginning, the prototype of media fandom.[37]

OPEN CHANNEL_D

Even the most dedicated fandom needs more than copiers, video tape recorders, and an occasional fan convention to survive and grow. Despite the ever-increasing number of fan works, particularly print fanzines like the long-running *Kuryakin File* which has produced thirty-three issues since 1985, *U.N.C.L.E.* remained a relatively small, boutique fandom. Then, in June

1995, a discussion mailing list called Channel _D was established on the computer of public librarian and longtime David McCallum fan, Ellen Druda. "Open Channel D" had been the famous line spoken into the communicators by the agents on the series whenever they wanted to contact U.N.C.L.E. headquarters in New York City.

There were ten people on the original list, recruited though chat rooms and through word of mouth spread at Media West Con, a multimedia fan writer's convention held every Memorial Day weekend in Lansing, Michigan. As moderator and list owner, Druda forwarded emails to each individual. When another fan, Linda Cornett, joined on July 4, 1995, her first post observed,

> Wow, this is service. Ellen mentioned in an AOL chat room Sunday evening that she might put together an *U.N.C.L.E.* list and ta-dah! Hi guys. Close your eyes. We are now sitting on the patio of the Holiday Inn in Lansing listening to the fountain and talking about our joint obsession. I, personally, am drinking Bailey's. We are talking about what? How about: any new news on the movie? . . . It's so nice to be with you all again.[38]

By mid-July, the number of members had doubled and another member, an IT professional named Rick Pavek, moved the list to a larger server which could distribute the posts by email to list members automatically. By March 1996, the group had grown to 115 members.

At that time, based on fanzine print runs and mailing lists, the size of *U.N.C.L.E.* fandom was estimated to be approximately 300 core fans, not all of them as yet online. The fandom was organized around several cluster points, including the aforementioned *U.N.C.L.E.* HQ, four newsletters (one from the United Kingdom), several regularly published fanzines, and the online Channel_D listserv. *U.N.C.L.E.* fans were connected to each other usually by two or more of these cluster points.[39]

A survey available through the email list and by traditional mail was conducted during three weeks in late February/early March of 1996. Questions asked respondents to rate their enthusiasm for *The Man from U.N.C.L.E.*, to name their favorite character and the reason they liked the show, and to count how many of their friends were also *U.N.C.L.E.* fans. They were also asked if they had ever engaged in twenty-eight separate fan activities in connection with *U.N.C.L.E.* A convenience sample of ninety-six responses—about one third of the core fandom—were then analyzed.[40]

Results from the survey found that ages ranged from teenage to the early sixties, with 80 percent of the fandom in their thirties and forties. The average age was thirty-nine. Seventy-two percent of fans had first become acquainted with *U.N.C.L.E.* during the original run of the series. Ninety-three percent rated their enthusiasm at a "four" or "five," while 40 percent said that what they liked most about the series was the characters.

In terms of gender, 73 percent of the active fandom was female, 27 percent was male. Further analysis revealed that, at least as far as *U.N.C.L.E.*, men and women had somewhat different fan experiences. The biggest difference was in their favorite characters. Forty-two percent of the fans who favored Napoleon Solo were male while 73 percent of the fans who favored Solo's Russian partner, Illya Kuryakin, were female.

There were other significant differences as well. While female fans liked the series for the characters, male fans pointed to the overall concept of the U.N.C.L.E. organization, along with the "cool" 1960s style and stories. Compared to male fans, female fans had nearly three times as many friends in the fandom, an average of eight compared to three. Although the level of enthusiasm was the same, female fans were more active and participated in different activities than men. Both male and female fans wanted to ultimately "possess" *U.N.C.L.E.* in some way. However, male fans did so by collecting information and memorabilia connected to the series. Female fans possessed the show by appropriating, reshaping, and reinterpreting—that is, "poaching" in Jenkins' terminology[41]—the source text, principally through writing fan fiction.

IN FROM THE COLD: *U.N.C.L.E.* FANDOM AS THIRD PLACE

In his 1989 book, *The Great Good Place*, Ray Oldenburg famously described what he called the "third place." It was not a work space and it wasn't home, except perhaps as a kind of home away from home. Third places were characterized as neutral, accessible and accommodating spaces where the mood was playful, there was a core community of "regulars" and the main activity was conversation.[42] A few years later, writing about one of the first virtual communities, The WELL, Howard Rheingold employed Oldenburg's conceptualization to explain the appeal of hanging out online.[43]

In various ways, the idea of a third place—Jenkins calls it the "weekend only world"—has also been used to characterize media fandom. That's not surprising because, as Gregory Benford predicted in his *Reason* article entitled "Alt.Fans,"[44] fandom would anticipate the activities, relationships, practices, and social aggregations that would prove so popular in cyberspace. Indeed, fandom and the Internet seemed made for one another.

Although the *U.N.C.L.E.* community arrived relatively early on the Web, *U.N.C.L.E.* fans were far from pioneers. Science fiction fandom was first, establishing an early discussion list on the old Arpanet in the late 1970s, and then transferring to the Unix Users Network or USENET, which had been originally established as a forum for computer support personnel to troubleshoot technology problems. Media fandom followed close behind. The rec. and alt. hierarchies became popular virtual meeting places for fans of various televisions series and films. But media fans, the vast majority of whom were women, often found themselves uncomfortable or unwelcome on USENET groups and bulletin boards where the general atmosphere was often aggressive, confrontational, and largely male.[45] When commercial services like General Electric's GEnie, Compuserve and Prodigy came online, the media fans spread out, establishing more private, members-only discussion groups with moderators and rules such as no flaming allowed. The first online mailing list for fans was established in 1992 for Forever Knight, a series about a vampire detective that ran from 1992 to 1996.[46]

As with other fan communities of the time period that were also reaching into cyberspace, what was happening virtually in *U.N.C.L.E.* fandom did not replace offline activity but largely promoted and extended it.[47] Some subscribers found out about the listserv through friends and fan connections, while others simply stumbled in after surfing the Internet. Academics, librarians, nurses, programmers, journalists, corporation managers, even a Canadian ex-spy subscribed, coming from the United States and Canada, but also from the United Kingdom, Europe, Australia, and eventually Japan. What they all had in common was access to a computer and for many, the sudden realization that they were not alone. A typical first "newbie" post read "My God! You people are here! I'm not the only one!" typed in large capital letters with lots of emoticons. The answering response on the listserv became something of a ritual. After being asked to introduce him or herself, the new "agent" was welcomed "in from the cold" by the other subscribers.[48]

The community continued to expand, increasing not only in the number of members, but in both interaction and creativity. During 1999, a year after the list moved to the Onelist site, over 7,000 messages were posted. In 2000, Onelist became Egroups and a year after that, Egroups finally became Yahoogroups. Despite these potentially disruptive migrations, until 2005, the annual total number of posts to the list averaged well over 5,000.[49]

Episode-related discussions about the plots, the characters, and the technology of the series were common, as well as debates over Hollywood's seemingly endless attempts at reviving *The Man from U.N.C.L.E.* Occasionally, conversations veered off-topic and became personal, as they did on September 11, 2001, when list members from across the United States and around the world updated each other on their experiences during the events of the day.[50]

Notably, the online community became an important place for announcements, reviews, advice, and support connected to fan-related activity. A second wave of fanzine writers were recruited and more amateur *U.N.C.L.E.* novels and short story collections from amateur presses were published. New Web sites and story archives were established, including the *Fans from U.N.C.L.E.*, *File 40*, and *David McCallum Fans Online*, all still in existence.[51] Attendance and activity at fan conventions also spiked, particularly at multi-fan cons like Eclecticon, held in a hotel near Newark Liberty Airport in New Jersey, and the aforementioned Media West Con in Lansing, Michigan. For example, during one four day stretch at the 2004 Media West Con, the *U.N.C.L.E.* fan community organized three panels and two parties, a play, a game, an *U.N.C.L.E.*-themed auction, shared several official dinners together, and submitted over a dozen fanzines in four categories for the con's annual Fan Q awards. There was also a suite filled with apparel and handmade merchandise emblazoned with the U.N.C.L.E. insignia to be shared and sold at cost. Well over fifty fans attended; con reports continued to be posted for days.[52]

FRAGMENTATION, MIGRATION, AND DIFFUSION

Online communities evolve through stages and the Channel_D listserv is no exception. Theorists have offered various life cycle models that usually incorporate some forms of inception, maturity, and death,[53] but Richard Millington introduced another phase after maturity: mitosis. That is, when a virtual community becomes mature and self-sustaining, it often begins to divide, giving birth to smaller sub-communities.[54]

Such fragmentation began even before the Channel_D list made the first move to Onelist. First, a sub-list, Channel_M, was established for David McCallum/Illya Kuryakin fans. Then, other sublists began to appear, devoted to Robert Vaughn/Napoleon Solo fans, to fan writers, to fan artists, to those fans who enjoyed slash stories,[55] even to off-topic discussions. Eventually, there were over a dozen separate lists.

New Web sites and archives, usually focused on one particular aspect of *The Man from U.N.C.L.E.*, continued to appear. Some offered episode guides, reviews or other nonfiction material, while others presented the fictional work of individual writers.

Still, the *U.N.C.L.E.* fandom remained fairly centralized until a group of younger fans created the MUNCLE Slash Community on a new social media site called LiveJournal in 2002. As a blogging community in which members set up and maintained their own individual journals, LiveJournal offered fans a new kind of experience. One could join a community and be part of a group, but at the same time, still maintain a personal space of one's own. Like Channel_D, MUNCLE led to more *U.N.C.L.E.*–related LiveJournal communities. Currently, there are twenty-two.

Some fans new to *U.N.C.L.E.* were recruited through LJ. Others migrated from the Yahoogroups list and found LiveJournal easier to navigate and manage with more individual privacy. "[On the lists] following conversations with people was too complicated," observed one fan. "And either you got too many e-mails daily, or, if you opted for the daily digest, you got one big e-mail where everything was jumbled together." Said another, "With listservs, even professional ones, my experience hasn't always been positive. Most of them suffer from flame wars and repetitive re-hashing of the same topics. There were also problems with members forwarding other members' posts to non-members without permission."[56]

Those fans who preferred LiveJournal agreed that interacting through linked blogs was "friendlier," and "more welcoming," encouraged more creativity and could accommodate graphics and various lengths of fan fiction. "I like LJ over Yahoogroups because it seems more personal," commented a fan who has had a personal LJ for six years.

> It's easier to post images, be playful. It seems more homey. Maybe it's the icons? And the format which makes it so easy to see everyone's responses at once. I love how threads start and at one glance it looks like a room full of friends having different side conversations, and coming back to the main conversation, or taking it off in a different direction. [57]

As *U.N.C.L.E.*-related LiveJournal communities began to proliferate and the social media site became more popular, particularly with fanfic writers and readers, Channel_W, the Yahoogroups' listserv for writers, suffered a severe drop in activity. In 2006, there were 2,378 posts. The following year, just 1,553. The year after that, only 197. So far, for 2014, there have only been forty-five posts.

Other discussion lists on Yahoogroups suffered similar drops in activity if not membership. For example, although the total of Channel_D members remains over 900, the actual activity has decreased from an annual high of 8,301 posts in 2006, to less than one-third of that total in recent years. In the case of Channel_D, however, the major migration has not been to LiveJournal, but to an even newer social media site, Facebook.

Starting around 2007, *U.N.C.L.E.* fans began establishing both groups and pages devoted to the series and various aspects of the series. As of the summer of 2014, there were seven *U.N.C.L.E.* groups and sixteen pages. The largest group, *The Man from U.N.C.L.E. Fan Club*, had over 1,000 members. The post popular page for *The Man from U.N.C.L.E. TV Show* had almost 5,000 likes.

THE EIGHTEEN YEARS LATER AFFAIR

Ironically, when producer Martin Fisher was searching for *U.N.C.L.E.* fans, the community itself was beginning to decentralize and disperse into a wider variety of cyberspaces. Fortunately, the *Fans from U.N.C.L.E.* site remained a popular and recognizable clearing house, ranking high in search engines. Fisher recruited over a half dozen fans, some of whom had, over the years, become entertainment professionals themselves, to appear in the ten hours of extras and to consult on the selection and editing of material as well as the marketing of the DVD boxed set. Fans even made it possible for now-retired associate producer George Lehr to be located and interviewed on camera.[58]

Released in November of 2007 by Time/Life, *The Man from U.N.C.L.E.* DVD boxed set was a success, although not quite as much as the *Get Smart* project. A year later, at the fifth annual TV DVD awards, it won Best 1950s/1960s Series but missed out to *Seinfeld* for Best of Show.[59]

U.N.C.L.E. fans were thrilled about the DVD set. In another survey of the *U.N.C.L.E.* fan community conducted for three weeks during the summer of 2014, 71 percent of the respondents said they had bought the set and another 14 percent planned to. Ninety-six percent rated it a "four" or "five" with 65 percent saying that they "absolutely loved it."[60]

The results from this survey, only the second aimed at the entire fandom in eighteen years, revealed what had changed and what had remained the same. The community had tripled in size to an estimated 1,000 active members and once more, about one third of the fans responded.[61]

As in 1996, over 90 percent of the convenience sample was enthusiastic about the series. There was still a wide age range from teens to senior citizens, but the vast majority—over 80 percent—of the respondents were still baby boomers. The average age was now fifty-six. Women were still in the majority at 62 percent, but the proportion of male fans had increased to 36 percent. Nevertheless, the cohort of those who had viewed *U.N.C.L.E.* during its original network run (now called within the fandom, "first cousins") still made up over three-quarters of the community.

In some familiar areas, male and female fans still had different tastes, opinions, and experiences. Ninety-two percent of the female fans said what they liked best about *U.N.C.L.E.* was the characters with the women overwhelmingly preferring Illya Kuryakin to Napoleon Solo 69 percent to 8 percent. The majority of male fans, some 69 percent, were split between liking the concept of the U.N.C.L.E. organization and the style, and plots and gadgets in the series. They preferred Solo to Kuryakin 33 percent to 16 percent, but a new category—liking Solo and Kuryakin equally—was checked by 28 percent of the fans, both male and female.

Eighteen years later, male fans were still more likely to collect *U.N.C.L.E.*-related articles and memorabilia than female fans, while women were more likely to read and write fan fiction compared to men five to one.

Interacting for almost two decades on social media also had affected the fandom. The most glaring change was the number of face-to-face fan friends. Now, there was no difference between the genders and the average had dropped from eight friends to three. Notably, over one third of the respondents—37 percent—said they had no face-to-face fan friends at all. Only a very small minority, some 7 percent, belonged to an *U.N.C.L.E.*-related club not online. Organizing or attending a fan convention, panels, or face-to-face gatherings were in severe decline.

Another difference was related to what social media site male and female fans preferred. Women were more active online overall, belonging to an average of six *U.N.C.L.E.*-related social media groups, compared to the men's average of three. This is probably related to which site each gender prefers. On Yahoogroups and Facebook, fans are members of an average of two groups and active male fans can be found on these two social media sites. However, on LiveJournal, the fandom is overwhelmingly female, perhaps because LJ is the preferred site for writing fan fiction. Also, on LiveJournal, fans are members of an average of six groups, with some ranging as high as nineteen.

Despite the fact that the overall profile of the fandom seemed to have hardly changed in eighteen years, the respondents of 2014 were not the same individuals as the respondents of 1996. Only 35 percent of the current fans were part of the fandom before the founding of the Channel_D listserve in 1995. The rest of the fans had joined after the community went online.

This is significant because it indicates that without social media, the *U.N.C.L.E.* fan community might have not only remained small, but might not have survived over five decades at all. "The Internet has been a boon to smaller fandoms like MFU," remarked one fan. "It allows them to survive by making it easy to find and stay in touch with like-minded people around the world."[62]

Another change is the relationship between offline and online interaction. *U.N.C.L.E.* is no longer what Lazar, Tsao, and Preece once called a hybrid "physic-virtual" community.[63] It is primarily online. Fans, of course, still collect memorabilia, but the search for, and trading of rare items has become far less challenging. Currently, on Ebay alone, there are over 1,000 *U.N.C.L.E.* related offerings.

It has also become easier to distribute, locate, and read fan fiction. Indeed, online archives continue to grow. Of the major *U.N.C.L.E.* collections, File 40 hosts 650 stories; the Chrome and Gunmetal Madhouse hosts 634, and the MFU Fan Fiction Archive has 341 stories with 614 links to more. On the multifandom sites, there are 538 *U.N.C.L.E.* stories in the Archive of Our Own (AO3) and the largest fan fiction repository, Fanfiction.net, holds 2.3K of *U.N.C.L.E.*-labeled amateur fiction. In the television category, *The Man from U.N.C.L.E.* ranks ninety-sixth. Only two other shows, *Star Trek: The Original Series*, and *Doctor Who* are as old.

Print fanzines and nonfiction newsletters are pretty much gone and because these were significant cluster points, bringing writers, editors, and readers together in social groups, it is a significant change. Regular issues of the *U.N.C.L.E. HQ* newsletter ended in 2007. The longest-running, best-selling print fanzine in the fandom, *Kuryakin File*, was published in 2014 only as a PDF. Although large, Hollywood-friendly events like Comic Con continue to attract record crowds,[64] smaller, fan-run cons are nearly extinct as well. Spy Con ended in 1998, Eclecticon in 2006, and MediaWest Con will held its final gathering on the 2015 Memorial Day weekend. For the first decade that the *U.N.C.L.E.* community was online, the virtual supported, extended, and promoted the real, the concrete, and the face-to-face. Now, interaction offline has been largely replaced.

Charting the journey of the *U.N.C.L.E.* fandom community over five decades, one can see how the community has migrated and fragmented even as it has grown larger; how personal and communal connections have been both lost and made; how practices and customs have changed and evolved, and most of all, how the virtual has slowly and surely replaced print media and face-to-face interaction. The community itself has also become more visible and mainstream. Echoing Benford, Henry Jenkins observed that while Media Fandom has influenced the mainstream, creating "affinity spaces" and "niche groups" everywhere, the mainstream has returned the favor. "There is no typical media consumer against which the cultural otherness of the fan can be located, "writes Jenkins. "Perhaps we are all fans or perhaps none of us is."[65]

And yet, perhaps because of the age of most *U.N.C.L.E.* fans, some remnants of the old ways survive. For example, an intimate *U.N.C.L.E.* gathering that attracts some two dozen fans continues to meet regularly every spring in the town of Arundel in the United Kingdom.[66] And in the fall of 2014, in order to celebrate *The Man from U.N.C.L.E.*'s fiftieth anniversary, 100 fans gathered in Los Angeles for "The Golden Anniversary Affair," a weekend of events, including a tour of the MGM /SONY Studios lot, featuring a peek inside of Stage 10 where U.N.C.L.E HQ once stood.[67] Since they wrote their first letters to Norman Felton as children, some fans have kept in touch with the professionals who worked on *U.N.C.L.E.*, both onscreen and off. At the Los Angeles gathering, fans shared dinners and panels with some of these folks including cinematographer Fred Koenekamp, director Joe Sargent, and associate producer George Lehr. Original props from the series, now owned by fans, were also on display. Observed Robert Short, one of the organizers of the event who also owns the refurbished *U.N.C.L.E.* car:

> I don't think there is any way we could have done this event without social media. If this was pre 1996 we would have had to spend a ton of money on advertising in whatever fanzines were available and even then it would have been hit and miss. Our biggest response was on Facebook by creating a page about the event and then specifically promoting it on the individual *U.N.C.L.E.* fan Facebook pages. That netted us responses from Argentina, Scotland, Japan, Chile, and all points across the United States.[68]

THE MAN FROM U.N.C.L.E. 2.0

"The Golden Anniversary Affair" may prove to be a last hurrah for a fandom that has weathered five decades of cultural and technological change because a new big-budget, feature length film version of the source text—consider it *U.N.C.L.E. 2.0*—is on the horizon. Directed by Guy Ritchie, the film will be a prequel, moving the action to before the U.N.C.L.E. organization was established. The reboot will star Henry Cavill as Napoleon Solo, Armie Hammer as Illya Kuryakin, and Hugh Grant as a younger, middle-aged Waverly.

Needless to say, older fans have been watching the film's progress with intense interest and not a little trepidation.[69] Few of these fans are active as yet on Twitter, Tumblr, and Instagram[70] but younger fans are, particularly those devoted to Cavill who post production reports and photos taken on location.[71] Perhaps, because of this activity, Warner Brothers has moved the release date from January 16, 2015 to a more attractive late summer date, August 14, 2015. Interestingly, a recent invitation to a test screening for the film posted online by Nielsen Movie View stipulated that the viewer and guest should be between the ages of fifteen to forty-four, which would eliminate a substantial proportion of the current fandom.[72]

These older fans today often discuss how the original series impacted their earlier lives as well. "I am a better person for my exposure to the *U.N.C.L.E.* universe, undoubtedly," one fan responded to an open-ended question on the 2014 survey, "and it prepared me to withstand the latter part of the 1960s and other tough times to come." Observed another: "I think of MFU as pivotal to my maturation from teenager to young adult. It helped shape my personal definitions of 'service' and honor within a broader definition of character." And yet another agreed: "The show changed my life. And meeting with other fans brought me out of my shell."[73]

In the 2014 survey, 65 percent of the respondents remembered playing *U.N.C.L.E.* in the backyard and playgrounds as a child and 45 percent admitted dressing as one of the characters at some point in their lives.[74] In some sense, because of social media, they never had to stop.

"It's been a grand old adventure," wrote one longtime fan in the survey. "Sometimes I wonder what a sixty-four-year old woman is doing being involved with such silliness, but I think it keeps me younger than my parents were at my age. And for that I am grateful."[75]

Will the new film mean an end to this "grand adventure" for the older fans, or the beginning of another adventure for younger ones? Will these newly arriving fans interact with the same enthusiasm on future social media sites, or perhaps, play *U.N.C.L.E.* in virtual playgrounds, acting through avatars in multi-player games? Will there be a 100th anniversary celebration some day? Only time will tell.

NOTES

1. David Lambert, Muppets, "*Get Smart* Reign at TV DVD Awards," TV Shows on DVD.com, http://www.tvshowsondvd.com/news/Site-News-4th-Annual-TV-DVD-Awards/84 32, accessed July 20, 2014. Also see Carl Birkmeyer, "The DVD Releases: The Old Produce the Best DVD Set of the Year Trick," WouldYouBelieve.com, http://www.wouldyoubelieve.com/dvd.html, accessed July 20, 2014.

2. Personal interviews with Jeff Peisch, November 13, 2008 and October 27, 2011.

3. Personal interview with Martin Fisher, January 15, 2008.

4. Henry Jenkins. *Convergence Culture: Where Old and New Media Collide* (New York: New York University Press, 2006).

5. Cynthia W. Walker, *Work/Text: Investigating the Man from U.N.C.L.E.* (New York: Hampton Press, 2013), 235–63; also Jon Heitland, *The Man from U.N.C.L.E. Book: The Behind-the-Scenes Story of a Television Classic.* (New York: St. Martin's Press, 1987).

6. *The Man from U.N.C.L.E., NBC, 09/22/64-01/15/68. (circa 1968).* Nielsen ratings summary, Norman Felton Collection, University of Iowa Library, Iowa City, Iowa.

7. "Great TV Spy Scramble," *Life*, October 1, 1965, 118–20. See also Cynthia W. Walker, "Man from/Girl from U.N.C.L.E." in *Encyclopedia of television* (2nd ed.) edited by Horace Newcomb. (Chicago, IL: Fitzroy Dearborn, 2004), 1404–406. Available online at http://www.museum.tv/eotv/manfromun.htm.

8. For a full listing, see Brian Paquette and Paul Howley, *The Toys from U.N.C.L.E.: Memorabilia and Collectors Guide* (Worchester, MA: Entertainment Publishing: 1990).

9. "The Greatest Thing Since Peanut Butter and Jelly," *TV Guide*, April 17, 1965, 6–9.

10. Leslie Raddatz, "The Mystic Cult of Millions: The People from U.N.C.L.E." *TV Guide*, March 19, 1966, 15–18.

11. Anthony Enns, "The Fans from UNCLE: The Marketing and Reception of the Swinging '60s Spy Phenomenon." *Journal of Popular Film and Television*, 28, (2000): 124–32. Also mentioned in "Fandemonium." *The Man from U.N.C.L.E.: The Complete Series DVD*. Bonus Disc 1. Time Life/Warner Bros. Entertainment Inc.

12. Memo from producer Norman Felton to Mort Werner, senior vice president for programming and talent, February 5, 1965.

13. Cynthia W. Walker, "Spy Programs," in *The Encyclopedia of Television*, (2nd ed.) edited by Horace Newcomb, (Chicago, Illinois: Fitzroy Dearborn, 2004), 2181–85. The reference was first made by Paul Johnson, "Sex, Snobbery and Sadism," *New Statesman*, April 5, 1958.

14. Paula Smith invented the term, "Mary Sue." See Cynthia W. Walker, "A Conversation with Paula Smith," *Transformative Works and Cultures*, 6 (2011). Available at http://journal.transformativeworks.org/index.php/twc/ article/view/243/205.

15. Heitland, *The Man from U.N.C.L.E. Book*, 53.

16. For a discussion of the devices and their contribution to U.N.C.L.E.'s "Wainscot Fantasy" see Cynthia W. Walker, "The Future Just Beyond the Coat Gook: Technology, Politics and the Postmodern Sensibility in the Man from U.N.C.L.E." *Channeling the Future: Essays in Science Fiction and Fantasy Television*, edited by Lincoln Geraghty (Plymouth, UK: Scarecrow Press, 2009), 41–55.

17. "Our Man with Uncle." *Secretariat*, 11 (February 16, 1966), 13.

18. Ian Markham-Smith, "Why Hoover Was Gunning for the Man from U.N.C.L.E." *Daily Mail*, (September 1, 2000), 27. Also noted in Danny Biederman, *The Incredible World of Spy-Fi: Wild and Crazy Spy Gadgets, Props and Artifacts from TV and the Movies.* (San Francisco, CA: Chronicle Books, 2004), 139.

19. Walker, *Work/Text*, 222.

20. Walker, *Work/Text*, 222–23.

21. See P. T. Smith, "The Fan from U.N.C.L.E." *Daredevils*, 10 (1984), 33–39; Bradley Sinor, "Collecting Paperbacks: The Man from U.N.C.L.E." *Baby Boomer Collectibles*, May (1994), 54–56.

22. Robert Vaughn, personal communication, February 14, 1997. Vaughn also recalled the event in "Fandemonium," in *The Man from U.N.C.L.E. The Complete Series DVD*. Bonus Disc 1.

23. For the full story of U.N.C.L.E.'s problems and cancellation, see chapter 9 of Walker, *Work/Text*, 235–70. Heitland, *The Man from U.N.C.L.E. Book*, blames the turn to camp in third season, 188–92.

24. Bob Thomas, "No It Wasn't Thrush That Did UNCLE in," Syndicated column, *Associated Press Features*, (1967). Retrieved from the Norman Felton Collection, University of Iowa Library, Iowa City, IA. Also see Lloyd Schearer, "Life and Death of the Man from U.N.C.L.E." *Parade*, (1968, March 17), 16–19.

25. Cynthia W. Walker, "The Gun as Star and the U.N.C.L.E. Special," in *Bang Bang, Shoot Shoot! Essays on Guns and Popular Culture* (2nd ed.) edited by Murray Pomerance and John Sakeris, (NY: Simon and Shuster, 2000), 203–13. Also see Walker, "The Future Just Beyond the Coathook," and Biederman, *The Incredible World of Spy-Fi*.

26. Smith, "The Fan from U.N.C.L.E." 35.

27. Sue Cole,"The History of U.N.C.L.E. HQ: A Brief Look at How This Newsletter Has Evolved and Progressed over the Years," *U.N.C.L.E. HQ Newsletter*, 193/194, (1995): 5–7.

28. Joan Marie Verba, *Boldly Writing: A Trekker Fan and Zine History, 1967–1987*. (Minnetonka, MN: FTL Publications, 1996), 39; 48.

29. Paula Smith, (1995, October). "The U.N.C.L.E. Diaspora and the Great In-Gatherings," *Z.I.N.E.S.*, 9, (1995), 20–21.

30. Herbert F. Solow and Robert H. Justman, *Inside Star Trek: The Real Story* (New York: Pocket Books: 1996).

31. Heitland, *The Man from U.N.C.L.E. Book*, 194.

32. Fans also point to U.N.C.L.E. tribute episode of *The A-Team* (1983–1987), on which Robert Vaughn was a regular. On October 31, 1986, General Hunt Stockwell, the character played by Vaughn, is reunited with his old partner Igor Trigorin (McCallum), only now they are enemies.

33. Articles on U.N.C.L.E. and the fandom appeared in various science fiction and cult magazines. A good example is "Collecting on Man from U.N.C.L.E. or What to Do When the Shock Treatments Fail Affair," by Charlie Kirby in *Daredevils*, 11 (1984), 24–28.

34. Sue Cole, personal communication. Also see Smith, "The U.N.C.L.E. Diaspora," 20. Craig Henderson's "U.N.C.L.E. Archive" on the *For Your Eyes Only* Web site is a good source of fan information. Accessed on September 1, 2014, http://www.for-your-eyes-only.com/Site/UNCLE_archive_1.html.

35. Henry Jenkins, *Textual Poachers: Television Fans and Participatory Culture* (NY: Routledge, 1992/2013), 277–81.

36. The efforts of Danny Biederman and Robert Short to revive U.N.C.L.E. in the late 1970s–early 1980s eventually led to the *Return* movie. Those efforts have been described a number of times, most notably in Heitland, *The Man from U.N.C.L.E. Book*, 194–201. Longtime fan Jon Burlingame produced the series of U.N.C.L.E. soundtrack CDs in the early 2000s for which fans were recruited to write the liner notes. See Burlingame's personal Web site, http://www.jonburlingame.com/cds/. Also see Craig Henderson's U.N.C.L.E. Archives online at http://www.for-your-eyes-only.com/Site/UNCLE_archive_2.html.

37. Francesca Coppa, "A Brief History of Media Fandom," In *Fan Fiction and Fan Communities in the Age of the Internet*, edited by Karen Hellekson and Kristina Busse (Jefferson, NC: McFarland & Co., 2006), 41–59. Also see Lev Grossman, "The Boy Who Lived Forever," *Time* (2011, July 7). Available at http://www.time.com/time/arts/article/0,8599,2081784,00.html.

38. Channel_D archive on Yahoogroups.com.

39. See Camile Bacon-Smith, *Enterprising women: Televison Fandom and the Creation of Popular Myth* (Philadelphia: University of Pennsylvania press, 1992) about the importance of "cluster points" in the organization of fandom.

40. The survey was conducted from February 25, 1996 to March 15, 1996. Results were reported in Cynthia W. Walker, "Coming in from the Cold: Gender, the Internet and Man from U.N.C.L.E. Fandom," presented at the 5th Annual Console-ing Passions: Television, Video and Feminism, April 25–28, 1996, University of Wisconsin-Madison.

41. Henry Jenkins, *Textual Poachers*.

42. Ray Oldenburg, *The Great Good Place* (Cambridge, MA: Da Capo Press, 1999). Also see, Charles Soukup, "Computer-Mediated Communication as a Virtual Third Place: Building Oldenburg's Great Good Places on the World Wide Web," *New Media and Society*, 8 (2006), 421–40.

43. Harold Rheingold. *The Virtual Community: Homesteading on the Electronic Frontier* (Cambridge, MA: MIT Press, 2000), 8–11.

44. Gregory Benford, "Alt.fans," *Reason*, 27 (1996): 43–44. Retrieved September 13, 2014 at http://reason.com/archives/1996/01/01/altfans.

45. Susan Clerc, "Estrogen Brigades and 'Big Tits' threads: Media Fandom Online and Off," *The Cybercultures Reader* edited by David Bell & Barbara M. Kennedy (New York: Routledge, 2000) 218–22.

46. Coppa, "A Brief History of Media Fandom." Also, see Francesca Coppa, "Pop Culture, Fans and Social Media," *Social Media Handbook*, edited by Jeremy Hunsinger and Theresa Senft, (New York: Routledge, 2014), 76–92.

47. Clerc, "Estrogen Brigades," 217; David Bell, "Cybersubcultures," *An Introduction to Cybercultures*, edited by David Bell (New York: Routledge, 2001), 168–69.

48. Walker, "Coming in from the Cold."

49. Channel_D archive on Yahoogroups.com.

50. Posts on Channel_D appeared all day, about three an hour.

51. *Fans from U.N.C.L.E.* at http://www.manfromuncle.org/; *File Forty* at http://file40.net/index.html and *David McCallum Fans Online* at http://www.davidmccallumfansonline.com/. Accessed September 25, 2014.

52. Channel_D archive, Yahoogroups.com, May and June, 2004.

53. Alicia Iriberri and Gondy Leroy, "A Life-Cycle Perspective on Online Community Success," *ACM Computing Surveys*, 41 (2009), article 11. DOI 10.1145/1459352.1459356 http://doi.acm.org/10.1145/1459352.1459356.

54. Richard Millington, *Buzzing Communities*, (Feverbee: 2012), 24–39.

55. For the uninitiated, *Fanlore* defines "slash" as "a type of fanwork in which two (or more) characters of the same sex or gender are placed in a sexual or romantic situation with each other. Slash more commonly refers to male/male pairings, with femslash being used more often to refer to female/female scenarios." Accessed September 29, 2014 at http://fanlore.org/wiki/Slash.

56. Personal communications, September 16–20, 2014.

57. Personal communication, September 16, 2014.

58. Personal interview with Martin Fisher, January 15, 2008.

59. David Lambert, "Seinfeld' Takes Home Best of Show at the Fifth Annual TV DVD," *TV Shows on DVD.com*, Accessed July 20, 2014 at http://tvshowsondvd.com/news/Site-News-5th-Annual-TV-DVD-Awards/10706#ixzz3DdNAgJYF.

60. Results are from an online survey using Survey Monkey, that ran from July 13, 2014 to August 8, 2014. There were six pages containing fifty-four questions, four of which were open-ended.

61. 358 fans responded anonymously.

62. Anonymous quotes are from replies to the open-ended questions on the 2014 summer survey.

63. Jonathan Lazar, Ronald Tsao, and Jennifer Preece, "One Foot in Cyberspace and the Other on the Ground," *Webnet Journal*, (July–September, 1999), 49–57.

64. Kwame Opam, "Camping in Hell: The Endless Lines of Comic-Con 2014," *The Verge*, http://www.theverge.com/2014/7/26/5940113/camping-in-hell-the-endless-lines-of-comic-con-2014. Accessed July 26, 2014.

65. Henry Jenkins, "Afterword: The Future of Fandom," in *Fandom: Identities and Communities in a Mediated World*, edited by Jonathan Gray, Cornel Sandvoss and C. Lee Harrington (New York: New York University Press, 2007), 364.

66. Information about the Arundel convention can be found at http://uncleconvention.webs.com/. Accessed on September 29, 2014.

67. The Golden Anniversary Affair's Web site is http://thegoldenanniversaryaffair.weebly.com/. Accessed on September 29, 2014.

68. Personal communication from Robert Short.
69. Bill Koenig, "U.N.C.L.E. Movie Gets Another Test Screening," *The HMSS Weblog*, http://hmssweblog.wordpress.com/2014/08/23/u-n-c-l-e-movie-gets-another-test-screening/. Accessed August 23, 2014.
70. According to the results from the summer 2014 survey, 85 percent of the respondents said they never use Twitter, 80 percent never use Tumblr and 98 percent never use Instagram.
71. Bill Koenig, "Cavill, Hammer Return to U.K. for U.N.C.L.E., Fan Site Says," *The HMSS Weblog*, http://hmssweblog.wordpress.com/2014/08/03/cavill-hammer-return-to-u-k-for-u-n-c-l-e-fan-site-says/. Accessed August 3, 2014.
72. Bill Koenig, "U.N.C.L.E. Movie Gets Another Test Screening." The actual online invitation could still be accessed on September 29, 2014 at https://secure.mymovieview.com/Panel/Invitation.aspx?StudyNumber=US14084579&passid=349&Display=RSVPHere&ReturnTo.
73. Anonymous quotes are from replies to the open-ended questions on the 2014 summer survey.
74. To the question, "Did you play U.N.C.L.E. as a child?" 64.5 percent of the respondents said yes and 35.5 percent said no. To the question, "Have you ever dressed as an U.N.C.L.E. character?" 30.8 percent of respondents said yes, as a child; 6.9 percent said yes, as an adult, and 10.8 said yes, as both child and adult.
75. Anonymous quote are from replies to the open-ended questions on the 2014 summer survey.

REFERENCES

Bacon-Smith, Camille. *Enterprising Women: Televison Fandom and the Creation of Popular Myth*. Philadelphia: University of Pennsylvania press, 1992.
Bell, David. "Cybersubcultures." In *An Introduction to Cybercultures*. Edited by David Bell, 168–69. New York: Routledge, 2001.
Benford, Gregory. "Alt.fans," Reason, 27(1996): 43–44. Retrieved September 13, 2014 at http://reason.com/archives/1996/01/01/altfans.
Biederman, Danny. *The Incredible World of Spy-Fi: Wild and Crazy Spy Gadgets, Props and Artifacts from TV and the Movies*. San Francisco, CA: Chronicle Books, 2004.
Birkmeyer, Carl. "The DVD Releases: The Old Produce the Best DVD Set of the Year Trick," WouldYouBelieve.com. Accessed July 1, 2014, http://www.wouldyoubelieve.com/dvd.html, 2.
Clerc, Susan. "Estrogen Brigades and 'Big Tits' Threads: Media Fandom Online and Off." In *The Cybercultures Reader*. Edited by David Bell & Barbara M. Kennedy, 218–22. New York: Routledge, 2000.
Cole, Sue. "The History of U.N.C.L.E. HQ: A Brief Look at How This Newsletter Has Evolved and Progressed Over the Years." *U.N.C.L.E. HQ Newsletter*, 193/194, (1995): 5–7.
Coppa, Francesca. "Pop Culture, Fans and Social Media." In *The Social Media Handbook*. Edited by Jeremy Hunsinger and Theresa Senft, 76–92. New York: Routledge, 2014.
———. "A Brief History of Media Fandom. In *Fan Fiction and Fan Communities in the Age of the Internet*. Edited by Karen. Hellekson and Kristina Busse, 41–59. Jefferson, NC: McFarland & Co., 2006.
Enns, Anthony. "The Fans from UNCLE: The Marketing and Reception of the Swinging '60s Spy Phenomenon." *Journal of Popular Film and Television*, 28 (3), (2000): 124–32.
Fisher, Martin (producer). "Fandemonium." *The Man from U.N.C.L.E.: The Complete Series DVD*. Bonus Disc 1. Time Life/Warner Bros. Entertainment Inc, 2007.
"The Greatest Thing Since Peanut Butter and Jelly," *TV Guide*, April 17, 1965, 6–9.
"Great TV Spy Scramble." *Life*, October 1, 1965, 118–120.
Grossman, Lev. "The Boy Who Lived Forever." *Time*. (2011, July 7). Accessed July 7, 2011 at http://www.time.com/time/arts/article/0,8599,2081784,00.html.
Heitland, Jon. *The Man from U.N.C.L.E. Book: The Behind the Scenes Story of a Television Classic*. New York: St. Martin's Press, 1987.

Henderson, Craig. "U.N.C.L.E. Archive." *For Your Eyes Only.* Accessed on September 1, 2014 at http://www.for-your-eyes-only.com/Site/UNCLE_archive_1.html.

Iriberri Alicia and Leroy, Gondy. "A Life-cycle Perspective on Online Community Success." *ACM Computing Surveys,* 41 (2009), article 11. DOI 10.1145/1459352.1459356 http://doi.acm.org/10.1145/1459352.1459356.

Jenkins, Henry. *Textual Poachers: Television Fans and Participatory Culture.* New York: Routledge, 1992/2013.

―――. "Afterword: The Future of Fandom." In *Fandom: Identities and Communities in a Mediated World.* Edited by Jonathan Gray, Cornel Sandvoss and C. Lee Harrington, 357–64. New York: New York University Press, 2007.

―――. *Convergence Culture: Where Old and New Media Collide.* New York: New York University Press, 2006.

Johnson, Paul. "Sex, Snobbery and Sadism." *New Statesman,* April 5, 1958.

Kirby, Charlie. "Collecting on Man from U.N.C.L.E. Or What to Do When the Shock Treatments Fail Affair." *Daredevils,* 11 (1984): 24–28.

Koenig, Bill. "U.N.C.L.E. Movie Gets Another Test Screening." The HMSS Weblog. Accessed August 23, 2014 at http://hmssweblog.wordpress.com/2014/08/23/u-n-c-l-e-movie-gets-another-test-screening/.

―――. "Cavill, Hammer Return to U.K. for U.N.C.L.E., Fan Site Says." The HMSS Weblog. Accessed August 3, 2014 at http://hmssweblog.wordpress.com/2014/08/03/cavill-hammer-return-to-u-k-for-u-n-c-l-e-fan-site-says/.

Lambert, David. "*Seinfeld* Takes Home Best of Show at the Fifth Annual TV DVD." TV Shows on DVD.com, October 14, 2008. Accessed July 20, 2014 at http://tvshowsondvd.com/news/Site-News-5th-Annual-TV-DVD-Awards/10706#ixzz3DdNAgJYF.

―――. Muppets, "'Get Smart' Reigns at TV DVD Awards." TV Shows on DVD.com, November 13, 2007. Accessed July 20, 2014, http://www.tvshowsondvd.com/news/Site-News-4th-Annual-TV-DVD-Awards/8432.

Lazar, Jonathan, Tsao, Ronald and Preece, Jennifer. "One Foot in Cyberspace and the Other on the Ground." *Webnet Journal* (July–September, 1999): 49–57.

The Man From U.N.C.L.E., NBC, 09/22/64-01/15/68. (circa 1968). Nielsen ratings summary.

Markham-Smith, Ian. "Why Hoover Was Gunning for the Man from U.N.C.L.E." *Daily Mail,* September 1, 2000: 27.

Millington, Richard. *Buzzing Communities.* Feverbee: 2012: 24–39.

Norman Felton Collection, University of Iowa Library, Iowa City, Iowa.

Oldenburg, Ray. *The Great Good Place.* Cambridge, MA: Da Capo Press, 1999.

Opam, Kwame. "Camping in Hell: The Endless Lines of Comic-Con 2014." *The Verge.* Accessed July 26, 2014. At http://www.theverge.com/2014/7/26/5940113/camping-in-hell-the-endless-lines-of-comic-con-2014.

Robinson, N. "Our man with Uncle." *Secretariat,* 11, February 16, 1966: 13.

Paquette, Brian and Howley, Paul. *The Toys from U.N.C.L.E.: Memorabilia and Collectors Guide.* Worcester, MA: Entertainment Publishing: 1990.

Raddatz, Leslie. "The Mystic Cult of Millions: The People from U.N.C.L.E." *TV Guide,* March 19, 1966: 15–18.

Rheingold, Harold. *The Virtual Community: Homesteading on the Electronic Frontier* (Rev. ed.). Cambridge, MA: MIT Press, 2000.

Schearer, Lloyd. "Life and Death of the Man From U.N.C.L.E." *Parade,* (1968): 16–19.

Sinor, Bradley. "Collecting Paperbacks: The Man from U.N.C.L.E." *Baby Boomer Collectibles,* (May 1994): 54–56, 58.

Smith, P.T. "The Fan from U.N.C.L.E." *Daredevils,* 10 (August, 1984): 33–39.

Smith, Paula. "The U.N.C.L.E. Diaspora and the Great In-Gatherings." *Z.I.N.E.S.,* 9, (1995): 20–21.

Solow, Herbert F. and Justman, Robert H. *Inside Star Trek: The Real Story.* New York: Pocket Books: 1996.

Soukup, Charles. "Computer-Mediated Communication as a Virtual Third Place: Building Oldenburg's Great Good Places on the World Wide Web." *New Media and Society,* 8 (2006): 421–40.

Thomas, B. "No It Wasn't Thrush That Did UNCLE in." Syndicated column, *Associated Press Features*. (1967). Retrieved from the Norman Felton Collection, University of Iowa Library, Iowa City, IA.

Verba, Joan Marie. *Boldly Writing: A Trekker Fan and Zine History, 1967–1987.* Minnetonka, MN: FTL Publications, 1996.

Walker, Cynthia W. *Work/Text: Investigating The Man From U.N.C.L.E.* New York: Hampton Press, 2013.

———. "A Conversation with Paula Smith." *Transformative Works and Cultures*, 6 (2011). Accessed September 1, 2014, at http://journal.transformativeworks.org/index.ph/twc/article/view/243/205.

———. "The Future Just Beyond the Coat Hook: Technology, Politics and the Postmodern Sensibility in the Man from U.N.C.L.E." In *Channeling the Future: Essays in Science Fiction and Fantasy Television.* Edited by Lincoln Geraghty, 41–55. Plymouth, UK: Scarecrow Press, 2009.

———. "Man from/Girl from U.N.C.L.E." In *Encyclopedia of Television* (2nd ed.). Edited by Horace Newcomb, 1404–406. Chicago, IL: Fitzroy Dearborn, 200. Available online at http://www.museum.tv/eotv/manfromun.htm.

———. "Spy Programs." In *Encyclopedia of Television*, (2nd ed.). Edited by Horace Newcomb, 2181–185. Chicago, Illinois: Fitzroy Dearborn, 2004.

———. "The Gun as Star and the U.N.C.L.E. Special." In *Bang Bang, Shoot Shoot! Essays on Guns and Popular Culture* (2nd ed.). Edited by Murray Pomerance and John Sakeris, 203–13. New York: Simon and Schuster, 2000.

———. "Coming in from the Cold: Gender, the Internet and Man from U.N.C.L.E. Fandom." Presented at the 5th Annual Console-ing Passions: Television, Video and Feminism, April 25–28, 1996, University of Wisconsin-Madison.

Chapter Twenty

Managing Multiscreen

Dan Faltesek

"FML, I'm dying here" remarks an account executive, hung out to dry, waiting for some answer to a client's multiscreen question.[1] The answer to "Do we do multiscreen?" is not to be found in this full page advertisement for Mixpo, an advertising software company. Framed as a conversation between two advertising agency employees through the image vernacular of an instant message conversation on an iPhone, the dialog wraps around the bottom side of the front cover to the bottom of the iPhone with its familiar home button. The resolution? The employee back at the agency had found an answer to the vexing multiscreen problem. Hopefully the client would accept the answer tomorrow.

Multiscreen advertising is the practice of coordinating the presentation of material for a campaign on multiple platforms, particularly referring to the coordination of user activity between social networking and television viewing. Mixpo assures us that, "Clients expect flawless execution across screens even if they don't explicitly ask for it. Are you ready?"[2] Social integration is no longer a feature for boutique firms to boast, but a requirement to be in the game. The anxiety for an advertiser would be understandable, the choice to spend money with any particular agency is a matter of taste and trust, a carefully crafted relationship. Media can be purchased by almost anyone and creatives abound. A firm lives or dies by its ability to prove to potential clients that their insights are worth paying for. Or to use Mixpo's copy, "Want some help?"

Multiscreen advertising is a new term for a relative old idea, the marketing mix. Coordinating a campaign across platforms has been an aspect of agency work for decades. Not planning for multiple platforms would be quite irregular. As a point of introduction, Mixpo's advertisement does a fine job of inspiring anxiety. Mixpo neither explains what multiscreen is nor how one could attempt to answer the question. Clients want multiscreen, and they may not even know it. Although there are clear similarities with the advertising products of the past, contemporary advertising discourse treats multiscreen as something separate, or at least as a concept distinct enough that it requires discussion as a thing in itself. Difference sells.

This chapter reads the break that advocates of multiscreen advertising technologies make between the advertising of the past and the adverting of the future. Understanding what fragmentation means today requires moving past the ever shrinking audience, toward a more complex notion of the fragmentation of the media system as a glut of choices and screens. First, this chapter argues that the fragmentation of the audience has been fully integrated as a feature of advertising discourses, and that campaigns today assume a fragmented audience. Second, inspired by strategies from media industry studies, this chapter takes corporate discourses seriously as a topic for internal critique. The nested theories and conceptions of the public, the market, and the user interface are read for the sort of media theory they present. Third, this chapter reads two distinct sets of primary texts: the first is a selection of industry sales texts for multiscreen products, particularly those produced by Microsoft Advertising. These are developers texts. The second primary text bank presented is a reading of stories and arguments about multiscreen advertising in the magazine of record for the industry, and the source of the Mixpo advertisement in this introduction, *Advertising Age*. Insights from these primary texts will be used to juxtapose multiscreen and second screen discourses, with theories of affective modulation and transmedia storytelling. What is at stake in the distinction between transitional advertising and multiscreen is not so much a new sort of interface, but a larger debate about the nature of fragmentation and metaphors for understanding the public. Multiscreen discourse is a testbed for ways of describing audience agency and commerce in a diverse world screens.

This research has implications for understanding social media and fan communities. Multiscreen and second screen advertising are presented to media organizations as the first line of strategic efforts to control the flow of conversation on social media, to seed, steer, foment, and silence fan communication. Instead of fan communities autonomously organizing themselves through a version of collective intelligence, multiscreen management offers the possibility for the strategic manipulation. Future research on fans and social media will need to take stock of the ways that strategic efforts shape the expressions of fandom online.

FRAGMENTATION AND ABUNDANCE IN THE MEDIA SYSTEM

Before turning to a reading of advertising texts, it is important to understand the state of the advertising market. To say that fragmentation is the condition of our times is routine, even banal. No longer do visions of a society unified by rationality, power, or absolute spirit carry any particular weight. Fragments, some less commensurable than others, are provisionally articulated into meaningful assemblages. Instead of unified a public watching broadcast channels and filling their reservoirs with pools of meaning, the proliferation of cable and online video sources has only fragmented the public sphere further. The sort of ratings that *Star Trek: The Next Generation* had twenty years ago are the stuff of top programs today.[3] The reasons for the collapse of the mass televisual public are several: more devices, more preferences, and more distribution channels.

The least offensive programing strategies of limited mid-century television systems provided the illusion of wholeness.[4] Ratings were high if only because options were limited. As long as an offering was not patently offensive, it could attract a large audience. Tightly focused advertisements would be difficult to schedule in this programming environment, as audiences were large and heterogeneous. Consider the 1980s classic Peter campaign for Folger's coffee.[5] Nothing beats Folger's for waking to find that an adult son has returned home for Christmas morning. Aside from the cultural specificity of Christmas, the emotional tonality of family and reunion should be accessible for many audiences. Schmaltz sells for a good reason.

Super Bowl commercials are the exception that proves the rule. Annually the prices of these advertisements, and their content, are a topic of consideration. Millions of dollars for seconds of airtime seems excessive or wasteful. The prices are clearly justified in that this moment is an island of stability in an otherwise stormy affective world. One evening in February is the outlier. Getting through to tens of millions of potential customers is much more difficult than it once was.

Fragmentation cannot continue without reaching a limit, which in the case of the media industries is generally the break-even point. At a certain point the capacity of distribution systems simply outstrips the size of the publics that might be served. Consider *Cutlery Corner*, a paid program on certain satellite television networks that exists to sell knives, scabbards, and swords.[6] Contrary to the vision of the operator of a network making important political decisions or engaging in meaningful deliberation to stage the content that would ideally be on a system, the *Cutlery Corner* provides the most flow of all. It is a flow, it demands attention as much as it is a highly personal. The presenters of the corner are so smooth, so polished, and so good at selling that they seem to be right there in your living room. Yet, the Knife channel is little more than a toss-in, a placeholder that could just as easily be replaced by a feed of your local city council meeting, or a live-feed from the zoo.

There is now such a supply of programming channels that they can no longer be filled in any meaningful way. The emptiness of the communication system is apparent. Listless Google + profiles. DirecTV placeholder screens. "Your advertisement here" banner ads. So many avenues for distribution, so little affective energy. In his Silicon Valley agenda setting tome, *The Long Tail*, Chris Anderson posed that there would be demand for even the most marginal media products.[7] Demand for niche products was supposed to extend forever. Deep bench products for sale are quite profitable, particularly where immaterial goods are concerned, although the limit for profitable sales is rather shallow. Although Anderson's model of artificial scarcity and media product markets makes a good point (there is plenty of business beyond the hits) the idea of unlimited demand is faulty. At a certain point there are too few viewers to sustain the cost of distribution, much less production. Given a large enough sample size, the demand drops to zero.[8] Noise music opera is a niche of a niche. An hour long album of glass breaking or infants crying may not sell. Finding and tuning advertising for publics associated with mid-level products is the key.

This glut begins to dissolve the sort of affective fabric of public culture. For Andrejevic, the glut of information is fundamentally conservative—answers to important questions are just a few facts away, with meaningful political categories suppressed by the capacity of the system of being over-full with information.[9] Post-modernism is not so much a critique as the status quo. Instead of subjects being increasingly thrilled with the number of attachments and articulations available, they become overwhelming, inspiring an ironic response.[10] The ironic attachment to mass media today is the inverse of the reaction to Folger's Peter campaign—instead of a warm feeling for the return of a family member we would consider why he would make such questionable fashion choices, why the family would not have some kind of alarm, and who would be so unhip as to name their child Peter. This becomes an inoculant. Remaining distant offers both an escape from a saturated environment, but also an opportunity to defer consumer choices. The expansion, and failure, of enhanced of distribution mechanisms contributes to a sort of affective flattening. More choices means less excitement. Rather than offering a stale critique of fragmentation, the infoglut (Andrejevic's titular term) centers our discussion on the affective crisis of choice and agency over the faux sentimentality of those hoping for a return to the network era of television news.

Richard Thaler, Cass Sunstein, and Jon Balz have argued that the proliferation of choices has become a time eating program.[11] Rather than liberating individuals by providing them additional choices, the rhetoric of agency in a neo-liberal society becomes cancerous with choice for its own sake becoming both means and ends. Instead of taking time to be an engaged, reflective person we are asked to make choices continuously. Choice becomes an onto-logical imperative—an insatiable command to select more things, more objects, and more experiences that can never be fulfilled. This facet of neoliberal culture takes on the characteristics of the Lacanian transition from desire to drive.[12] Instead of making choices that could result in happiness or at least some satisfaction choices become an end itself. Searching and analyzing options would consume all affective possibilities. It is not simply that one is choosing between undifferentiated things, as Zizek would frame as the choice between Coke and Pepsi,[13] but that choice has become the only choice. Fetishistic individuality, as a form of radically personal fragmentation has a long history in advertising, as Thomas Frank described in the cultural fascination with Burger King's slogan about breaking rules, little did he know their next slogan would be "Have It Your Way." Fragmentation calls for an ever-increasing number of choices between the fragments and identities, and that choice calls for constant research that eventually consumes the entire time and attention of the public. Unlike money or material that can be created, time is impossible to manufacture. The flood of choices that it demands cast a long shadow on public culture in the early twenty-first century. Choice proliferation is a critical aspect of fragmentation in the multiscreen environment, fragmentation will increasingly depend not only on small groups of eyes, but clicks peppering a server.

Advertising in the early twenty-first century is in a fraught position: small audiences, low conversion rates, and ever increasing competition for dollars from any number of online impression sources. The advertising industry has been given exactly what it asked for, targeted access to pure demographics and customers making a myriad of measurable micro-choices. This revolution in measurement was made possible by a revolution in distribution. Instead of the large undifferentiated audiences that sustained the industry through the twentieth century, the early twenty-first century has provided the sorts of niches that advertisers had always wanted. Be careful what you ask for, you may get it.

In contrast, the social media seemingly offers new sites where attention may be pooling. Hundreds of millions of users logon to Facebook, Twitter, and Instagram each day. These users are both providing a new well of attention to sell, as well as valuable cross-promotion opportunities. The idea of a large scale social network as an island of stability in the otherwise tumultuous seas of fragmentation resonates with the primary texts form the industry. In this context, multiscreen advertising can be understood as yet another adaptive strategy for managing the risk posed by changes in the industry.

READING MULTISCREEN

From the outset it should be clear that this reading of multiscreen advertising is concerned with the ways that advertisers talk about multiscreen advertising, not the impact of that advertising or the texts circulated through multiscreen schemes. This distinction is important as it sets both the kinds of evidence and the practice of reading.[14] By reading the texts that promote various schemes we can gain leverage on the sorts of logic of the ideal operation in those industries and business plans. Susan Buck-Morss reading of Walter Benjamin's *Arcades Project* is an inspiration for this approach to reading the idea designs and the dream images of what media technologies could be always include their own undoing.[15] By reading the plan and the appeal of the technology we can understand the past and the future. As a practice for critical research reading dialectical images for their ideal figures, building plans, and failure points allows for different approaches to research to come together in a single project.

There are two kinds of texts read in this chapter, those from major multiscreen marketing firms describing their technology, and those from the trade magazine *Advertising Age* discussing best practices and research regarding multiscreen advertising. *Advertising Age* is the magazine of record for the advertising industry. As a trade magazine it serves a different role than a traditional newspaper, instead of the ideal of objectivity or the lesser substitute balance, trade magazines negotiate positions and refine industry discourses. Trades provide a wealth of signs for the normally internal thought processes of organizations. Materials from major advertisers will be read in much the same way. This is distinct from a reading of industry lore, the every talk and narratives that make sense of an industry for participants. These stories offer important insights into management and labor while the stories and plans read in this research are more concerned with deliberate strategy, the public face.

This reading of multiscreen advertising depends on the idea of a public sphere, particularly the idea that publics form in circulation. The emphasis of this chapter is on the ways in which strategic actors work to seed the public sphere with content, or to use Habermasian terms to colonize the lifeworld. Advertising industry talk often theorizes the public sphere in as much as it describes effects, flows, and pools. A number of authors have undertaken readings of industry discourses in public magazines, most notably Amanda Lotz, my difference in approach here is minor.[16] By paying attention to their supposed ideal models of public interaction we can learn about the trajectories of their companies, rather than where their companies have been. Instead of focusing on risk management as the heuristic driving decisions, this view sees the evaluation of potential as the site for analysis. To understand the discourse of multiscreen we will start with a reading of the discourses used to sell platforms, then we will turn to the discussion of those platforms in the trade literature.

SELLING MULTISCREEN

Multiscreen advertising promises to connect narrative across platforms. As a primary advertising format, Microsoft features multiscreen alongside in ad apps, mobile, exchanges, and a number of other relevant campaign styles. The key details that set a multiscreen campaign apart, aside from utilizing multiple devices, are enhanced targeting, creative sequencing, cross-screen frequency management, and device specific insights. [17] Of these affordances, targeting, cross-screen inventory management, and device based accounting are old hat. What stands out is the idea of creative sequencing. To develop the potential of cross-screen frequency Microsoft turns not to analytics, but to Carl Jung. Mike Shields writing in the *Wall Street Journal*, has argued that the idea of multiscreen offers Microsoft a way to stay in the advertising game. [18]

Microsoft's Jungian typology is no mere rehash of the Meyers-Briggs (another industrial reading of Jung), but a new set of "archetypes" that have been validated through "qualitative interviews, ethnographies, and quantitative data" to more fully map the psyche of the potential user. [19] The archetypes validated by Microsoft in partnership with BBDO and Ipsos OTX are: the TV as the Everyman, the Gaming Console as the Jester, the Computer as the Sage, and the Table as the Explorer, the e-reader as the Dreamer, and Mobile as the Lover. [20] Each archetype poses a figure not of a user, but of a use. The computer as sage is a reflection of the idea that the computer would be used for more complex reading and writing tasks, as well as for work. The jester for a console gaming system, the e-reader for imagination, and the phone as an intimate make are quite reasonable. If anything the language of Microsoft's archetypes seems to be adding a level of affective sophistication to established patterns in usability research. It is not that people tend to use devices in particular ways randomly, but that these choices are the reflection of their desires and fears.

Understandably, Microsoft has another aspect of its suite of advertising products matched with each archetype. Multiscreen is a metaphor for the larger campaign and that campaign is supposed to be coordinated with the various psychic dimensions of the audience. The rhetoric of sequencing provides Microsoft a way of making the implementation more than the sum of its parts. Multiscreen redefines the roles for the traditional advertising products offered while offering important insights into corporate psychoanalysis.

"Life is but a screen" sung by a doo-wop group and images of people using various devices in everyday life through an old-TV post-production

filter frames the thesis: multiscreen advertising is an adaptive layer that enhances television advertising.[21] With investors including Cox and Samsung,
and a wide partner list including many consumer product firms, Collective
(an advertising agency) theorizes multiscreen differently. Collective (another
multiscreen adverting product developer) brings a playbook for multiscreen
beginning with a large off-set quote from Kal Liebowitz, "TV remains the
single most efficient way to create critical mass among consumers with the
emotional power of sight, sound, and emotion that only video can deliver."[22]
Multiscreen advertising is a way of adding complimentary video access
across new platforms that would have been scary in the past. Collective's
points of emphasis are likewise televisual: ratings graphics and day-part
analyses.

The use of charts and graphs is quite normal for advertising. Nielsen
ratings are still a key aspect of the sensibility of any advertising technology.
Day-parts is the term that refers to the different times when audiences are
watching television. The day-parts are traditionally broken into blocks: early
morning, daytime, early fringe, prime access, primetime, late news, late
fringe, and post–late fringe. Different audiences watch television at these
times. Daytime typically has a very different audience and commercial
spread than prime time. Soap operas are thus named because of the logic of
advertising soap when homemakers may be watching. Day-parting is a key
aspect of the campaign for a recent convert from television advertising as it is
something they are already familiar with.

Collective also emphasizes the idea of sequencing advertising. The sequence when deployed during the proper day-parts with well-designed advertising can be a meaningful experience for the audience and effectively advertise the product. It is in the conclusion of their day-parting documents when
Collective poses that there may be something other than passive viewing of
single devices at play—one does not simply stare at their computer for an
entire workday, they are almost always inputting or outputting something, or
at least sending business Tweets. This alternative is known as second screen.
Seconding screening will receive a more detailed treatment in its own section
as it is distinct from, yet very similar to multiscreening.

IAB also took traditional advertising as a starting point by pairing day-
parts, genres, and recall studies to support the idea that a multiscreen campaign might be complimentary with existing advertising.[23] Instead of emphasizing the distinctiveness of multiscreen advertising IAB emphasizes the continuity with existing advertising strategy. It is not that multiscreen strategy is
new, but that it is the status quo. Mapping the world of multiscreen depends
on isolating the number and purpose of screens that the user interacts with.
Relative engagement with these screens and what that might mean for the
audience is not defined. The alternative to meaningful cultivation of multiscreen strategies is right in line with existing media industry anxieties: dis-

traction. IAB relies on a trope of inevitability. If viewers are already distracted, why not distract them with your own content?

The research facilitated by IAB is critical as the partnership between IAB, Ipsos, and Nielsen provides the basic research that confirms that users are engaged with multiple screens, this research when diffused across partnerships informs the entire world of multiscreen strategies. Each way of theorizing multiscreen advertising adds layers of value to the initial proposal for multiscreen research and further defines the stakes. Advertising depends on measurement.

YuMe, on commission from Nielsen, has introduced a multiscreen reach analysis tool.[24] The tool takes the established baseline engagement statistics and poses how they might change with different marketing mixes. Marketing mix analysis is nothing new: it has been a key idea in strategy for decades.[25] It should be no surprise that shifting some television advertising spend increases reach, but that is the purpose of advertising in the first place, emphasizing the continuity of approaches. A more radical shift in discourse, closer to Microsoft's would appear too risky.

DEBATING MULTISCREEN

Multiscreen discourse first appeared in *Advertising Age* as an aspect of the Newfront, a version of the traditional advertising Upfront for non-television media. Prior uses of multiscreen referred not to the idea of a mixed, sequential campaign, but to the literal use of multiple screens in an installation. In the traditional advertising market, the spring features an event known as Upfronts where television networks pitch their programming to potential advertisers to secure advertising. The vast majority ad inventory is sold during Upfronts, while some is sold later during the scatter markets. The Newfronts are an attempt to harness the same energy for the sale of multiscreen advertising.[26] Starting coverage of a new advertising concept at Upfronts is telling, as Michael Learmonth reported, "But if the digital 'NewFront' is about positioning web video as an alternative to TV, at least one big owner of TV networks is hoping not to beat them, but join them." The prospect of a Newfront, or at least digital integration into the upfront process was reported as early as 2006.[27] Multiscreen advertising has similar goals to all other advertising: to increase sales of a product. These Newfronts and the destabilizing effect of alternative program streams has had an effect on the market as well as some firms are beginning to hold money back from the Upfronts for future scatter and digital purchases.[28]

Not all reporting on multiscreen advertising has been positive, particularly when described through the language of the second screen the concept is already tired to the point that Ad Age suggested retiring the term as a cliché at the end of 2013. Mark Bergen reported that multiscreen seemingly impossible to measure.[29] Edmund Lee went further to suggest that the central issue with multiscreen advertising was that it could not escape the event horizon of distraction and the cognitive limits of the audience.[30] Advertisers take the limited cognitive resources of the public as a given. Time and attention are the metaphysical boundaries that circumscribe all advertising logic. As a decision science advertising is the study of allocation of articulation in a limited world. Lee's answer to the concern about cognitive capacity is that the audience has become more savvy, that instead of being exposed to many streams with equal intensity that the public has embraced a "hunter-gatherer" mentality for selecting the content they want to embrace. For advertisers, this sort of logic fits neatly into the larger taxonomic logic of multiscreen as the best material would need to be compelling enough to be retained for later engagement.

Jeanine Poggi reported in an in-depth article on the ongoing success of CBS that multiscreen efforts are at the end of the day unimportant to their strategy on the whole, "NCIS may not have social buzz, but it's the most-watched,' Mr. Moonves said. 'Call us least-sexy, that's fine. Give me what's measurable."[31] For all the anecdotal storytelling of the evidence that decides debates in the media industries, it still comes down to a combination of induction and verification. A cute story isn't good enough, there needs to be some concrete evidence. As Brian Steinberg reported, "But in a sign that Big TV still knows where the bulk of its revenue comes from, Mr. Moonves put everything in its place: 'The first screen comes first, and there's no second screen without it,' he said."[32] Moonves has an important point, if multiscreen engagement increases attention without sales it is a waste of money. Lost revenue clarifies many perspectives. In as much as a CEO can speak to the intention of a design, Moonves has made it clear that multiscreen practices are first and foremost, commercial.

In the span of a week, two rival executives at Coca-Cola played out both the sides of this debate concisely. On Monday, March 13, 2013 Eric Schmidt, senior manager-marketing strategy at Coca-Cola presented research that concluded that social buzz had a 0.01 percent impact on sales.[33] This is a very small impact even in an industry that measures small impacts. The most effective advertising came from television, followed by digital display, and search. Schmidt qualified his argument with details about the difficulty of conducting the research in the first place—automated sentiment analysis is not accurate enough, and the results of the study are not well suited to generalization across industries. Headlines from this research provoked enough of a response to force Coca-cola to clarify on Thursday. Wendy Clark, senior vice president of integrated marketing at Coca-Cola, argued in a response on a Coca-Cola Web site that was republished in *Advertising Age* that, "no single medium is as strong as the *combination* of media. We see this first-hand in our campaigns that integrate TV and social."[34] Clark presents minimal support for the claim that the marketing mix is a superior option. Clark's argument hinges on the idea that multiscreen might mark a distinction that matters, that there is something different enough about the marketing mix that it could either evade measurement or at least call for further consideration. Multiscreen advertising is the way that we read the larger complex of advertising relationships and their potentials, even if those potentials are unclear at best.

It seems likely that the Coca-Cola debate is symptomatic of larger problems in the market segment. Cola sales are falling, coffee is rising: the lack of buzz may be the result of a preference for beverages that do not fizz.[35] The debate over the impact of social buzz would matter little if Cola is post-peak. It should be no surprise that the coverage of the YuMe marketing mix tool in *Advertising Age* was sponsored by YuMe. Although some reports are willing to carry the language of the break between old and new advertising, for example using a quote from Charlie Hinton, an analytics executive at AT&T, "It's so much different than TV," the emphasis in the writing is often that multiscreen is an extension of existing advertising logics.[36] It seems that rhetoric of the break can only sustain itself so long against the inexorable demand to measure.

SECOND SCREENING

Less enchanting for advertisers already skittish about new platforms is the idea that a campaign might be designed around the audience not watching the commercials so that they might Tweet about them instead. Why pay for ads to run during the ads you already paid for? Advertising is, after all, strategic communication that should increase the awareness of, or likelihood that the audience might purchase a product. Second screening appeared earlier in *Advertising Age*, and does not appear at all as a key term in many sales documents for multiscreen technology.

It should concern multiscreen marketers and planners that the medium to this point has been dominated by Twitter.[37] Jennifer Healan, group director-integrated marketing at Coca-Cola remarked about a second screen campaign, "We had 9 million streams and grew our Twitter following by 22 percent, but then the conversation just fell from there. We didn't continue the conversation or capitalize on all those people talking about our brand."[38] Matt VanDyke from Lincoln Global emphasized the point that the second screen space was crowded, "'Everybody's got the second screen going on, and [the game] is going to be a really crowded space,' VanDyke continued 'People [watching] are going to be overwhelmed.'"[39] Aside from claims from advertising agencies that second screen was going to be dominant that season, especially in 2011, the tone of the trade press on second screen interaction was not particularly energetic. Perhaps IAB's inevitability of distraction argument from the review of multiscreen sales pitches provides a preemptive answer to this argument as it gives the advertiser permission to distract the audience. The debate about second screen has one side: second screens distract users.

If truly effective multiscreen advertising depends on coordinating archetypes, use patterns, and actually existing use practices, the idea that the second screen user is tied to a very limited device/platform is a real problem. Twitter is no rich-media gathering place. Twitter is temporally bounded. Conversations are only sequenced as long as they remain ahead of other material. The flow of time on Twitter is a raging river, not a carefully tended coy pond. Microsoft and others pushing their own proprietary multiscreen locations would offer a meaningful place of retreat—they can offer the sort of screen engagement that Twitter cannot.

As a discourse, second screen is distinct from multiscreen as it seems to point to overtly social campaigns that lack the sort of coordination, control, and thought of multiscreen. The shift to multiscreen as a key term reflects disenchantment with the weak returns of second screen. This shifts the register of the campaign by emphasizing the agency of the users and their capacity for engagement into conversational practice. It should be clear at this point that the publics organized around hashtags through Twitter are distinct from those sorts of passive publics that might experience affective engagement. Efforts at cultivating deeply engaged publics through Twitter have to this point been less than successful. Rarely a week goes by without a failed sponsored hashtag for questions and answers. Republican partisans claimed to be "winning Twitter" during the 2012 election because of superior numbers of mentions on Twitter with little attention to the substance of those messages.[40] No matter how many Twitter mentions one might have elections are decided on the basis of which candidate has the most votes.

When social publics are at their best, they are engaged in reflexive circulation and community building, they are assemblages of people who through their attention have constituted a public sphere. To become engaged in the second screen public sphere, a marketer would need to shift the structure of circulation, to institute a center to an otherwise diffused network. In his reading of Warner's counter publics through the life of Kierkegaard, David Wittenberg argued that the act of going public, of entering the debate changes the structure of the debate at hand.[41] For Wittenberg, when Kierkegaard became a celebrity philosopher, he gave up the position that allowed him to engage in public discourse, he was no longer a subject but an object.[42] For a brand or firm to enter into public discourse would require that they surrender their status as subject, surrendering their political capacity. It is no surprise that emerging best practices in Twitter public relations advise companies to only respond to Tweets to which they might actually have capacity to address and to allow other Tweets to pass as mere noise—to use the vernacular of social media: don't feed the trolls. The logic is clear: once the more trollish members of the public know they have found the company and not the idea of the company, they attack *en masse*. Some marketers and reality television producers have effectively managed this capacity by wrapping the social into the television experience. By placing useful content from engaged users into the stream of the narrative the program can become more engaging. Instead of a second screen customer service experience, they work to carefully craft an experience on multiple screens. [43] Marketers are learning the lessons of failed second screen campaigns and are reticent to become an object, opting instead to remain the subject manipulating social publics.

Microsoft would avoid recommending a heavy Twitter presence as a part of the marketing mix for more than the goal of not aiding a competitor, but also because the sort of affective engagement it represents is not a meaningful part of the multiscreen experience. Microsoft wants control of the emotional scene. The Jungian archetypes posed by Microsoft do not seem to offer space for a contentious public debate, or the Arguer. Further, the emotional tonality of a Twitter conversation is not neatly tuned. Rather than the sort of careful psychic tuning posed in the research, Twitter, as it actually exists, is populated by public relations hacks, trolls, psychopaths, and journalists. The second screen is distinct from the multiscreen, rarely do the two concepts appear together.

AFFECTIVE MODULATION

Microsoft's typology of affective modes goes a step further than simply claiming to move with the audience across devices—it supposes that there are distinct affective modulations for different media devices. Although this seems obvious it bears repeating that different sorts of media texts look and feel different. If one were to claim that the series *Breaking Bad* had the same emotional tonality and resonance as an episode of *Grey's Anatomy*, he or she would not be taken seriously. There is nothing quite as unsettling as a co-viewer YouTube viewer of *Sesame Street* clips with a small child as to when violent or obscene advertisements run between lessons from Cookie Monster, Big Bird, and Elmo. M is not just for mother and monster.

Raymond Williams's idea of flow emphasizes the continuity of the affective temporality of media, the schedule and sequence of events matters. [44] What Microsoft's approach to archetypes does so effectively is to ask for a more sophisticated reading of flow that depends on the enfolding of the users in their own virtual space. [45] This is where Microsoft's approach to theorizing multiscreen differs from those that read for a more traditional marketing mix.

For Coca-Cola and partners, the starting point is brand engagement with the product. For Collective and IAB, the emphasis point is the connection with traditional media pricing schemes. For Microsoft, the point of departure is affective. What flow presumes, and multiscreen campaign discourses depend on, is the idea that the attention of the user can never be made whole. Attention exists in fleeting moments and provisional articulations that could become something greater. It can never be made whole.

These theories of affective modulation frame the key to the effort as being engagement with the possible viewer and this exists in feeling. Devices, content, and users are a complex enfolding that requires careful affective read. Joshua Gunn's reading of the jouissance of the iPod operates along similar lines with the gadget serving as a joint between multiple worlds.[46] The iPod user equipped with their own personal soundtrack moves through the world their own affective envelope. Although this could be used by fan communities or other groups to produce their own affective spaces within dominant space, the possible use of this approach by strategic actors should be clear. Appropriate affective modulation is critical for meaningful public communication. If style and content are not carefully matched, the basis for a meaningful political movement or brand campaign sublimates. The rhetoric of consistency presents a rival theory of affective matching, instead of matching the affective potential of a device to a user at a particular time, the consistency theory suggests that a brand would need to be stable across the platforms at different times. This move toward consistency reduces the granularity of the campaign, but retains the idea of the individual mood and discursive constitution as the sine qua non of the advertising campaign. Affective modulation through multiscreen engagement challenges the underlying coordinates of the advertising ecosystem and in doing so forces a more fundamental reevaluation of the nature of the interface between public culture and the system. Not only should a discourse meet with the genre and style needs of the system, it also needs to be tuned to the same affective note. It is not simply that one should present arguments on the basis of the rationality within a sphere, but that they also need to operate in the affective register of that sphere. At times, protest movements may successfully deploy incongruent affective styles, but as our studies of the advertising industry have indicated, tightly coupled and planned affective engagement is far more effective than not.

TRANSMEDIA, AVOIDED

One particularly compelling theory that does not appear in the industry discourse on multiscreen: transmedia. This is particularly surprising given the similarity of these approaches. Henry Jenkins's definition of transmedia storytelling is strong here, "Transmedia storytelling represents a process where integral elements of a fiction get dispersed systematically across multiple delivery channels for the purpose of creating a unified and coordinated entertainment experience."[47] This is quite similar to the idea of creating an advertising campaign that maps onto different devices with different affective styles. One major reason for avoiding the interaction of transmedia and multiscreen comes through the question of control. Transmedia has often been the point where participatory culture and audience agency comes into contact with planning. Jenkins has noted in his research that this question of control is critical as the loss of control for major firms is almost impossible to justify. They simply cannot lose control. Second screening has a lukewarm response for the same reason. If it gives control of the experience way, or at least appears that way, firms will be reticent to engage. Multiscreen advertising strategy is typically far more programmatic and deliberate than the open-ended associations that typify transmedia. A first line adaptive strategy for firms could come in the way that the options involved in transmedia are foreclosed. Instead of allowing users to create fan fiction or engage in independent creative work the multiscreen advertiser poses a form of treatment that would affectively engage the audience without asking to produce new symbols. The range of interactive options (and the risk of creativity) are greatly reduced by limiting the audience to clicking share.

Making a distinction between simply selecting or sharing, and entering text and making powerful creative decisions are important both for criticism and for the corporate use of multiple screens. Multiscreening provides a theory of advertising that avoids the agency of the audience, it simply plugs into more of their supposed ports until they are matched with the correct affective content. Microsoft's theory of affective modulation is limited as it can only be expressed in the imaginary terms of already established types, and this is substantially more advanced than many other approaches.

It is useful to defend the idea of transmedia storytelling from reduction to multiscreen, but it seems difficult to see any future for the term in an industry where the traditional vocabulary has been remapped to assume the function of the advertisement in multiple domains. The newness of multiscreen and the depth of attachment are central to the idea of transmedia. Instead of asking users to participate and create, multiscreen campaigns tell them how to feel. This works across the purpose of producing synergy, the additive effect supposed by so many advocates of multiscreen. If multiscreening depends on both a limit on the creativity of the user base as well as each element being similar in total reach and design to a television advertisement the capacity for the campaign to be different is limited.

Microsoft mapped the archetypes to the devices and not to the users. Once mapped back to the user, the discourse of multiscreening can only be sensible as an aspect of the traditional advertising industry or through the language of radical agency and creativity of transmedia. Either of those options would not meet the needs of the multiscreen manager. Everyone knows that the affect of the individual user is the central question: it is simply too painful a question to ask. Multiscreen theories of social media community formation eschew transmedia because of the promise of the idea. The public sphere theory propagated by multiscreen advertisers is that of a mass public that is directly manipulated by affective modulation. Social media in this theory of the public is evacuated of its creative potential, being rendered yet another means of commodifying the audience.

The underlying engine of Jenkins's convergence and transmedia theories was a robust conception of collective intelligence. Additive comprehension and fan culture could work synergistically to increase fan engagement with the text, and thus revenues, by giving up some control of the circulation of discourse to those fans. Affective common places that would seemingly animate these networks are those of tactical and provisional association—this is the vulnerability in the theory of circulation inherent in fan culture that multiscreen advertisers exploit. Strategy can come full circle, configuring the terrain in such a way that tactics may be selected by advertiser on the basis of their potentially beneficial results.

CONCLUSION

Multiscreen advertising is a strategy to control community formation on social network sites. At the highest level, multiscreen discourse poses a theory of a modulated psyche where publics are contacted in different emotional scenes for different purposes. This chapter argued that the multiscreen discourse within the advertising industry has been structured to maintain continuity with existing advertising discourses of attachment, psyche, control, and publicity. These discourses pose novel theories of differential affective investment of individuals into different devices at different times. Advertisers dream that multiscreen approaches offer a way to more completely interface with the gestalt of the public, to finally reach a resonance that could accomplish all the goals that advertising has had for a century through the digital extensions of their psyche. Yet, for all the sophistication and research inherent in the discourse, it seems possible, if not likely, that the very emphasis on control that makes programmatic multiscreen advertising seem possible will be its undoing. Within industry discourse new approaches to advertising must take a dual role, claiming to be unique and yet familiar. The potential for synergy between screens gave Clark, defending a lack of measurable results in terms of soda sold by multiscreen advertising for Coca-Cola, the leverage to refute the idea that buzz was irrelevant. It did not matter that more soda was not sold, the multiscreen effort surely shaped the affective landscape. Device-centric affective theorizing allowed Microsoft to theorize feeling and time differently. Lurking in these discourses is the agency of the audience, that prickly matter that in a fragmented media environment the users/audiences/publics are asked to make more choices all the time, and that these choices have an affective dimension in their own right.

Social media research on fan communities must address these ongoing efforts to strategically form and manipulate publics. Although fan creativity may be stimulated by corporate interaction, the reality of capital structuring that public is unavoidable. What we find through reading of industry discourse about multiscreen advertising strategy is that there are theories both regarding talk on social media as well as theories about the affective modulation of devices that fan communities might use. As this trend of seeding social content continues, the agency of fan communities will be further constrained as strategic actors find ways to reward communities for acting to further their campaigns. The most troubling aspect of this sort of advertising campaign is not the overt discourse of control that might come from the industry, but the lack of a discussion of audience creativity and engagement. Advertising discourses do not treat fan publics as a future site for creative engagement, but as a resource to be effectively utilized. Multiscreen advertising techniques promise to usher in a new era of careful affective modulation. The future is not one of cooperation and participation via social media, but of increased control. The ways in which fan communities resist, reverse, and revise these efforts at control will be a major topic of research for years to come.

NOTES

1. This advertisement is a wrapper on the cover of the print edition of Ad Age, June 2013.
2. Ibid.
3. Ratings during the final season of *The Next Generation* were high, with nearly twenty million viewers per episode. Compared with the most recent season of American Idol, checking in with less than half the total viewership. Svetkey, Benjamin. "Star Trek: *The Next Generation* Readies for Last Episode." *Entertainment Weekly*, May 6, 1994. http://www.ew.com/ew/article/0,,20661787_302144,00.html; Yahr, Emily. "As 'American Idol' Ratings Plummet Further, Let's Stop Trying to Reinvent This Once-Hot Show." *Washington Post*, May 22 2014. http://www.washingtonpost.com/blogs/style-blog/wp/2014/05/22/as-american-idol-ratings-plummet-further-lets-stop-trying-to-reinvent-this-once-hot-show/.
4. There are a number of cases that could be called to argue that macro-public seemingly hailed into existence by network television has been overstated.
5. "The Ads of Christmas Past: Folgers (1986) 'Peter Comes Home'—an Advertising Blog by Mascola Group." Accessed July 25, 2014. http://mascola.com/insights/the-ads-of-christmas-past-folgers-1986-peter-comes-home/.
6. The best example of this type of programming is Cutlery Corner, this is their homepage: http://www.cutlerycorner.net/.
7. Anderson, Chris. *The Long Tail*. 2nd ed. New York: Hyperion, 2008.
8. This is not a particularly new idea, since Anderson's initial publication there have been a number of arguments that the idea of the tail, particularly in music, eventually are exhausted. A

strong corollary here is the idea of inelastic demand. Normally this refers to the idea that changing price may not effect the quantity of a thing demanded, cheaper toilets may not compel homeowners to install more bathrooms for example. The long tail really is an argument about substantially increasing the supply of goods in the marketplace, the total demand level in the market remains unchanged. This would also explain how disappointing sales are in the tail: absent demand, supply goes unsold. Mulligan, Mark. "The Death of the Long Tail." *Music Industry Blog.* Accessed July 28, 2014. http://musicindustryblog.wordpress.com/2014/03/04/the-death-of-the-long-tail/.

9. Andrejevic, Mark. *Infoglut: How Too Much Information Is Changing the Way We Think and Know.* New York: Routledge, 2013.

10. Weiland, Matt and Thomas Frank, eds. *Commodify Your Dissent: Salvos from The Baffler.* 1st ed. New York: W. W. Norton & Company, 1997.

11. Thaler, Richard H., Cass R. Sunstein, and John P. Balz. *Choice Architecture.* SSRN Scholarly Paper. Rochester, NY: Social Science Research Network, April 2, 2010. http://papers.ssrn.com/abstract=1583509.

12. Zizek, Slavoj, *How to Read Lacan.* W. W. Norton & Company Incorporated, 2006.

13. Zizek, Slavoj, *Did Somebody Say Totalitarianism?: Five Interventions in the (mis)use of a Notion.* Verso, 2002, 241.

14. This approach has been well done by Amanda Lotz. The key difference in my approach is that the reading of circulating discourses is considered architectonically and affectively as a point for political critique. As a sort of surreal method for media studies this approach takes everyday industry discourses seriously as an expression of the affective conditions of possibility for communication infrastructures. An example of this sort of industrial reading method without the surrealist turn: Lotz, Amanda. *The Television Will Be Revolutionized.* New York: New York University Press, 2007.

15. The idea of the dialectical image is described quite clearly in the work of Buck-Morss. Buck-Morss, S. *The Dialectics of Seeing: Walter Benjamin and the Arcades Project.* MIT Press, 1991.

16. Ibid.

17. "Multi-Screen Advertising." *Microsoft Advertising.* Accessed July 28, 2014. http://advertising.microsoft.com/en/multi-screen.

18. Shields, Mike. "How Microsoft Is Helping Brands Like Lexus Advertise Across Three Screens." *Wall Street Journal,* July 7, 2014. http://blogs.wsj.com/cmo/2014/07/07/how-microsoft-is-helping-brands-like-lexus-advertise-across-three-screens/.

19. Stromberg, Joseph. "Why the Myers-Briggs Test Is Totally Meaningless."*Vox,* July 15, 2014. http://www.vox.com/2014/7/15/5881947/myers-briggs-personality-test-meaningless.

20. "Meet the Screens Reveals Multi-Screen Insights." *Microsoft Advertising.* Accessed July 28, 2014. http://advertising.microsoft.com/en/cl/630/meet-the-screens-us-whitepaper-multi-screen-user-research.

21. The full commercial video is accessible at this link: http://vimeo.com/69798572.

22. Both the daypart and regular playbook sales documents are hosted as inlays on this page. "Collective | Multi-Screen Advertising." Accessed July 28, 2014. http://collective.com/audience-network/.

23. Interactive Advertising Bureau, "The Multi-Screen Marketer," whitepaper, May 2012. http://www.iab.net/media/file/The_Multiscreen_Marketer.pdf.

24. The YuMe tool can be accessed at: http://www.yumecalculator.com/calculator.html. This is a fascinating interactive advertising that seems to offer projections without any discussion of what is under the hood, or why a campaign with 75 percent of spend shifted to digital would not be desirable.

25. Marketing mix analysis is intuitive and has been gaining in popularity steadily over the past fifty years as per Google nGram.

26. Learmonth, Michael. "TV Power NBC Universal Joins Web Video 'Upfront.'"*Advertising Age,* March 8, 2012. http://adage.com/article/special-report-tv-upfront/tv-power-nbc-universal-joins-web-video-upfront/233167/.

27. Atkinson, Claire. "Ready for Prime Time?" *Advertising Age,* May 8, 2006.

28. Poggi, Jeanine. "Think CBS Isn't Sexy? Moonves Doesn't Care." *Advertising Age*, May 13, 2013.

29. Bergen, Mark. "Welcome to the New First Screen: Your Phone." *Advertising Age*, March 17, 2014. http://adage.com/article/digital/millward-brown-study-shows-mobile-outpacing-tv/292183/.

30. Lee, Edmund. "How Social Media Stole Your Mind, Took Advertising with It." *Advertising Age*, February 27, 2011. http://adage.com/article/digital/social-media-stole-mind-advertising/149120/.

31. Poggi, "Think CBS Isn't Sexy? Moonves Doesn't Care."

32. Steinberg, Brian. "Plethora of Promos Will Soon Mob Your TV Screen; Tolerance of 'Messy' Media Prompts Testing of Top-to-Bottom Ads ." *Advertising Age*, August 10, 2009.

33. Neff, Jack. "Buzzkill: Coca-Cola Finds No Sales Lift from Online Chatter." *Advertising Age*, March 18, 2013. http://adage.com/cmo-strategy/coca-cola-sees-sales-impact-online-buzz-digital-display-effective-tv/240409/.

34. Clark, Wendy. "Coca-Cola's Wendy Clark Defends 'Crucial' Social Media."*Advertising Age*, March 21, 2013. http://adage.com/article/cmo-strategy/social-media-matter-marketing-coca-cola/240444/.

35. The larger transition in the beverage industry is quite real. Cola drinks seem to be in a transition. Zmuda, Natalie. "Coca-Cola Boosts Media Spending as Demand Slows." *Advertising Age*, February 18, 2014. http://adage.com/article/news/coca-cola-boosts-marketing-spend-demand-slows/291747/.

36. Bergen, Mark. "Why Marketers Just Can't Crack Mobile." *Advertising Age*, May 19, 2014. http://adage.com/article/news/marketers-crack-mobile/293249/.

37. Peterson, Tim. "Shazam Devises Ad-Retargeting Program with Facebook, Makes Its App Stickier." *Advertising Age*, February 3, 2014.

38. Zumda, Natalie. "Behind the Scenes of Coca-Cola's Super Bowl and 2013 Ad Plans." *Advertising Age*, February 4, 2013. http://adage.com/article/cmo-strategy/scenes-coca-cola-s-super-bowl-2013-ad-plans/239568/.

39. Delo, Cotton. "Social Networks Aim to Land More Super Bowl Ad Dollars."*Advertising Age*, January 14, 2013. http://adage.com/article/special-report-super-bowl/social-networks-aim-land-super-bowl-ad-dollars/239148/.

40. This is a great overview of Powerline's claim that Romney was winning social media, as well as the debunking of that claim. Wilhelm, Alex. "Romney Winning the Social Media War? Nope." *The Next Web*, September 7, 2012. http://thenextweb.com/socialmedia/2012/09/07/romney-winning-social-media-war-nope/.

41. Wittenberg, David. "Going out in Public: Visibility and Anonymity in Michael Warner's 'Publics and Counterpublics.'" *Quarterly Journal of Speech* 88, no. 4 (2002): 426–33.

42. Ibid, 432.

43. Cunningham, Sean. "How Media Agencies Can Stay Relevant in a Programmatic Age (Despite CMO's Plans). *Advertising Age,* December 3, 2013. http://adage.com/article/media/media-agencies-stay-central-a-programmatic-age/245496/.

44. Williams, Raymond. *Television*. 3rd ed. New York: Routledge, 2003.

45. Microsoft's shift to Jung is an enfolding is fascinating in this respect.

46. Gunn, Joshua, and Hall, Mirko. "Stick It in Your Ear: The Psychodynamics of iPod Enjoyment." *Communication and Critical/Cultural Studies* 5 (2008): 135–57.

47. Jenkins, Henry. "Transmedia 202: Further Reflections."*Confessions of an Aca-Fan*, August 1, 2011. http://henryjenkins.org/2011/08/defining_transmedia_further_re.html.

REFERENCES

Anderson, Chris. *The Long Tail*. 2nd ed. New York: Hyperion, 2008.
Andrejevic, Mark. *Infoglut: How Too Much Information Is Changing the Way We Think and Know*. New York: Routledge, 2013.
Atkinson, Claire. "Ready for Prime Time?" *Advertising Age*, May 8, 2006.

Bergen, Mark. "Welcome to the New First Screen: Your Phone." *Advertising Age*, March 17, 2014. http://adage.com/article/digital/millward-brown-study-shows-mobile-outpacing-tv/29 2183/.

———. "Why Marketers Just Can't Crack Mobile." *Advertising Age*, May 19, 2014. http://adage.com/article/news/marketers-crack-mobile/293249/.

Clark, Wendy. "Coca-Cola's Wendy Clark Defends 'Crucial' Social Media." *Advertising Age*, March 21, 2013. http://adage.com/article/cmo-strategy/social-media-matter-marketing-coca-cola/240444/.

"Collective | Multi-Screen Advertising." Accessed July 28, 2014. http://collective.com/audience-network/.

Cunningham, Sean. "How Media Agencies Can Stay Relevant in a Programmatic Age (Despite CMOs' Plans)." *Advertising Age*, December 12–13, 2013. http://adage.com/article/media/media-agencies-stay-central-a-programmatic-age/245496/.

Delo, Cotton. "Social Networks Aim to Land More Super Bowl Ad Dollars." *Advertising Age*, January 14, 2013. http://adage.com/article/special-report-super-bowl/social-networks-aim-land-super-bowl-ad-dollars/239148/.

Gunn, Joshua, and Mirko Hall. "Stick It in Your Ear: The Psychodynamics of iPod Enjoyment." *Communication* and 5 (2008): 135–57.

Interactive Advertising Bureau. "The Multi-Screen Marketer." *Interactive Advertising Bureau*, May 2012. http://www.iab.net/media/file/The_Multiscreen_Marketer.pdf.

Jenkins, Henry. "Transmedia 202: Further Reflections." Confessions of an Aca-Fan, August 1, 2011. http://henryjenkins.org/2011/08/defining_transmedia_further_re.html.

Kittler, Fredrich. *Gramaphone, Film, Typewriter*. Translated by Winthrop-Young, Geoffrey and Wutz, Michael. Stanford, CA: Stanford University Press, 1999.

Learmonth, Michael. "TV Power NBC Universal Joins Web Video 'Upfront.'" *Advertising Age*, March 8, 2012. http://adage.com/article/special-report-tv-upfront/tv-power-nbc-universal-joins-web-video-upfront/233167/.

Lee, Edmund. "How Social Media Stole Your Mind, Took Advertising with It." *Advertising Age*, February 27, 2011. http://adage.com/article/digital/social-media-stole-mind-advertising/149120/.

Lotz, Amanda. *The Television Will Be Revolutionized*. New York: New York University Press, 2007.

"Meet the Screens Reveals Multi-Screen Insights." *Microsoft Advertising*. Accessed July 28, 2014. http://advertising.microsoft.com/en/cl/630/meet-the-screens-us-whitepaper-multi-screen-user-research.

Mulligan, Mark. "The Death of the Long Tail." Music Industry Blog. Accessed July 28, 2014. http://musicindustryblog.wordpress.com/2014/03/04/the-death-of-the-long-tail/.

"Multi-Screen Advertising." *Microsoft Advertising*. Accessed July 28, 2014. http://advertising.microsoft.com/en/multi-screen.

Neff, Jack. "Buzzkill: Coca-Cola Finds No Sales Lift from Online Chatter." *Advertising Age*, March 18, 2013. http://adage.com/article/cmo-strategy/coca-cola-sees-sales-impact-online-buzz-digital-display-effective-tv/240409/.

Peterson, Tim. "Shazam Devises Ad-Retargeting Program with Facebook, Makes Its App Stickier." *Advertising Age*, February 3, 2014.

Poggi, Jeanine. "Think CBS Isn't Sexy? Moonves Doesn't Care." *Advertising Age*, May 13, 2013.

Shields, Mike. "How Microsoft Is Helping Brands Like Lexus Advertise Across Three Screens." *Wall Street Journal*, July 7, 2014. http://blogs.wsj.com/cmo/2014/07/07/how-microsoft-is-helping-brands-like-lexus-advertise-across-three-screens/.

Steinberg, Brian. "Plethora of Promos Will Soon Mob Your TV Screen; Tolerance of 'Messy' Media Prompts Testing of Top-to-Bottom Ads." *Advertising Age*, August 10, 2009.

Stromberg, Joseph. "Why the Myers-Briggs Test Is Totally Meaningless." *Vox*, July 15, 2014. http://www.vox.com/2014/7/15/5881947/myers-briggs-personality-test-meaningless.

Svetkey , Benjamin. "Star Trek: The Next Generation Readies for Last Episode." *Entertainment Weekly*, May 6, 1994. http://www.ew.com/ew/article/0,,20661787_302144,00.html.

Thaler, Richard H., Cass R. Sunstein, and John P. Balz. *Choice Architecture. SSRN Scholarly Paper*. Rochester, NY: Social Science Research Network, April 2, 2010. http://papers.ssrn.com/abstract=1583509.

Weiland, Matt. *Commodify Your Dissent: Salvos from The Baffler*. 1st ed. New York: W. W. Norton & Company, 1997.

Wilhelm, Alex. "Romney Winning the Social Media War? Nope." *The Next Web*, September 7, 2012. http://thenextweb.com/socialmedia/2012/09/07/romney-winning-social-media-war-nope/.

Williams, Raymond. *Television*. 3rd ed. New York: Routledge, 2003.

Wittenberg, David. "Going Out in Public: Visibility and Anonymity in Michael Warner's 'Publics and Counterpublics.'" *Quarterly Journal of Speech* 88, no. 4 (2002): 426–33.

Yahr, Emily. "As 'American Idol' Ratings Plummet Further, Let's Stop Trying to Reinvent This Once-Hot Show." *Washington Post*, May 22. http://www.washingtonpost.com/blogs/style-blog/wp/2014/05/22/as-american-idol-ratings-plummet-further-lets-stop-trying-to-reinvent-this-once-hot-show/.

Yue, Michelle. "The Ads of Christmas Past: Folgers (1986) 'Peter Comes Home'—an Advertising Blog by Mascola Group." Accessed July 25, 2014. http://mascola.com/insights/the-ads-of-christmas-past-folgers-1986-peter-comes-home/.

Zizek, Slavoj. *Did Somebody Say Totalitarianism?: Five Interventions in the (Mis)use of a Notion*. Verso, 2002.

———. *How to Read Lacan*. W W Norton & Company Incorporated, 2006.

Zmuda, Natalie. "Coca-Cola Boosts Media Spending as Demand Slows." *Advertising Age*, February 18, 2014. http://adage.com/article/news/coca-cola-boosts-marketing-spend-demand-slows/291747/.

———. "Behind the Scenes of Coca-Cola's Super Bowl and 2013 Ad Plans." *Advertising Age*, February 4, 2013. http://adage.com/article/cmo-strategy/scenes-coca-cola-s-super-bowl-2013-ad-plans/239568/.

Index

A&E, 15, 147–148, 150, 151, 152, 153
AMC, 7, 33, 135, 183, 185, 190, 192, 202
Army Wives, 235–236, 237, 239, 240, 241,
    242–243, 244, 246, 247, 248, 250, 251,
    252, 253, 254

The Big Bang Theory, 119
Blog(s), 31, 36, 63, 64, 66, 67, 69, 70, 72,
    74, 75, 77, 78, 81, 82, 83, 84, 129, 134,
    135, 138, 162, 167, 168, 170, 171, 172,
    173, 183, 189, 196, 223, 226, 243, 278,
    285, 321, 340, 361, 362
Breaking Bad, 31, 202, 207, 209, 210, 213,
    385
Buffy the Vampire Slayer, 185, 192, 202,
    208, 211, 212, 213, 238, 336

Community, 109, 110, 114, 115, 116, 117,
    119, 120–121, 122, 202, 205, 209,
    211–212

Doctor Who, 211, 279, 295, 297, 298, 300,
    301, 302, 304, 305, 306, 307, 308, 309,
    317, 318, 321, 324, 327, 357, 364
Duck Dynasty, 147–148, 150–157, 262

Facebook, 3–4, 9, 28, 35–46, 95, 136, 139,
    148, 150–153, 155–157, 162, 165–167,
    173–177, 183–184, 187–190, 192, 196,
    201–203, 209, 223–224, 261–263,

268–272, 277–278, 281, 302, 317–319,
    321–322, 326–328, 362–365, 377
Firefly, 202, 205, 206, 209, 210, 212–213,
    214, 317, 327
Food Network, 261–262, 264, 265, 266,
    267, 268, 270–271, 271
Foursquare, 3, 46, 343, 344

Game of Thrones, 31–32, 34, 36, 127, 132,
    135, 317–318, 321–322, 325–327
General Hospital, 221, 222, 224
Grey's Anatomy, 241, 344, 385

Here Comes Honey Boo Boo, 148, 150,
    167, 262, 263
HBO, 31, 127, 128, 129, 130, 132, 136,
    137, 139, 192, 277, 282, 288, 318, 325

Jenkins Henry, 109–114, 133, 137, 204,
    214, 219, 319, 321, 357, 359, 365,
    386–388

Late Night with Jimmy Fallon, 7, 48
Lost, 31, 113

The Newsroom, 277, 282

One Life to Live, 221, 223

parasociality, 183, 185, 187, 197
Parks & Recreation, 24

396

Index

Pinterest, 1, 3, 4, 7, 295, 304, 305, 306, 307, 308, 309, 344

The Real Housewives, 150
reality television, 6, 148, 149, 150, 161, 162–165, 176, 177, 384
Reddit, 43, 49, 110, 121, 201, 203, 205, 206, 207, 208, 210, 211, 212, 213

The Sopranos, 31

Toddlers & Tiaras, 161–162, 166–177
True Blood, 136, 192
Tumblr, 10, 14, 161, 166, 167, 167–170, 171–172, 173, 176, 177, 306, 317, 318, 320, 322, 366
Twin Peaks, 335, 338, 339, 340, 341, 342, 344

Twitter, 3–4, 9–10, 15–16, 28–29, 32–36, 45–46, 52–53, 63–64, 67–68, 70–76, 79–81, 83–84, 90, 92, 95, 99–100, 136, 162, 165–167, 173, 175–177, 183–184, 187–192, 196, 201–203, 207–209, 219–220, 223–228, 263, 271, 277–282, 301–302, 317, 321, 335–338, 340–341, 343–346, 366, 377, 383, 385

Uses and gratification theory, 25

Veronica Mars, 201
The Voice, 205

The Walking Dead, 33, 183, 185, 188–190, 193–196

The X-Files, 29, 30, 339

About the Contributors

Benjamin Brojakowski (PhD student and teaching assistant, Bowling Green State University) is a full-time Boston Red Sox fan with a passion for research. When he isn't watching baseball, he works as an instructor/teaching assistant and PhD student at Bowling Green State University. His research primarily focuses on identity and race issues in social media, online impression management strategies, and media effects. His goal is to earn a full-time professor position after the completion of his degree.

Jason Roy Burnett (MA, MSEd) is a doctoral student in media and communication at Bowling Green State University. He has been researching narrative, identity, and intertextuality since 2004, working in literature, counseling, and now communication. A theory junkie, Jason's primary interests lie in the intersections of meaning that are utilized in performative identity. He has presented original work in a variety of academic contexts and published work in the *Review of Social Science, Law, and Psychology*. Jason lives in the Midwest with a very patient wife and a very demanding terrier.

Ryan Cassella (MA, Fairfield University) is a media professional with research interests that often intersect with his day-to-day work. He is especially fascinated by the changing role of the modern media maker in the ever evolving digital age. When he isn't consumed with thoughts of glowing screens and endless sound-bytes, Ryan enjoys the great outdoors and other assorted adventures with his friends and family.

Garret Castleberry (MA, University of North Texas) is a PhD candidate and the director of forensics for the Department of Communications at the University of Oklahoma. His ongoing research and publications investigate polyvalent critical/cultural themes as well as socioeconomic and mythic narratives embedded within contemporary transmedia texts and popular culture. With strategic emphasis in television studies and the post-network televisual mediascape, Garret previously published in special issues of *Cultural Studies, Critical Methodologies*, and *International Review for Qualitative Research* with upcoming contributions in *The ESPN Effect* (2015) and *Communication Basics for Millennials* (2015). Recent publications combine genre studies, autoethnography, and ideological criticism to explore the rhetorical work performed by televisual artifacts.

Matthew Collins is an alumnus of Christopher Newport University where he earned dual degrees in communication and history. His research involves television in a new media landscape as well as studies in fandom. Currently he works as a junior sales associate for the Daily Press Media Group in Newport News, Virginia, helping local businesses increase their Web presence through online marketing.

Ted M. Dickinson (PhD candidate, the Ohio State University) is one of those "nontraditional" students. He is fortunate enough to have a wife, Jessica, with a degree in social psychology who can understand his mutterings about research. His research interests include fandoms and advertising in video games, and he has previously been published in the *Journal of Communication* for something completely unrelated to either of those.

Marsha Ducey is assistant professor of journalism at the College at Brockport, State University of New York. After a nearly ten-year career in newspaper journalism, she returned to school and earned her master's degree in media studies at Syracuse University and doctorate in communication at the State University of New York at Buffalo. Ducey serves as faculty adviser to the College at Brockport's student newspaper and Society of Professional Journalists chapter in addition to teaching journalism classes. Ducey's research interests center on two areas: journalism and the role of gender in media. She authored the "Women in Journalism" entry in the *Encyclopedia of Journalism*; worked as a research assistant for Project Daytime, a long-term research project focused on the study of cultural indicators in daytime serials; and has presented numerous papers at academic conferences including the Popular Culture Association and the Association for Education in Journalism and Mass Communication.

Dan Faltesek is an assistant professor of social media at Oregon State University. His research explores the intersection of economic and legal arguments with the design of communication networks. Critical research on the rhetoric of a market for data, analytics, and visualization platforms will be a major theme in his work over the next several years.

Krystal Fogle (MA, communication) enjoys both watching and analyzing television. She has presented scholarship on *Sherlock* and *Doctor Who*, and has researched and written about various pop culture artifacts including You-Tube and Pinterest. She will be pursuing a PhD in rhetoric and media.

Brian Geltzeiler (MBA, American University) is a vice president and financial advisor at a major Wall Street firm. He also founded the NBA-themed Web site http://www.hoopscritic.com in 2010 and is a regular radio guest on sports talk shows throughout the country. He is currently a talk show host on *Sirius XM NBA Radio* hosting five shows per week on the network. He recently was brought onboard as a contributor on *Medium*'s new sport's writing site called *The Cauldron*.

Dedria Givens-Carroll (PhD, University of Southern Mississippi) is an associate professor at the University of Louisiana at Lafayette, joining academe after twenty-five years as a public relations professional. She teaches public relations courses and researches religious communication and public relations.

Steve Granelli (PhD student, Ohio University) is interested in the study of the creation and management of fan communities as well as rhetoric of popular culture. He is a former co-host at SiriusXM and faculty member at both Syracuse University and the State University of New York at Oswego, where he served as the co-director of the Hollywood POV program. Steve's study of popular culture allows him to combine research interests with his unbridled fandom of *Game of Thrones, The Walking Dead*, and professional wrestling.

Michel M. Haigh (PhD, University of Oklahoma) is an associate professor in the College of Communications at Pennsylvania State University. She teaches courses in the public relations sequence. Her research interests are mass media influence and strategic communication. She has co-authored more than thirty-five conference presentations, six of which have been recognized with a "Top Paper" award. In 2014, she was recognized on the NerdScholar list of "40 Under 40: Professors Who Inspire." Haigh has published more than thirty articles in journals such as *Journalism & Mass Communication Quarterly, Journal of Broadcasting & Electronic Media, Communication Monographs, Communication Research, Journal of Social and Personal Relationships, Newspaper Research Journal, Corporate Communications: An International Journal*, and *Communication Quarterly*.

Leandra H. Hernandez (PhD, Texas A&M University) is an associate faculty member in the Department of Arts and Humanities at National University. Her research interests fall under two main categories: health communication with a focus on women's reproductive health and media representations of gender. Thus, she enjoys researching reproductive politics; Hispanic/Latina health experiences; patient-provider communication; women's pregnancy experiences; media and news coverage of reproductive politics; and media representations of masculinities and femininities, particularly in reality television shows. She has two book chapters about gender performances in *Toddlers & Tiaras* and *Duck Dynasty* in *Reality Television: Oddities of Culture* (2014) edited by Alison Slade, Amber J. Narro, and Burt Buchanan. She enjoys teaching courses about health communication, gender and the media, and feminist theories.

Laura Kane is a PhD student at Oregon State University in the School of Design and Human Environment. Her research interests include the study of cosplay (the act of dressing up as characters from popular media) and cosplay culture as it relates to internet behavior, psychological and sociological aspects of apparel design, and the apparel design process. After she completes her PhD, Laura plans to teach sewing and apparel design related courses at the college level.

Shaughan A. Keaton (BA, MA, University of Central Florida; PhD, Louisiana State University) is assistant professor of communication studies at Young Harris College. Shaughan grew up in Ohio and follows the Browns, Cavaliers, and Buckeyes, although he also cheers for his alma maters, UCF and LSU. He is an ardent fantasy sport participant, having been in the same league for eleven years (three-time champion). Shaughan has co-edited *The Influence of Communication in Physiology and Health* (2014), in which he authored a chapter dealing with the physiological effects of sport fandom. He is also a musician, having performed in over three hundred shows.

Jenny Ungbha Korn is a scholar of race, gender, and online identity with academic training from Princeton, Harvard, Northwestern, and the University of Illinois at Chicago. She has appeared in video, online, radio, and print stories revolving around race, gender, and online identity, including NPR, CNN, SXSW, Colorlines, and others. She has published in *Contexts*, *Our Voices*, *Multicultural America*, *Hashtag Publics*, *Intersectional Internet*, *Journal of Economics and Statistics*, *Encyclopedia of Asian American Culture*, and more. Her research examines race and gender online within Facebook, Twitter, YouTube, Tumblr, Chatroulette, blogs, email, and other computer-mediated communication.

Julia E. Largent (PhD student, Bowling Green State University) comes to media and communication studies with a peace studies background. She studies how media is used to create communities, to bring people together, and to enact social change. She focuses primarily on documentary studies, fan studies, social media and virtual pacifism, and nonviolence in video games. Julia has presented her research at several academic conferences and looks forward to many more.

Corey Jay Liberman (PhD, Rutgers University) is an assistant professor in the Department of Communication Arts at Marymount Manhattan College. His research spans the interpersonal communication, group communication, and organizational communication worlds and he recently co-authored a textbook dealing with organizational communication (*Organizational Communication: Strategies for Success*, which was published in 2013) and edited a case study book dealing with persuasion (*Casing Persuasive Communication*, which was published in 2014). He is currently working on his next two book projects, both of which deal with risk and crisis communication, which have tentative publication dates of May 2016. Currently, he is most interested in the social practices of dissent within organizations, specifically the antecedents, processes, and effects associated with effective employee dissent communication.

William E. Loges (PhD, University of Southern California) is assistant professor of new media communications and sociology at Oregon State University. His research focuses on the social roles of mass media. Dr. Loges has published in *Communication Research, Journal of Communication, Journalism Quarterly, Journal of Social Issues, Prehospital and Disaster Medicine, Journal of Applied Communication*, and he co-authored the entry on values in the *Encyclopedia of Sociology* and the entry on mass media and crime in the *Encyclopedia of Crime and Justice*.

Kathryn L. Lookadoo (MA, the Ohio State University) is a third year PhD candidate at the University of Oklahoma. Her research interests include entertainment media and health, parasocial relationships, and fan-celebrity online interactions. When she is not pursuing those interests, you can find her in front of the television "researching" her next project.

Christopher A. Medjesky (PhD, Bowling Green State University) is an assistant professor of communication studies at Defiance College. His research focuses on the relationship between rhetoric and media, primarily through television studies. This scholarship often focuses on the significance of humor, intertextuality, and interactivity in rhetoric.

Darcey Morris is a doctoral candidate at Georgia State University writing a dissertation titled "The Rise and Fall of Cable Branding: A Case Study of Time Warner," which analyzes cable network branding and promotional strategies in the post-network era. Prior publications include an article in *Communication Review* titled "The Sex Wars Continue: *Hung*'s Postfeminist Debate" and a chapter on cable branding in the reader *Media Economies*. Darcey is currently a lecturer in the communication studies department at Towson University.

Amber J. Narro (PhD, University of Southern Mississippi) is an associate professor of communication and obtained her doctorate of mass communication from the University of Southern Mississippi. She specializes in multiplatform journalism and researches political news coverage, journalism trends, and communication for nonprofit organizations. With professional experience in both journalism and public relations, she has practical knowledge to add to her courses in journalism, public relations, and public communication. Dr. Narro is the coordinator for the England study abroad program at Southeastern, and she has published articles in national and international journals. She has a bachelor's degree in mass communication and journalism and a master's degree in organizational communication, both of which she obtained from Southeastern Louisiana University.

Sabrina K. Pasztor (MA, Cornell University) is completing her doctoral work in June 2015 in communication/media studies with a concentration in gender/women's studies at the University of Illinois at Chicago (UIC). She was formerly an instructor at UIC and adjunct faculty in business and professional communication at DePaul University. She joined the University of Southern California (USC) as an assistant professor of clinical management communication in August 2015. Her research focuses on mass media (television), social media and popular culture, media framing (gender wage gap), media economics, and gender studies. Sabrina has presented at twelve communication, culture studies, and public policy conferences, and published in *Media Report to Women* and in an upcoming anthology focusing on the gendered representation of technology usage in Disney original movies. Her guilty pleasures are reality television, traveling, reading spy and mystery novels, anything *Harry Potter*, and dance music. Before returning to academia, she was a management consultant for ten years specializing in communication strategy, organizational change management, business process reengineering, and curriculum development.

Mike Plugh (MA, Fordham University) is a lecturer in the Department of Communication and Media Studies at Fordham University. His research interests include media ecology, technology and sociocultural change, and issues of education, community, and identity. Over the years, Plugh has also contributed writing on the subject of Japanese baseball to a number of publications, including *Baseball Prospectus*, *Sports Illustrated*, and *Medium*. He is a long-suffering, but enthusiastic supporter of the New York Knickerbockers. He is currently finishing his PhD at Temple University in Philadelphia.

Alane Presswood is currently pursuing a PhD at Ohio University. She enjoys researching the effects of social media on relationship construction and maintenance. When not nose-deep in scholarly books and articles, she prefers to spend her time puttering about in the kitchen or getting lost in the deeper annals of Netflix.

Brody J. Ruihley is an assistant professor of sport administration in the School of Human Services at the University of Cincinnati (Ohio). Ruihley's educational background consists of a bachelor's degree in communication from the University of Kentucky (2005), a master's degree in sport administration from the University of Louisville (2006), and a doctorate degree in sport management from the University of Tennessee (2010). Ruihley's primary research interests lie in the areas of fantasy sport and public relations in sport. Ruihley is the co-author of *The Fantasy Sport Industry: Games Within Games* (2014).

Alison F. Slade (PhD, University of Southern Mississippi) is currently a stay-at-home mom with five children. In her spare time, she continues with her academic research interests, which include reality television, social media, and fan culture. For the past three years, Dr. Slade has hosted the nationally syndicated radio program *The Alison Slade Show*, focusing on political discourse from an independent conservative view. Dr. Slade has appeared as a media expert on the *Redding News Review*, *America's Morning News*, and a recent NBC Universal documentary on *Duck Dynasty*. She was also a contributor in the award-winning book *Rock Brands: Selling Sound in a Media Saturated Culture* (Dr. Elizabeth Christian, ed.) and co-editor of *Mediated Images of the South: The Portrayal of Dixie in Popular Culture* and *Reality Television: Oddities of Culture*.

Danielle M. Stern is an associate professor in Christopher Newport University's Department of Communication. Her research engages the role of feminism in transforming popular culture and pedagogy. She teaches courses in critical/cultural theory, media studies, and gender. She is co-editor of *Lucky Strikes and Three-Martini Lunch: Thinking About Television's Mad Men*. Her nearly twenty scholarly articles have been published in journals such as

Text and Performance Quarterly, and the *Communication Review*, as well as in various edited books.

Cynthia W. Walker (PhD, Rutgers University) is an associate professor and chair of the Department of Communication and Media Culture at St. Peter's University in Jersey City, New Jersey, where she teaches courses in public relations and marketing, media literacy, gender and communication, film, animation, broadcast studies, scriptwriting, and research writing. A working journalist for over thirty-five years, she currently covers professional regional theater in New Jersey. A fan of the *Man from U.N.C.L.E.* series, Dr. Walker has written a number of U.N.C.L.E., spy and fandom-related articles including entries for the Museum of Broadcast Communication's *Encyclopedia of Television*, edited by Horace Newcomb. She delivers onscreen commentary in the extras of the *Man from U.N.C.L.E.: The Complete Series* DVD set, currently available from Time/Life and Warner Home Video. Her book, *Work/Text: Investigating The Man from U.N.C.L.E.*, proposes a new dialogic model of mass media and was published in the summer of 2013.

Nicholas Watanabe (PhD, University of Illinois) is assistant teaching professor in the Department of Parks, Recreation, and Tourism at the University of Missouri, working in the sport management emphasis. His main research interests are sport management, economics, finance, communications, and development. He has published in journals including: *International Journal of Sport Finance*, *International Journal of Sport Communication*, *International Journal of Sport Management and Marketing*, and *Sport Marketing Quarterly*.

Shelley Wigley (PhD, University of Oklahoma) is an associate professor in the Department of Communication at the University of Texas–Arlington. She teaches undergraduate courses in the public relations sequence as well as theories of persuasion and crisis management courses at the graduate level. This past spring 2015, Wigley taught the university's first strategic social media communication course, which is part of the Department's Emerging Media Certificate. Wigley's research interests include crisis communication, social media, and media relations. She has published articles in several journals including *Communication Monographs*, *Communication Research*, *Public Relations Review*, *Public Relations Journal*, and *Corporate Reputation Review*. Wigley is a member of the Greater Fort Worth Chapter of the Public Relations Society of America and the Social Media Club of Fort Worth.

CPSIA information can be obtained at www.ICGtesting.com
Printed in the USA
BVOW02*1621051115

425743BV00001B/1/P